Advanced Wireless Communication and Sensor Networks

This book covers wireless communication, security issues, advanced wireless sensor networks (WSNs), routing protocols of WSNs with cross-layer solutions, emerging trends in advanced WSNs, power management, distributed sensing and data gathering techniques for WSNs, WSNs security, applications, research of advanced WSNs with simulation results, and simulation tools for WSNs.

Features:

- Covers technologies supporting advanced wireless communication systems, sensor networks, and the conceptual development of the subject.
- Discusses advanced data gathering and sharing/distributed sensing techniques with its business applicability.
- Includes numerous worked-out mathematical equations and formulas, as well as essential principles including figures, illustrations, algorithms, and flow charts.
- Provides pervasive background knowledge including both wireless communications and WSNs.
- Covers wireless networks as well as sensor network models in detail.

This book is aimed at graduate students, researchers, and academics working in the field of computer science, wireless communication technology, and advanced WSNs.

Advanced Wireless Communication and Sensor Networks

Applications and Simulations

Edited by
Ashish Bagwari, Geetam Singh Tomar,
Jyotshana Bagwari, Jorge Luis Victória Barbosa,
and Musti K.S. Sastry

CRC Press is an imprint of the
Taylor & Francis Group, an **informa** business
A CHAPMAN & HALL BOOK

Designed cover image: © Shutterstock

First edition published 2023
by CRC Press
6000 Broken Sound Parkway NW, Suite 300, Boca Raton, FL 33487-2742

and by CRC Press
4 Park Square, Milton Park, Abingdon, Oxon, OX14 4RN

CRC Press is an imprint of Taylor & Francis Group, LLC

© 2023 selection and editorial matter, Ashish Bagwari, Geetam Singh Tomar, Jyotshana Bagwari, Jorge Luis Victória Barbosa and Musti K.S. Sastry; individual chapters, the contributors

Reasonable efforts have been made to publish reliable data and information, but the author and publisher cannot assume responsibility for the validity of all materials or the consequences of their use. The authors and publishers have attempted to trace the copyright holders of all material reproduced in this publication and apologize to copyright holders if permission to publish in this form has not been obtained. If any copyright material has not been acknowledged please write and let us know so we may rectify in any future reprint.

Except as permitted under U.S. Copyright Law, no part of this book may be reprinted, reproduced, transmitted, or utilized in any form by any electronic, mechanical, or other means, now known or hereafter invented, including photocopying, microfilming, and recording, or in any information storage or retrieval system, without written permission from the publishers.

For permission to photocopy or use material electronically from this work, access www.copyright.com or contact the Copyright Clearance Center, Inc. (CCC), 222 Rosewood Drive, Danvers, MA 01923, 978-750-8400. For works that are not available on CCC please contact mpkbookspermissions@tandf.co.uk

Trademark notice: Product or corporate names may be trademarks or registered trademarks and are used only for identification and explanation without intent to infringe.

Library of Congress Cataloging-in-Publication Data
Names: Bagwari, Ashish, editor. | Tomar, Geetam, editor. |
Bagwari, Jyotshana, editor. | Barbosa, Jorge Luis Victória, editor. | Sastry, M. K. S., editor.
Title: Advanced wireless communication and sensor networks : applications and simulations /
edited by Ashish Bagwari, Geetam Singh Tomar, Jyotshana Bagwari, Jorge Luis Victória Barbosa, M.K.S. Sastry.
Description: First edition. | Boca Raton : Chapman & Hall/CRC Press, [2023] |
Includes bibliographical references and index.
Identifiers: LCCN 2022055752 (print) | LCCN 2022055753 (ebook) |
ISBN 9781032347189 (hbk) | ISBN 9781032352916 (pbk) |
ISBN 9781003326205 (ebk)
Subjects: LCSH: Wireless sensor networks.
Classification: LCC TK7872.D48 A2965 2023 (print) |
LCC TK7872.D48 (ebook) | DDC 006.2/5–dc23/eng/20230106
LC record available at https://lccn.loc.gov/2022055752
LC ebook record available at https://lccn.loc.gov/2022055753

ISBN: 9781032347189 (hbk)
ISBN: 9781032352916 (pbk)
ISBN: 9781003326205 (ebk)

DOI: 10.1201/9781003326205

Typeset in Palatino
by Newgen Publishing UK

Dedicated to our real God "Parents," Baba Kedarnath, and Anshul, Anushka, Agrim, Adesh, & Shivanya.

Ashish Bagwari, Jyotshana Bagwari

Contents

Acknowledgments..xi
Editor Biographies..xiii
List of Contributors..xvii

Part I Advanced Wireless Communication: Overview, Challenges, and Security Issues

1. Wireless Communication: Overview and Fundamentals........................3
   Rajib Biswas

2. Introduction to Wireless Communication and Its Applications................11
   Kapil Jain, Neeti Khandekar, Prajakta Kulkarni, and Vivek Singh Kushwah

3. Power and Information Transfer Using IoT with NOMA-based GA-LPTS FBMC for Advanced Wireless and Sensor Networks..................21
   K. Ayappasam, Ganesan Nagarajan, and Ashish Bagwari

4. 5G-NR Wideband MIMO Antenna Design Using Stepped Radiators for Wireless Communication...39
   Mantar Singh Mandloi, Ajay Parmar, Karan Gehlod, Ashish Shakya, Priyanshi Malviya, and Leeladhar Malviya

5. Advanced Wireless Communication and Sensor Networks: Applications and Simulations...57
   Ciro Rodríguez and Isabel Moscol

6. Advanced Wireless Communication: Technology Overview, Challenges, and Security Issues...65
   Amit Kumar, Adesh Kumar, Geetam Singh Tomar, and Aakanksha Devrari

Part II Advanced Wireless Sensor Networks: Architecture, Consensus, and Future Trends

7. Advanced Wireless Sensor Networks: Introduction and Challenges............83
   Pooja Joshi, Somil Kumar Gupta, Kapil Joshi, and Jyotshana Bagwari

8. Wireless Sensor Networks: Routing Protocols and Cross-Layer Solutions......97
   N. Thirupathi Rao, Eali Stephen Neal Joshua, and Debnath Bhattacharyya

9. Social Impacts of Technology with the Emergence of IoT, 5G, and
   Artificial Intelligence.. 119
   Ghazanfar Latif, Jaafar Alghazo, and Sherif E. Abdelhamid

Part III Advanced Wireless Sensor Networks: Power, Data Gathering Techniques, and Security

10. Power Management Strategies in Wireless Sensor Networks................... 137
    Senthil Kumaran Rajendran and Ganesan Nagarajan

11. Power Management in Wireless Sensor Networks........................... 151
    Prasanta Pratim Bairagi, Kanojia Sindhuben Babulal, and Mala Dutta

12. Wireless Sensor Networks: Power Management............................ 173
    Ciro Rodríguez and Isabel Moscol

13. Security Enabling for IoT and Wireless Sensor Network Based Data
    Communication.. 181
    Ghazanfar Latif, Jaafar Alghazo, and Zafar Kazmi

14. Wireless Sensor Network Security 197
    Ciro Rodríguez and Isabel Moscol

Part IV Advanced Wireless Sensor Networks: Applications, Opportunities, Challenges, and Simulation Results

15. Advanced Wireless Sensor Networks: Applications and Challenges 215
    Vandana Roy, Shyam S. Gupta, and Binod Kumar Soni

16. A Novel Heuristic for Maximizing Lifetime of Target Coverage
    in Wireless Sensor Networks.. 227
    Pooja Chaturvedi, A.K. Daniel, and Vipul Narayan

17. Network Recovery in Dense and Emergency Areas Using a Temporary
    Base Station and an Unmanned Aerial Vehicle 243
    Sayanti Ghosh, Sanjay Dhar Roy, and Sumit Kundu

18. Wireless Sensor Networks with the Internet of Things 259
    *Ashish Bagwari, Geetam Singh Tomar, Jyotshana Bagwari,
    Jorge Luis Victória Barbosa, K.S. Sastry, and Manish Dixit*

19. Wireless Sensor Networks for Energy, E-Health, Building Maintenance
    and Agriculture Areas, and Simulation Results............................. 273
    Nakka Marline Joys, N. Thirupathi Rao, and Debnath Bhattacharyya

**20. A Survey on Opportunities and Challenges for Next Generation
Wireless Sensor Networks** 293
T. Perarasi, M. Leeban Moses, and K. Shoukath Ali

21. Various Simulation Tools for Wireless Sensor Networks 313
Hakan Koyuncu and Ashish Bagwari

Index 331

Acknowledgments

We express our deep gratitude to our parents Mr. Pradeep Kumar Bagwari and Mrs. Saraswati Bagwari, who have been encouraging and supportive throughout the tenure of our research work, and also thanks for Baba Kedarnath for their blessings to achieve this goal. We also wish to acknowledge our family members, i.e. personal and professional (WIT), for their continuous support, help and encouragement, without which this research work would not have been completed in time. Last but not least, special thanks to our lovely nephew Anshul Semwal, nieces Anushka Gaur and Shivanya Painoli, and our sons Agrim Bagwari and Adesh Bagwari for their true love.

Dr. Ashish Bagwari, India
Dr. Jyotshana Bagwari, India

Editor Biographies

Ashish Bagwari received B. Tech. (with Honors), M. Tech. (with Honors and Gold Medalist), and Ph.D. degrees in Electronics and Communication Engineering in 2007, 2011, and 2016 respectively. He is currently working as Head for the Department of Electronics and Communication Engineering at Women Institute of Technology (WIT) (Institute of State Government) Dehradun, Affiliating Institution of Uttarakhand Technical University (State Government Technical University), Dehradun, India. He has more than 12 years of experience in industry, academics, and research. He received the Best WIT Faculty award in 2013 and 2015 and Best Project Guide Award in 2015. Dr. Bagwari has been awarded the Corps of Electrical and Mechanical Engineers Prize from the Institution of Engineers, India (IEI) in December 2015 for his research work and was named in Who's Who in the World 2016 (33rd Edition) and 2017 (34th Edition). He also received an Outstanding Scientist Award 2021 from VDGOOD Technology, Chennai, India in November 2021, and Dr. A.P.J. Abdul Kalam Life Time Achievement National Award 2022 from the National Institute for Socio Economic Development (NISED), Bangalore, India in June 2022.

Dr. Bagwari has published more than 125 research articles in various international journals (including SCI, Scopus, and ISI indexed) that also include IEEE international conferences. His areas of interest are cognitive radio networks, mobile communication, sensor networks, wireless, and 5G communication, digital communication, mobile ad-hoc networks, etc. Dr. Bagwari is an active member of various professional societies like the Institute of Electrical and Electronics Engineers (IEEE, USA) as a Senior Member, the Institute of Electronics and Telecommunication Engineers (IETE, India) as a Fellow, professional member of ACM (Association for Computing Machinery), member of the Machine Intelligence Research Laboratory Society, member of the International Association of Engineers, etc. He has also been an Academic Editor of *Mobile Information Systems* (Hindawi, SCI indexed journal), Academic Editor of International Journal of *Wireless Communications and Mobile Computing*, (Wiley-Hindawi, SCI indexed journal), an advisor and reviewer for several well-known international journals published by IEEE, Taylor & Francis, Springer, Elsevier, IJCCN, ISTP, IJATER, JREEE, JCSR, and JNMS, CICN-2011, 2013, 2015, 2016, 2017, 2018, 2019, 2020, 2021, 2022, CSNT-2014, 2015, 2016, 2017, 2018, 2019, 2020, 2021, 2022, ICMWOC-2014, I4CT'2014, and ICEPIT-2014. Dr. Bagwari is a member of the Board of Studies of Dr. B.A. Technological University, Lonere, Maharashtra, Uttarakhand Technical University (UTU), Dehradun, and Uttaranchal University (UIT), Dehradun, etc. for the departments of Electronics & Communication Engineering. Dr. Bagwari has organized 4 international conferences, 20+ workshops/symposiums, 8 Short Term Course Programs (STCP) / Faculty Development Programs (FDP), 12 expert lectures, and has been invited as a key speaker and session chair at various workshops/conferences and delivered 17 expert lectures at different institutions and universities within India. Dr. Ashish visited China in the year 2007 when he associated with Reliance Industries, Mumbai, India. Dr. Ashish has filed 2 Indian patents and is currently pursuing various research projects/consultancies funded by the Government of India and private organizations. Dr. Ashish has assessed and evaluated more than 7 post-graduate students' projects (at M.Tech.

Course Level) and 16 under-graduate students' projects (at B.Tech. Course Level), guided one Ph.D. student, is currently guiding 3 Ph.D. students, and has evaluated/reviewed 9 Ph.D. theses as an external examiner. He has written 5 books and more than 10 book chapters for reputed international publishers. Dr. Ashish is also "Lead Guest Editor" of the special issues "Radio Networks for New Disruptive Digital Services in the Fourth Industrial Revolution," *Computers, Materials & Continua* (SCIE Q1 indexed journal with impact factor 3.772); "Applications of Cognitive Radio in Emerging Technologies," *International Journal of Distributed Sensor Networks* (SCIE Q1 indexed journal with impact factor 1.640); and "Cognitive Radio as Futuristic Technology for Cloud Computing, Mobile Computations, Big Data, and Blockchain Techniques," *Frontiers In Computer Science* (ESCI indexed journal). He is the "Guest Editor" for a special issue "AI and Security Application in Green Energy and Renewable Power" of *Applied Sciences* (SCI indexed journal, impact factor 2.679) for the year 2021–22.

He organized an International Conference on Wireless Technologies, Networks and Science (2022), in association with Al-Balqa University, Jordan, as General Chair. (https://icwtns.aairlab.com)

Visit Google Scholar: https://scholar.google.co.in/citations?hl=en&user=6kYT6h0AAAAJ

Geetam Singh Tomar received his UG, PG, and Ph.D. degrees in Electronics Engineering from The Institute of Engineers, Calcutta, MNREC (NIT) Allahabad, and State Technology University Bhopal, respectively, and was a post-doctoral fellow at the University of Kent, Canterbury, UK.

Currently he is the director of REC Sonbhadra, and is also the director of Machine Intelligence Research Labs, Gwalior, India. He is associated with professional societies like IEEE as Senior Member and IETE and IE (I) as Fellow. He is Chair of IEEE MP Sub-section and Chair of IETE MP. He is also ex-director of BIAS Bhimtal, and THDC Institute of Hydropower Engineering and Technology, Tehri, India. Prior to this he worked as a visiting professor at the School of Computing, University of Kent, and faculty at the University of the West Indies, Trinidad and Tobago. Apart from this, he has served in IIITM Gwalior, MITS Gwalior and other institutes in India. He worked in the Indian Air Force for 17 years prior to joining academia, having total experience of more than 35 years.

He received an International Pluto Award for Academic Achievements in 2009 from IBC, Cambridge, UK. He is IEEE Senior Member, Fellow IETE and IE (I) member CSI, ACM and ISTE. He is chief editor of five international journals (and eight in the past). Has published 4 patents, and more than 200 research papers in international journals/conferences. Has written 11 books from Springer, CRC, IGI, and Indian publishers, organized more than 30 IEEE international conferences in 7 countries, and delivered more than 20 keynotes addresses abroad, and regularly visits two universities abroad as research consultant. He has completed 5 sponsored projects for DRDO and DST, including two major projects. He is a member of two IEE/ISO working groups and member of technical committees for IEEE SMC and TCCC. He is presently a member of the Board of Governors, IIM Kahsipur, Chairman for IETE Gwalior for two terms, and Chairman IEEE MP Sub Section.

Visit Google Scholar: https://scholar.google.com/citations?user=SVAxkl8AAAAJ&hl=hi

Editor Biographies

Jyotshana Bagwari received B. Tech. (with Honors), M. Tech. (GATE qualified in 2012), and Ph.D. degrees in Computer Science and Engineering. Presently, she is Director of the Advanced and Innovative Research Laboratory (AAIR Lab), India. Earlier, she served Robotronix Engineering Tech Pvt. Ltd, Indore, India as a Project Engineer in the department of Research & Development. She also served at the Department of Computer Science and Engineering, Birla Institute of Applied Sciences (BIAS), Bhimtal, Uttarakhand, as an Assistant Professor. She has more than four years of experience in industry and academia. Dr. Bagwari has published more than 30 research papers in various international journals (including SCI/ISI/Scopus indexed) and IEEE international conferences. Her current research interests include networking, cryptography, cognitive radio networks, etc. She is a member of IAENG, Global Member of the Internet Society, member of the Institute of Science Index, member of MIR Labs, member of the International Association of Computer Science and Information Technology, and member of IDES. She has also been an editor and reviewer for several well-known international journals: *IJIEAC, IJRTM, International Journal of Frontiers of Mechatronical Engineering, IJITCS, IJCMEEE (COMPEL), IJCN, IET Communications, Ain Shams Engineering Journal, IEEE CSNT* (2019, 2020), and *IEEE CICN* (2020).

Dr. Bagwari successfully completed two NPTEL on-line certified courses on the topic "Cryptography and Network Security" with 83% in the year 2019, and the topic "Design and analysis of algorithms" with 70% in the year 2017. Dr. Bagwari has filed one patent and has written three books and more than seven book chapters in reputed publications.

Dr. Bagwari is also a "Lead Guest Editor" of special issues "Applications of Cognitive Radio in Emerging Technologies," *International Journal of Distributed Sensor Networks* (SCIE Q1 indexed journal with impact factor 1.640); and "Cognitive Radio as Futuristic Technology for Cloud Computing, Mobile Computations, Big Data, and Blockchain Techniques," *Frontiers In Computer Science* (ESCI indexed journal).

Visit Google Scholar: https://scholar.google.co.in/citations?hl=en&user=FKraujAAAAAJ

Jorge Luis Victória Barbosa received his B.S. degrees in Data Processing (1990) and Electrical Engineering (1991) from the Catholic University of Pelotas, Brazil. He obtained his M.S. and Ph.D. degrees in Computer Science from the Federal University of Rio Grande do Sul, Brazil, in 1996 and 2002, respectively. He conducted post-doctoral studies at Sungkyunkwan University (Suwon, South Korea, 2016) and University of California Irvine (USA, 2020). Nowadays, he is a full professor in the Applied Computing Graduate Program (PPGCA) at the University of Vale do Rio dos Sinos (UNISINOS), São Leopoldo, Brazil. Additionally, he is a researcher of productivity at CNPq (the Brazilian Council for Scientific and Technological Development) and head of the Mobile Computing Laboratory (MobiLab/UNISINOS). His research interests are ubiquitous computing, ambient intelligence, big data, Internet of Things (IoT), machine learning, data analysis, games development, context histories, context prediction mainly through similarity and pattern analysis, applied computing in health, accessibility, learning, industry and agriculture. He is a member of the Brazilian Computer Society (SBC).

Visit Google Scholar: https://scholar.google.co.in/citations?hl=en&user=JSBL9uEAAAAJ

Musti K.S. Sastry received his B.Tech. in Electrical Engineering from JNTU, Kakinada (A.P.) in 1990, M.Tech. and Ph.D. from NIT, Warangal in the years 1996 and 2002, respectively, and is an accomplished academic with more than 30 years of professional experience. He is currently with the Namibia University of Science and Technology as an Associate Professor in the Department of Electrical and Computer Engineering. He is involved with several projects in the areas of big data, enterprise information systems, and demand side management. He has more than 30 years' experience in industry, academia, and research and has worked in different parts of the world. Earlier he was with NIT Warangal, India as a lecturer; Global Energy Consulting Engineers, Hyderabad, India as Assistant General Manager; and then served the University of the West Indies in the Caribbean. Dr. Sastry has a significant interest in core and multidisciplinary research and has published more than 50 research articles and book chapters in the extended areas of energy systems, database and information systems, big data, and soft computing. He won the IEEE travel award for his paper at Kobe, Japan in *IEEE Circuits and Systems* in 2005. He has organized 4 Short Term Course Programs/Faculty Development Programs, several expert lectures, and has been invited as a key speaker and session chair to various workshops/conferences. Dr. Sastry has visited various universities across the globe to serve in various capacities. He is a visiting professor with ESIGELEC, Rouen, France (data sciences); Sophia University, Japan (data sciences), University of Turabo, Puerto Rico (energy systems), and KL University, India. He is on the panel of accreditors of IEEE/ABET, USA, and visited a few universities as part of the ABET accreditation process. Dr. Sastry is a senior member of IEEE, USA, a member of IET-UK, and a life member of IE, India.

Visit Google Scholar: https://scholar.google.co.in/citations?hl=en&user=9MQJ-P0AAAAJ

Contributors

Sherif E. Abdelhamid
Virginia Military Institute
Lexington, Virginia, USA

Jaafar Alghazo
Virginia Military Institute
Lexington, Virginia, USA

K. Shoukath Ali
Bannari Amman Institute of Technology
Erode, Tamilnadu, India

K. Ayappasam
Raak College of Engineering and
   Technology
Puducherry, India

Kanojia Sindhuben Babulal
Central University of Jharkhand
Ranchi, Jharkhand, India

Ashish Bagwari
Women Institute of Technology and
   Uttarakhand Technical University
Dehradun, India

Jyotshana Bagwari
Advanced and Innovative Research
   Laboratory (AAIR Lab)
Dehradun, India

Prasanta Pratim Bairagi
Assam Downtown University
Guwahati, Assam, India

Jorge Luis Victória Barbosa
University of Vale do Rio dos Sinos
   (Unisinos),
São Leopoldo, RS, Brazil

Debnath Bhattacharyya
K L Deemed to be University (KLEF)
Guntur, Andhra Pradesh, India

Rajib Biswas
Tezpur University
Assam, India

Pooja Chaturvedi
Institute of Technology, Nirma University
Ahmedabad, Gujrat, India

A.K. Daniel
Madan Mohan Malaviya University of
   Technology
Gorakhpur, Uttar Pradesh, India

Aakanksha Devrari
University of Petroleum & Energy Studies
Dehradun, India

Manish Dixit
Madhav Institute of Technology and
   Science
Gwalior, Madhya Pradesh, India

Mala Dutta
Assam Downtown University
Guwahati, Assam, India

Karan Gehlod
Azure Software
India

Sayanti Ghosh
National Institute of Technology
Durgapur, India

Shyam S. Gupta
Amity University
Gwalior, Madhya Pradesh, India

Somil Kumar Gupta
Uttaranchal University
Dehradun, India

xvii

Kapil Jain
Amity University
Gwalior, Madhya Pradesh, India

Kapil Joshi
Uttaranchal Institute of Technology,
  Uttaranchal University
Dehradun, India

Pooja Joshi
Uttaranchal Institute of Technology,
  Uttaranchal University
Dehradun, India

Eali Stephen Neal Joshua
Vignan's Institute of Information
  Technology
Visakhapatnam, Andhra Pradesh, India

Nakka Marline Joys
Anil Neerukonda Institute of Technology
  & Sciences
Visakhapatnam, India

Zafar Kazmi
Prince Mohammad Bin Fahd University
Saudi Arabia

Neeti Khandekar
ITM Group of Institutions
Gwalior, Madhya Pradesh, India

Hakan Koyuncu
Altinbas University
Istanbul, Turkey

Prajakta Kulkarni
ITM Group of Institutions
Gwalior, Madhya Pradesh, India

Adesh Kumar
University of Petroleum & Energy Studies
Dehradun, India

Amit Kumar
Veer Madho Singh Bhandari Uttarakhand
  Technical University
Dehradun, India

Sumit Kundu
National Institute of Technology
Durgapur, India

Vivek Singh Kushwah
Amity University
Gwalior, Madhya Pradesh, India

Ghazanfar Latif
Prince Mohammad Bin Fahd University
Saudi Arabia

Leeladhar Malviya
Shri G.S. Institute of Technology and
  Science
Indore, Madhya Pradesh, India

Priyanshi Malviya
Madhav Institute of Science and
  Technology
Gwalior, Madhya Pradesh, India

Mantar Singh Mandloi
Rewa Engineering College
Rewa, Madhya Pradesh, India

Isabel Moscol
Universidad de Piura
Piura, Peru

M. Leeban Moses
Bannari Amman Institute of Technology
Erode, Tamilnadu, India

Ganesan Nagarajan
Puducherry Technological University
  (Erstwhile Pondicherry Engineering
  College)
Puducherry, India

Vipul Narayan
Galgotias University
Greater Noida, India

Ajay Parmar
Shri G.S. Institute of Technology and
  Science
Indore, Madhya Pradesh, India

T. Perarasi
Bannari Amman Institute of Technology
Erode, Tamilnadu, India

Senthil Kumaran Rajendran
IFET College of Engineering
Villupuram, India

N. Thirupathi Rao
Vignan's Institute of Information
  Technology (Autonomous),
Visakhapatnam, Andhra Pradesh, India

Ciro Rodríguez
Universidad Nacional Mayor de Marcos
Lima, Peru

Sanjay Dhar Roy
National Institute of Technology
Durgapur, India

Vandana Roy
Gyan Ganga Institute of Technology and
  Sciences
Jabalpur, Madhya Pradesh, India.

K.S. Sastry
Namibia University of Science and
  Technology
Namibia

Ashish Shakya
Azure Software
India

Binod Kumar Soni
Doece, UIT-RGPV
Bhopal, Madhya Pradesh, India

Geetam Singh Tomar
Shonbhadra Engineering College
Shonbhadra, Uttar Pradesh, India

Part I

Advanced Wireless Communication: Overview, Challenges, and Security Issues

1
Wireless Communication: Overview and Fundamentals

Rajib Biswas

CONTENTS
1.1 Introduction 3
1.2 Basic Outline of Communication Systems 4
1.3 Wireless Technologies 5
    1.3.1 Wi-Fi 802.11 5
    1.3.2 Bluetooth 6
    1.3.3 WiMax 802.16 6
    1.3.4 Radio-frequency ID (RFID) 7
1.4 Signaling and Fading 8
1.5 Concluding Remarks 8
References 9

1.1 Introduction

Wireless communication is one of the dominant communication systems in the current era. It has permeated into every sphere of our modern day-to-day activities and the industrial sector. If we look at the history of wireless communication, it dates back to 1897. That year saw the historic demonstration of wireless telegraphy by Guglielmo Marconi. Four years later, the world saw the successful installation of radio reception across the Atlantic Ocean. One can easily apprehend that wireless technology has undergone tremendous developments in the subsequent years [1–3]. In the later years, most wireless systems have become optional. Sometimes, wired technology outsmarts the wireless. For example, optical fiber technology is gaining a good foothold. At the same time, cellular technology is gradually replacing the existing wired one. All these are testimony to the fact that there always exists a choice in such substitutional replacement of technology. There are certain benefits of using wireless communications. The notable merits include cost, mobility, ease of installation, reliability, and disaster recovery. Starting with cost, it is quite apparent that wireless systems are very cost-effective in nature as they do not require any installation of cables or allied accessories. Apart from this, wireless communication remains the preferable choice in buildings of historical importance. Because of absence of wires, it always provides better mobility within its range. During natural calamities, the damage that can happen to these system is found to be minimal. Meanwhile, new paradigm shifts have also revolutionized wireless communication. Some adaptations such as Bluetooth, Wi-Fi 802.11, WiMax 802.16 need mention here. Brimming with a multitude of merits, these standards have improved wireless communications systems to a great extent. This chapter highlights the key aspects of wireless communications. In essence, this chapter attempts to give a brief appraisal of fundamentals and applications.

1.2 Basic Outline of Communication Systems

As shown in Figure 1.1a, the fundamental blocks of a communication system comprise a transmitter, a communication channel, and a receiver. Alongside this, Figure 1.1b portrays another block diagram in extended mode. The transmitter is followed by a band pass filter where the modulated signal is bandlimited prior to amplification. The amplified modulated signal then enters the demodulator part where the signal is detached from the carrier. The demodulated signal after getting amplified passes through the low pass filter before being fully recovered at the destination. The components shown here are simply representative of the whole communication set-up. With different wireless technology, additional components may creep in, executing specific functions. As for example, the transmitter itself generally possesses several basic constituents such as source codes, channel coder, modulator, power amplifier, local oscillators, and antenna. The first component source coder encodes the message to check redundancy. The channel codes then encode the input—thereby ensuring channel-induced error-free transmission. The local oscillator is responsible for generation of the carrier which is then embedded with an encoded message by the modulator [4–8]. The power amplifier boosts the signal to a desired level before it is transmitted by the antenna. Likewise, the receiver contains identical components with opposite functions, as for instance a source decoder which decodes the encoded message, followed by a channel decoder. Afterwards, the modulated signal enters the down converter where the radio frequencies are down converted to intermediate frequencies. The down converted signal then enters the demodulator where the carrier is suppressed. In the

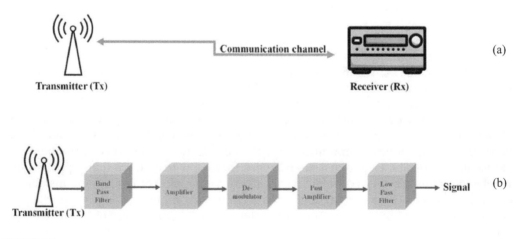

FIGURE 1.1
Schematic layout of communications systems: (a) abridged (b) extended.

TABLE 1.1

Fundamental technical terms used in wireless communication

Fundamental terms	Description
Station	As the name implies, it embraces all those devices which are under the Wireless LAN. It may be mobile or fixed. It has to be 80.2.11 compliant MAC as well as holding a PHY interface to the wireless medium.
Access point	This generates WLAN, thereby delivering connection to distribution services via a wireless medium. Basically, it establishes wired connection to Wi-Fi clients and routers. In essence, the Wi-Fi is enabled by the switching hub through the ethernet cable via access points. While enabling the Wi-Fi, the access points transcend the signal acquired from ethernet down to 2.4–5.0 GHz.
Client	Device that connects to a Wi-Fi (wireless) network.
Basic service set (BSS)	Refers to an assembly of stations tuned for 802.11 standard. It may comprise one access point with a couple of clients.
Basic service set identifier (BSSID)	Explicit in the name, it functions as identifier for the basic service sets. Basically, it is a 48-bit media access control address. More specifically, it possesses layer2 identifier of each individual basic service set.
Distribution system (DS)	A system that interconnects a set of basic service sets and integrated Local Area Networks (LANs) to create an Extended Service Set (ESS). It is used to extend wireless network coverage.
Extended service set (ESS)	This set of services comprises multiple access points with their corresponding clients.
Independent basic service set (IBSS)	As per name, it allows client to execute peer-to-peer communications without the need of an access point.
SSID/ESSID:	Known as a service set identifier (SSID), it enables connection with a chosen network in presence of multiple networks existing in the same physical area.
Roaming	Feature which ensures a seamless connection between a mobile client in the same extended service set.

subsequent stage, the demodulated signal is amplified by a low noise amplifier and the message is retrieved. Depending on the various configurations of wireless communications, the detailed components may vary and contain some additional parts. Table 1.1 lists some of the basic technical terms frequently used in wireless communications.

1.3 Wireless Technologies

There is a multitude of wireless technologies such as cellular phones, cordless receivers, Wi-Fi, Bluetooth and many more. Some of them are in turn replaced by wired technology. For example, we can cite the case of growing demand for optical fiber communication. In the following section, the major developments in wireless technology spanning Wi-Fi 802.11, WiMAX 802.16, RFID are dwelt upon.

1.3.1 Wi-Fi 802.11

802.11 is a wireless standard. It builds up the communication interface between base station and the clients. In general 802.11 comprises various versions which differ from one

TABLE 1.2

Various Wi-Fi standards at [1]

Wi-Fi standards	Frequency of operation (GHz)	Rate of data transmission
802.11 b	2.4	1–11 Mbps
802.11a	5	54 Mbps
802.11g	2.4	54 Mbps
802.11n	2.4, 5	600 Mbps
802.11ac	Same as above	3.5 Gbps
802.11ax	Same as above	9.6 Gbps

TABLE 1.3

Classes of Bluetooth linkage at [3]

Class	Power	Transmission loss	Range
Class 1	100 mW	20 dBm	100 m
Class 2	25 mW	4 dBm	10 m
Class 3	1 mW	0 dBm	1 m

another in terms of operating frequency, data rates, as well as coding techniques. The general data rates range from 1 Mbps to 9.6 Gbps. Accordingly, the different versions of it can be listed as 802.11b, 802.11a, 802.11g, 802.11n, etc. Table 1.2 summarizes the properties of these versions.

1.3.2 Bluetooth

This wireless technology basically implements low power radio receiver and communication protocol. It has limited range, more precisely short range. The speed of communication among connected devices remains within Mbps. Comprising three classes (listed in Table 1.3), it can hop in between 79 channels—each of them possessing a bandwidth of 1 MHz resulting in reduced interference. Apart from this, Bluetooth technology is not limited to "line-of-sight" communication. It endorses dynamic creation of private networks.

The linked devices under Bluetooth continuously change their channels so as to make themselves immune from the interference of other devices. More specifically, known as frequency-hopping spread spectrum (FHSS), it was invented by Hedy Lamaar in the year 1942. During normal conditions, the hopping is efficient at 1,600 times per second. Under specific conditions, it is changed up to 3,200 times per second.

1.3.3 WiMax 802.16

This is an advanced version of 802.11 standard. In the year 1999, IEEE introduced this advanced wireless technology. Their main objective was to deploy Broadband Wireless Metropolitan Area Networks via Broadband Wireless Access standards. Precisely, WiMAX is the enhanced version of 802.16e known as Worldwide Interoperability for Microwave Access. In general, it endorses two aspects: the physical layer and media access control layer (MAC). The transmission is executed by scalable orthogonal frequency division multiple access modulation protocol. This is supported by bandwidths ranging from 1.25

MHz to 20 MHz with a carrier capacity of 2,048. WIMAX has robust provisions for all weather conditions. Suppose the weather is not good—then WiMAX resorts to a prudent coding mechanism such as binary phase shift keying. However, very efficient 64 quadrature amplitude modulation is adopted whenever the weather conditions are favorable. Meanwhile, 16 QAM or quadrature phase shift keying is applied. It is noteworthy that WiMAX uses MIMO configuration purposefully to achieve non-line-of-sight. Likewise, hybrid automatic request ensures proper checking of errors. The media access control layer is equipped with sublayers accommodating extra layers such as asynchronous transfer mode or internet protocol on air interface. Use of data encryption standard or advanced encryption standard ensures secure information exchange, thereby activating authentication as well as encryption. WiMAX also allow users to operate in idle mode or sleep mode. This prevents excessive power usage and increases power saving. Above all, WiMAX supports connection orientation. Data transmission occurs only when there is allocation of communication channel by bare station. Apart from this, quality of service (QoS) is another striking feature of WiMAX which generates efficient transportation of application data.

1.3.4 Radio-frequency ID (RFID)

RFID is another development in the field of wireless communication systems. It is basically an automated data collection strategy, utilizing radio frequencies for transferring data between reader and mobile unit with the objective of identification, tracking, and categorization. Being a contactless system, it is very handy. The utilization of inexpensive components make the whole process cost-effective. The assigning of unique identification and subsequent background interaction makes RFID a versatile tool for several applications. If we look at the composition of an RFID system, it starts with RFID tags, an antenna (mostly microchip), and an RFID reader connected to a server or data collection network [2–4].

The RFID tags may be of two types: passive and active. The former is in no need of power as it can extract it from an interrogator board. Sometimes, it contains a condenser which delivers power via discharging when there is no power. Having less memory, these tags cannot be rewritten—being written once and read many times. Apart from this, they possess less range. On the contrary, active tags drain power from an active source. Being endowed with a long range, bits can be rewritten. With a larger memory content, they cost more. Coming to the RFID reader, it is in general capable of establishing bi-directional data links. They also create inventory tags and communicate with the dedicated network server. Notably, they can handle around 100–300 tags per second. These readers (also known as interrogators) may be fixed or mobile units which can be transported/carried as required.

Being a versatile wireless technology, it has proved immensely helpful in many spheres of application. Shopping malls, smart groceries, etc. are adequately using this RFID technology. Big companies such as Walmart, Amazon, and Procter & Gamble are exhaustively utilizing this technology for suppliers as well as outlets. The health sector is also using RFID to keep track of admitted patients.

Now with the proliferating use of RFID, there arise a couple of security concerns. If care is not taken, it may lead to eavesdropping. For instance, one could cite cases of muggers retrieving information from users to find out valuable items are possessed by the user, resulting in theft or snatching. Likewise, unauthorized use of RFID tags may help in creating user profiles to execute malicious intent. Maybe stronger encryption can solve such problems—however, this comes with a bigger price tag, which is another hurdle.

1.4 Signaling and Fading

As discussed in the previous sections, the communication channel plays a vital role in the implementation of wireless communications. Over this communication channel, there is transmission of signals. The effective transmission of signal is a very important aspect as it will ensure a successful recovery at the receiving end [4–8]. During signaling, baud rate and bit error rate are of prime importance. The former is directly linked to rate of transmission of signaling element while the latter refers to the rate at which errors may emerge per number of bits being transmitted. In order to avoid inter symbol interference, it is necessary that robust signal coding techniques be applied. In such cases, quadrature phase shift keying or bipolar phase shift keying can be advocated. In addition, the error detection procedures have to be equally prudent so that noise- and error-free reception can be realized.

Equally important is the frequency impact. Needless to say, variation of channel strength over time and frequency is inevitable. These variations give rise to a technical term, fading. Fading basically refers to decline of input signal to a receiver. The physical variables defining the radio path are often influenced by weather conditions as well as geometrical configuration of the path. Heavy ground fog or extremely freezing air over warm ground may also result in substantial fading. The occurrence of fading has a tendency to increase with repeater spacing and rise in frequency. The main categories are listed below.

Large-scale fading: As the signal encounters objects such as high-rise buildings or hills during its passage, it undergoes path loss. This path loss varies directly with distance and shadowing by the obstacles that come directly in the propagation path of the signal. Supposing the handheld wireless device traverses a distance equaling the size of the cell, this fading becomes dominant. Importantly, it has no dependence on frequency.

Small-scale fading: This fading depends on frequency. Whenever there are multiple signals propagating between the transmitter and receiver, it is quite common that interference occurring in both modes, spanning destructive as well as constructive, leads to this kind of path loss. Apart from being frequency dependent, small-scale fading arises within the scale of carrier wavelength. Large-scale fading becomes significant in cell-site planning whereas small-scale fading is impactful during the design and implementation of consistent communications.

1.5 Concluding Remarks

In this chapter, the fundamentals of wireless communication are overviewed. The basic layout is detailed. The recent wireless technologies are also appraised here. Signaling and fading constitute a sizable portion of wireless communication. Deterioration in signal might occur if the fading factors are not properly accounted for. Although they are circumstantial, they become cumulative if not treated. Being endowed with a multitude of merits, it has become a norm in today's communication world. However, certain factors such as interference and security affect wireless communications. As the signals are transmitted in free space, there is a fair chance of intrusion by an interceptor who can retrieve valuable information therefrom. In addition, interference in signals is another intriguing factor. With some wireless technologies, the signals used by them often interfere because of similar frequency allocations, which is further compounded by free space utilization. Although nominal, the radiation hazard from radio energy as used in wireless communication is another aspect which needs to be managed with care. In view of these, it is believed that advanced wireless communication will solve these critical issues via adoption of encryption as well as radiation shielding receivers. Moreover, the interference can be taken care of through precise spread spectrum protocols [8].

References

[1] "IEEE Standard for Information technology—Telecommunications and information exchange between systems Local and metropolitan area networks—Specific requirements-Part 11: Wireless LAN Medium Access Control (MAC) and Physical Layer (PHY) Specifications," in IEEE Std 802.11-2016 (Revision of IEEE Std 802.11-2012), pp.1–3534, 2016.

[2] J. Bray, C.F. Sturman. Bluetooth: connect without cables. Prentice Hall, 2000.

[3] M. Miller. Discovering Bluetooth. Sybex, 2001.

[4] C. She, P. Cheng, A. Li, Y. Li. Grand challenges in signal processing for communications. Frontiers in Signal Processing, 1, 2021.

[5] R. Biswas. A brief appraisal of machine learning in industrial sensing probes. In: D. Shubhabrata, J. Paulo Davim, editors. Machine learning in industry. Springer, 2021

[6] R. Biswas. A brief appraisal on blockchain assisted secured healthcare. In: M.D. Dutta, R.M. Visconti, G.C. Deka, editors. Blockchain in digital healthcare. Chapman & Hall, 2021.

[7] R. Biswas. A brief overview of IoT architecture and relevant security protocols. In: S. Velliangiri, S.A.P. Kumar, P. Karthikeyan, editors. Internet of Things: integration and security challenges. CRC Press, 2021.

[8] R. Biswas. Spectrum sensing techniques: an overview. In: A. Bagwari, J. Bagwari, G.S. Tomar, editors. Sensing techniques for next generation cognitive radio networks. IGI Global, 2018.

2

Introduction to Wireless Communication and Its Applications

Kapil Jain, Neeti Khandekar, Prajakta Kulkarni, and Vivek Singh Kushwah

CONTENTS
2.1 Introduction 11
2.2 5G 12
2.3 Dual Band 12
2.4 Frequency below 6GHz 13
    2.4.1 Advantages of Sub-6GHz 13
2.5 Speed of 5G 14
2.6 Comparison of 5G Technology with Other Generations 14
2.7 Microstrip Patch Antenna for 5G 14
2.8 Simulation Results and Analysis 16
    2.8.1 Gain 16
    2.8.2 Voltage Standing Wave Ratio (VSWR) 17
    2.8.3 Far Field for Antenna 17
    2.8.4 Radiation Pattern 17
    2.8.5 Rectangular Shaped Antenna Design (Simulation Results) 17
2.9 Conclusion and Future Work 18
Acknowledgments 20
References 20

2.1 Introduction

In history, communication through telephone was introduced by Alexander Graham Bell in 1876. And from that day to now, we have seen many new changes in this technology. As the number of generations increases, the speed of networks related to them also increases.

Wireless communication is the type of data communication which transfers information between two points or more without the use of any physical device or medium like an electrical conductor, optical fiber, transmission lines or other media for the transfer that are continuous and guided. The most common wireless technologies use radio waves. In this type of communication of data, the data is delivered wirelessly. Wireless communication has several applications in this modern era. A few of the applications are:

1. Cellular telephony
2. Internet
3. Improved security systems
4. Broadcast information
5. Satellite communication
6. GPS (Global Positioning System)
7. Ad-hoc network
8. Bluetooth
9. Wi-Fi
10. Antenna
11. Drone via antenna

Here, we are going to take the antenna applications of wireless communication. Like most devices, it advances day by day according to the requirements. Antennas are designed for many purposes and requirements. For communication purposes, wireless communication is the easiest way to communicate with someone. As the name suggests, it is communication without using wires or electrical conductors. It is the transfer of messages between two or more points which use the antenna for transmitting and receiving purposes. The last two decades brought major technological changes in the wireless communication field. To meet the demands of users, services like FM, Internet, live video calling, live television and many more have been revolutionized with the progress in coding techniques, networking protocols, high speed data transmission and interference [1]. Applications of dual band antennas for communication purposes include drones, mobile base stations, Wi-Fi, satellite communication, infrared communication, microwave communication and many more.

Dual band is also a good option for increasing the speed and advancement of antennas.

2.2 5G

5G technology stands for the fifth generation. It is expected that this technology will change the way people live and work. 5G technology is the latest cellular technology and is increasing the wireless network rapidly all over the world. Moreover, it is designed in such a way that it connects the whole world virtually, including all machines, objects or devices [2].

With an increase in user demand, 4G can now be easily replaced with 5G with its advanced technology, Beam Division Multiple Access (BDMA), or Filter Bank Multicarrier (FBMC). It has so many applications that make it different from other technologies, such as delivery of high data rates, low latency, high reliability or availability, high performance user experience and more bandwidth which increases the data transmission rate over the wireless system through advanced antennas [3].

2.3 Dual Band

Dual band expresses that it is something which is made up of two things: it contains two bands that are used to provide better performance. It means that by this device two wireless signals can be transmitted simultaneously. And comparing dual band routers is far easier than the single band router. In this chapter we use two frequencies for designing the antenna by using dual band technique for which the frequencies we use are 5.0 GHz and 5.4 GHz for microstrip patch antenna.

2.4 Frequency below 6 GHz

GSM or LTE networks are not capable of supporting higher data rates, while 5G technology is made to deliver high data rates. Our mobile gets the coverage connected with the network by the antennas or towers placed around us. At predefined frequencies these towers radiate the signals which are referred to as frequency or bands. 5G technology divides frequency bands into two categories, sub-6 GHz and mmWave. Figure 1 shows 5G technology frequency bands which is categorized into mmWave and sub-6GHz. Here sub-6 is the frequency below 6 GHz and mmWave is the higher frequency radio bands.

Sub-6 GHz is also known as Frequency Range-1: like the 5G frequency band, it is also categorized into two parts, low-band and mid-band, in which low-band ranges from 600 MHz to 2.4 GHz. Due to massive MIMO, we get better speed than 4G. In a low-band network we get at least 100 Mbps. In mid-band networks it ranges from 3 GHz to 6GHz while some countries increasingly range up to 7 GHz. In it we get speeds higher than 200 Mbps and up to 900 Mbps. It is said that nowadays India is aiming for ranges 3.3 to 3.6 GHz and 5.2 to 5.9 GHz. Sub-6 GHz band is more than enough for consumer use and can support higher bandwidth than the LTE frequency band [4].

2.4.1 Advantages of Sub-6 GHz

It is less complex, easy to develop, test and deploy, it can work on 4G architecture, and it is less expensive than mmWave network. Whereas the frequency above 6 GHz has a higher

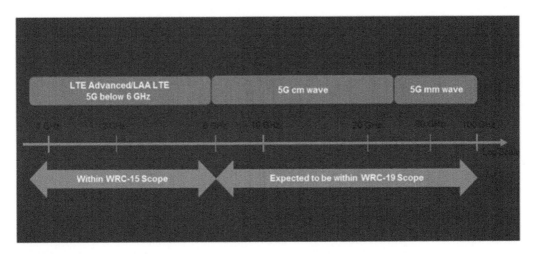

FIGURE 2.1
Frequency bands.

frequency range like 85GHz, and therefore increases complexity and losses. It also requires more repeaters and the length of the antenna too becomes shorter.

2.5 Speed of 5G

According to wireless industries, the speed of 5G is 10 times faster than 4G LTE as expected. It is enough to download a movie in just a few seconds. 5G has high data rates, and is designed to give peak rates up to 20 Gbps. In addition to this, it provides maximum network capacity by expanding the spectrums. It also provides immediate response due to its low latency which also helps to keep the data rate high consistently [5].

2.6 Comparison of 5G Technology with Other Generations

See Table 2.1 for comparison between 3G, 4G and 5G in terms of selected parameters.

2.7 Microstrip Patch Antenna for 5G

Microstrip antenna has a very important role in wireless communications. Its important components are a thin metallic patch made up of copper or gold, dielectric substrate and ground plane. The patch and the ground plane are separated due to the dielectric

TABLE 2.1

Comparison of 5G technology with other generations

Parameter	3G	4G	5G
Origin	2004–2005	2009	2018
Bandwidth	5–20 MHz	5–20 MHz	>1 Gbps
Data rate	Up to 2Mbps	Up to 20 Mbps	Up to 100 Mbps
Speed	384Kbps to 2Mbps	100 Mbps	1–10 Gbps
Multiplexing	CDMA	OFDM	OFDM
Frequency bandwidth	1.8–2.5 GHz	2–8 GHz	24–86 GHz
Latency	100–500ms	30–50ms	1–21ms
Network structure	Wide area cell-based	LTE or eUTRAN	DCs with radio access networks
Switching	Circuit/Packet	Packet	All packet
Access technologies	Wideband CDMA	Multi-carrier-CDMA or OFDM(TDMA)	CDMA and BDMA
Internet service	Broadband	Ultra-broadband	Wireless WWW
IP	Multiple versions	Multiple versions	Multiple versions
Core network	Packet network	LTE	Internet, 5G-NI
Service	Integrated high-quality audio, Video and data	HD quality, IP telephony, gaming, 3D TV	Dynamic information access

substrate. However, microstrip patch antennas have many types, such as rectangular, triangular, circular, elliptical and more. The feed line and radiating patch of microstrip patch antennas are usually photo-etched on the dielectric substrate. Rectangular and circular have many applications in different fields like 5G. The proposed work is to make a dual band microstrip patch antenna with 5.0–5.4 GHz using rectangular shape [6].

The design equation to analyse the length (L) and width (W) of antenna is

$$W = c/2f \times \sqrt{2/(\varepsilon_r+1)}$$

Here, ε_r is dielectric constant of substrate.

In rectangular patch, the radiation is generated with two equivalent slots which have two edges. There are also two opposing edges that are parted by **W** width and whose feedline is at the centre of the radiating edges. Due to that, they do not radiate far enough. Therefore, we can conclude that radiating patch can be displayed by two slots which are separated by a transmission line. And also, each slot is represented by susceptance **X** and conductance **G** which is in the form of a parallel circuit [7].

Antenna fabrication includes selection of frequency on which it is going to operate, thickness of the substrate used and also which substrate is being used, shape of antenna (here it is rectangular), calculation of dimensions, and last but not least simulation in software.

And then the real fabrication starts. With the help of dimensions of patch and the resonant frequency used, we can easily design a microstrip antenna [8].

Rectangular microstrip patch antennas are very advantageous as compared to any other type of antenna. The benefits of using this type of antenna are cost-effectiveness and ease of fabrication and installation. Integration of these antennas is easier than conventional antennas. Figure 2.2 shows the microstrip patch antenna of rectangular shape. Patch(L×W) and ground plane is separated by dielectric substrate. A Feed is connected to the patch.

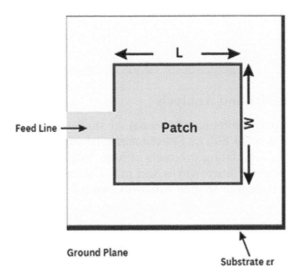

FIGURE 2.2
Rectangular microstrip patch antenna.

TABLE 2.2

Microstrip rectangular shaped patch antenna: parameters and their dimensions

Parameter	Dimension (mm)
W_s	60
L_s	70
L_1	11.7
L_2	33.5
L_3	5
W_1	23
W_2	3
L_{g1}	8
W_{g1}	31
G	2
D	2

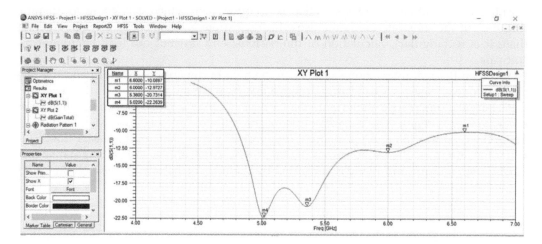

FIGURE 2.3
S-parameter (polar plot).

2.8 Simulation Results and Analysis

The given microstrip patch antenna of rectangular shape is simulated and designed on HFSS. HFSS software is a 3D EM, i.e., electromagnetic simulation software that is used for designing and then simulating products of high frequency such as antenna arrays, antennas, RF components of microwaves and many more. See Table 2.2 for the design aspect of microstrip rectangular shaped patch antenna using 5G application.

2.8.1 Gain

A patch antenna that is single provides a directive gain, i.e. maximum, of 6–9 dB. Gain of an antenna can be defined as the ratio between the maximum intensity of radiation in a given direction and the maximum intensity of radiation from a reference antenna which is in the same direction for 5.0–5.4 GHz. Figure 2.3 shows the gain of the designed antenna.

Wireless Communication and Its Applications 17

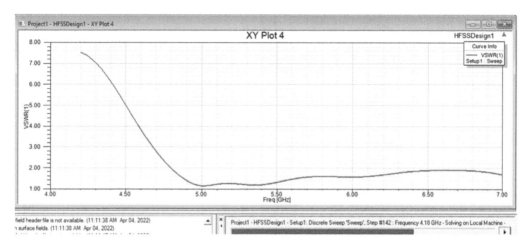

FIGURE 2.4
VSWR of antenna.

The Y-axis here shows the magnitude of max value, i.e. 1. Moreover, the X-axis shows the frequency range in GHz in the s-plane.

2.8.2 Voltage Standing Wave Ratio (VSWR)

The voltage standing wave ratio or VSWR is also known as standing wave ratio. The range for VSWR is 1 to ∞. The value under 2 of VSWR is suitable for antenna applications. That means, the ratio of microstrip patch antennas for all the 5G uses should be within the range of 1⩽VSWR⩽2 [9]. Figure 2.4 shows the range for VSWR is 1 to ∞. The value under 2 of VSWR is suitable for antenna applications. That means, the ratio of microstrip patch antennas for all the 5G uses should within the range of 1<VSWR<2.

2.8.3 Far Field for Antenna

Far field as the name suggests is the area or region which is far from the antenna. It can also be called by a different name, the radiation field, because of the high radiation in this field. In this region the outgoing wavefront is planar and the radiation pattern becomes independent from the distance of the antenna. Figure 2.5 shows the far field pattern of antenna. In this region the outgoing wavefront is planar and the radiation pattern becomes independent from the distance of the antenna.

2.8.4 Radiation Pattern

Figure 2.6 shows a diagrammatic representation of the antenna and its radiation properties, which means that it shows how much energy will be radiated (or received) by the antenna as a function of space. Here we can find the radiation pattern of a dual band microstrip antenna with 5.0–5.4 GHz frequency.

2.8.5 Rectangular Shaped Antenna Design (Simulation Results)

Figures 2.7 show the simulated results of a dual band rectangular microstrip patch antenna on HFSS software. Frequency range is 5.0-5.4 GHz. Figure 2.8 shows the front view of

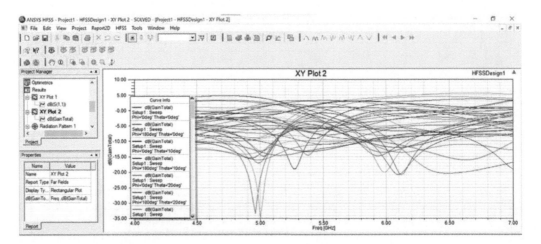

FIGURE 2.5
Far field pattern of antenna.

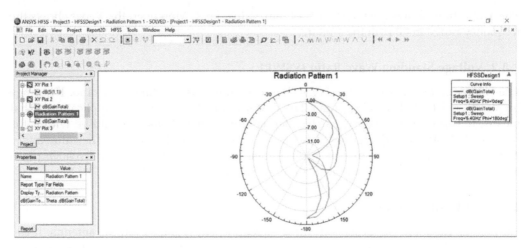

FIGURE 2.6
Radiation pattern of the antenna.

simulated results of dual band rectangular microstrip patch Antenna for 5G on HFSS software. Frequency range is 5.0-5.4 GHz.

2.9 Conclusion and Future Work

This chapter gives an overview of 5G with its applications because its high performance, greater reliability and low latency make it different from the rest. We use antennas for wireless communication that works on radio frequency. They are used to provide good

Wireless Communication and Its Applications 19

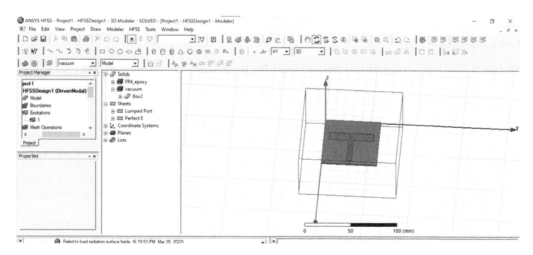

FIGURE 2.7
Simulated diagram of 5.0–5.4 GHz dual band rectangular microstrip patch antenna.

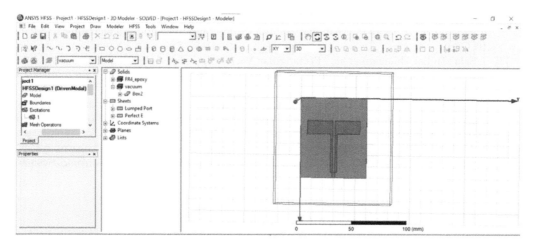

FIGURE 2.8
Front view of the simulated diagram.

quality communication services in urban and suburban areas. We have antennas around us, for better connectivity. Some parameters such as latency, bandwidth, data rate, speed, etc. are the main comparison parameters of 5G, 4G, and 3G as discussed above. Sub-6 frequency bands transmit frequency below 6 GHz and the massive MIMO is responsible for 5G's high speed. A microstrip rectangular patch antenna is used for effective wireless communication. It consists of a rectangular patch, feed line, dielectric substrate and ground plane. Here the dual band plays the vital role in defining the antenna parameters. At last, all the results were made using HFSS software for antenna design with the operating frequency of 5.0–5.4 GHz for the 5G antenna.

Acknowledgments

We thank our teachers, mentors and parents for their support and motivation.

References

[1] Jibendu Sekhar Roy, "Multiple-Antenna Techniques in Wireless Communication – Technical Aspects," Volume 4, Issue 1, 2018, International Journal of Engineering Research & Technology (IJERT).

[2] Khushneet Kour et al., "A Review Paper on 5G Wireless Networks," Volume 4, Issue 32, 2018, International Journal of Engineering Research & Technology (IJERT).

[3] Guanchong Niu et al., "Block Diagonal Hybrid Precoding and Power Allocation for QoS-Aware BDMA Downlink Transmissions," Volume 4, Issue 16, 2020, Sensors.

[4] Emil Björnson et al., "Massive MIMO in Sub-6 GHz and mmWave: Physical, Practical, and Use-Case Differences," Volume 26, Issue 02, 2019, IEEE Wireless Communications.

[5] Mohsen Attaran, "The Impact of 5G on the Evolution of Intelligent Automation and Industry Digitization," 2021, Journal of Ambient Intelligence and Humanized Computing.

[6] John Colaco et al., "Design and Implementation of Microstrip Patch Antenna for 5G applications," 2020, 5th International Conference on Communication and Electronics Systems (ICCES).

[7] J. Salai Thillai Thilagam et al., "Rectangular Microstrip Patch Antenna at ISM Band," 2018, Second International Conference on Computing Methodologies and Communication (ICCMC).

[8] S.Krishna Priya et al., "Design and Research of Rectangular, Circular and Triangular Microstrip Patch Antenna," Volume 8, Issue 12S, October 2019, International Journal of Innovative Technology and Exploring Engineering (IJITEE).

[9] John Colaco et al., "Design and Implementation of Microstrip Patch Antenna for 5G applications," 2020, 5th International Conference on Communication and Electronics Systems (ICCES).

3

Power and Information Transfer Using IoT with NOMA-based GA-LPTS FBMC for Advanced Wireless and Sensor Networks

K. Ayappasam, Ganesan Nagarajan, and Ashish Bagwari

CONTENTS
3.1 Introduction 22
3.2 Problems in the Existing Systems 22
    3.2.1 Peak to Average Power Ratio 22
    3.2.2 Power Spectral Density (PSD) 23
    3.2.3 Signal to Noise Ratio 23
    3.2.4 Spectral Efficiency 23
    3.2.5 Computational Complexity 23
3.3 Related Works 24
3.4 Proposed Scheme 24
    3.4.1 Downlink Phase 24
        3.4.1.1 Information Receiver 26
        3.4.1.2 Energy Receiver 26
    3.4.2 Uplink Phase 27
3.5 Proposed NOMA based Genetic Algorithm in Layered PTS FBMC–OQAM Scheme 27
    3.5.1 Genetic Algorithm in Layered PTS 28
3.6 Simulation Results and Analysis 30
    3.6.1 Comparison of PAPR 30
    3.6.2 Comparison of Spectral Efficiency 30
    3.6.3 Computational Complexity 30
    3.6.4 Golden Section Search Method (Convergence Curve) 33
    3.6.5 Performance of the Optimized Resource Allocation Scheme 33
3.7 Conclusion and Future Work 33
Acknowledgements 36
References 36

DOI: 10.1201/9781003326205-4

3.1 Introduction

Wireless networks pave the way for cutting-edge geographies like smart cities, and houses and automobiles with increased security. Furthermore, cellular mobile data solicitations such as high definition video (HDV), Internet of Things (IoT), machine to machine (M2M) communication, and social networking, need rapid broadband access. The goal of multicarrier modulation (MCM) is to divide the bandwidth of the signals into corresponding subcarriers of the entire band of signals. The innovative fifth generation (5G) MCM, such as FBMC, have a significant role. The major purposes of these waveforms are to enhance spectral efficiency and versatility through filtering. There are two types of multi access systems. They are orthogonal and non-orthogonal multiple access (OMA and NOMA) schemes. Several access solutions in telecommunication networks enable several users to stake restricted network assets. The novel 5G based NOMA system is implemented with IoT to access a large number of users. Unlike traditional orthogonal multicarrier techniques, NOMA uses non-orthogonal spread on the transmission side which intentionally produces both inter-cell and intra-cell interferences. Concurrent power and information transfer (CPIT-IoT) structure is developed using NOMA with genetic algorithm based layered partial transmit sequence (GA-LPTS) FBMC system [1–2] to solve the drawbacks of OFDM (orthogonal frequency division multiplexing) and also the energy-constrained lifespan of Internet of Things (IoT). This method proposes that user nodes (UNs) be capable of both harvesting of energy and decoding of information using different antennas [3]. To cancel the inherent interference, an iterative interference cancellation method is used at both base stations and UNs.

3.2 Problems in the Existing Systems

To preserve orthogonally across distinct sub-channels, the grant-based synchronization cyclic-prefix (CP) is required in the case of OFDM; also it suffers due to deprived OOB emission, which causes unadorned interference in asynchronous settings. Asynchronous transmissions have proven that OFDM systems perform poorly. In addition, low spectrum efficiency, high peak to average power ratio, high computational complexity and deprived SNR are important challenges in 4G-based orthogonal multiple access systems.

3.2.1 Peak to Average Power Ratio

When the incoming signals are added coherently at the transmitter, the peak power abruptly improves to N times the average values of all N subcarriers, disrupting the normal operation of the Q point of the power amplifier's final stage. As a result, the PAPR has been discovered. The PAPR is defined as the ratio of the peak and average power characteristics of all N subcarriers. A lesser value of PAPR offers better performance. The real and imaginary input symbols with m_1^{th} symbol on the k_1^{th} subcarrier are given as

$$x_k(m) = a_k(m) + j\,(b_k(m)) \qquad (1)$$

where, k varies from 0 to K-1 and m varies from 0 to M_1-1. Mathematically, the PAPR for s(t) in the rth interval is defined as

$$PAPR_r = 10log_{10} \frac{max_{rT \le t \le (P+1)T} |s(t)|^2}{P_{avg}}, r = 0,1,2...M1+\alpha-1 \qquad (2)$$

In this estimation, signals are separated to the summation of M and α intervals with T lengths and center powers P.

3.2.2 Power Spectral Density (PSD)

The PSD of the transmitted FBMC-OQAM signals can be calculated as follows:

$$s(t) = \sum_{k=0}^{k=1} s_b(t) \qquad (3)$$

where, $s_b(t)$ is base band modulated signal. The unit of power spectral efficiency is dB/Hz.

3.2.3 Signal to Noise Ratio

It is the greatest signal power divided by the noise power. SNR is measured in decibels (dB). It is denoted mathematically as

$$SNR = \frac{P_{Signal}}{P_{Noise}} \qquad (4)$$

P_{Signal} – Signal Power

P_{Noise} – Noise Power

3.2.4 Spectral Efficiency

It is defined as the rate at which information can be transmitted over a given bandwidth. It is generally assessed in bits per second per Hz. It is measured in bits/second/Hz/site for cellular geographic locations. The following equation may be used to compute the spectral efficiency of the FBMC-OQAM system.

$$\eta = \alpha \zeta \left(1 - P_{\sqrt{M}}\right) v_T v_F \qquad (5)$$

where, α denotes the rate of codings which is done at channel, ζ denotes amount of bits/subcarrier = $log_2 M$ and the time–frequency efficiency is denoted as $v_T v_F$.

3.2.5 Computational Complexity

It is limited by the quantity of addition and multiplication operations. The computational complexity reduction ratio is abbreviated as CCRR, which is distinct as

$$\text{CCRR} = \left(1 - \frac{Complexity\ of\ Proposed\ method}{Complexity\ of\ Existing\ method}\right) \times 100\% \qquad (6)$$

3.3 Related Works

A lot of research has been done in recent years to estimate the concert of CPIT in IoT based networks [4–6]. The majority of physical layer research has focused on combining CPIT with methods such as cooperative communications, full-duplex and multicarrier waveforms. Mahama et al.[7] examined two optimized challenges added to the goal of determining the ideal ratio of PS to advance transmission power the source. References [8–10] explore PIT for sensor based IoT networks which has several transmission and reception devices in order to manage alternation of energy and information through the PS at the reception nodule. It has also been discovered to decrease overall transmission power through optimized information topologies and beam forming methods [11–12]. The recital of CPIT-related structures of multicarrier signals is studied in [13–15]. Reference [16] investigates an OFDM with PIT-related structure with double MA and EH methods. As a first, the time division multiple access (TDMA) transmission scheme with a TS-created reception is described. Secondly, the orthogonal frequency division multiple access (OFDMA) transmitter added PS-based reception is described. References [17–19] suggest an OFDM-based SWIPT structure and the separated as two subcarrier clusters, for ID and EH. Furthermore, to optimize the gathered energy, a combined subcarrier and power allocation issue is constructed [20–21]. Reference [22] presents a similar model in which the aim is to exploit the customers' rate of sum by power allocation and channel. The receiving signals are divided into two clusters such as EH and ID, and the receiver does not require PS or TS [23]. The vast majority of previous work on PIT multicarrier waveforms is based on OFDM [24–25].

3.4 Proposed Scheme

The proposal for an FBMC with PIT based IoT with genetic algorithm based layered PTS system is explored in depth. The suggested system, depicted in Figure 3.1, is composed of a one antenna base station and K user networks. But each UN is a double antenna system, one for receiving and transmitting information and the other for receiving and transmitting energy. The UNs will transmit and receive data via TDMA.

3.4.1 Downlink Phase

During the downlink, the user networks conduct ID and EH as shown in Figure 3.1. The data on even-numbered subcarriers is proven to produce minor interferences to other even based subcarriers. These are due to the fact that the FBMC scheme uses prototype filters which are strongly restricted in the frequency field, limiting out of band (OOB) leakage to nearby subcarriers. As a result, OQAM data is lonely placed on even based subcarriers during downlink to lower the degree of inherent interference at the UN. The condensed interferences at the user network can minimize the complexity and the

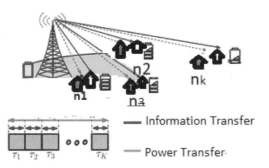

(a). First Phase of FBMC-OQAM with CPIT System

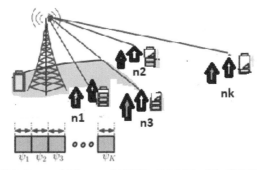

(b). Second Phase of FBMC- OQAM with CPIT System

FIGURE 3.1
FBMC-OQAM system model with CPIT system.

ingesting of the information receiver. The values of n-th transmission symbols vectors from the BS to the kth user network at the time of its designated time slots k are expressed in this configuration as

$$X_{n,k}^{dl} = G_{n,k} \Phi a_{n,k}^{dl} \tag{7}$$

where, $x^{dl}_{n,k} = [x^k_{0,n}, x^k_{1,n}, ..., xk\ L-1,n]$ are the vectors of the signal which is transmitted and the term L = O × M with O is the overlapping factor. The symbol Φ denotes the M × M inverse fast Fourier transform (IFFT) and the row of ith and column of jth are denoted as

$$\sqrt{\frac{1}{M}} \exp(\sqrt{-1}\frac{2\pi ij}{M}) \tag{8}$$

where, M is the subcarriers assigned and $a_{n,k}^{dl}$ denotes transmitted data, as

$$a_{n,k}^{dl} = \left\{ \begin{cases} a_{n,m}^{e,dl} & for\ m \in \{0,2...M-2\} \\ 0\ for\ m \in \{1,2,...M-1\} \end{cases} \right\} \tag{9}$$

$G_{n,k}$ denotes the L×M filter matrix and the l-th diagonal is assumed as

$$g_{n,k}[l] = g[l - nM] \tag{10}$$

where, $g[i]$ is the coefficient of filter.

3.4.1.1 Information Receiver

The signals received are filtered and down sampled. Then, the signals which are received are transmitted through the FFT block in the information receiver. The filtered signal is denoted by

$$y_{n,k}^{ID,dl} = G_{n,k}^H H_k^{dl} x_{n,k}^{dl} + G_{n,k}^H z_k^{dl}$$

$$= G_{n,k}^H H_k^{dl} x_{n,k}^{dl} + \Phi_{n,k}^{dl} + G_{n,k}^H z_k^{dl}$$

$$= \tilde{H}_k^{dl} a_{n,k}^{dl} + G_{n,k}^H z_k^{dl} \tag{11}$$

where, $\tilde{H}_k^{dl} = G_{n,k}^H H_k^{dl} \Phi_{nk}^{dl} z_k^{dl}$ is denoted as L×1 additive white Gaussian noise (AWGN) vectors at UN_k and H_k^{dl} is denoted as the L×L matrix of multi-path fading channel between the base station and the user networks. The values of lth diagonal of H_k^{dl} characterizes the complex channel gain of real and imaginary values of the lth sub-channel of kth user network and can be modeled as

$$h_{k,l} = \sqrt{A}\left(\frac{\tilde{d}}{d_k}\right)\sqrt{g_t g_r}.exp(j2\pi\frac{d_k}{\lambda} \tag{12}$$

where, d_k denotes the remoteness between the base station and k-d^{th} user network which is the orientation remoteness for path loss; the parameter A is the attenuation, k denotes the path loss exponent, g_t denotes gain of the base station antenna and g_r denotes gain of user network antenna. Then, the λ signifies RF wavelength with center frequency f_c. The frequency field signal after down sampling and after fast Fourier transform is denoted as

$$y_{n,k}^{dl} = a_{n,k}^{dl} \Phi_k^H y_{n,k}^{IDdl}$$

$$= \Phi_k^H H_k^{dl} a_{n,k}^{dl} + \Phi_{n,k}^H + G_{n,k}^H z_k^{dl}$$

$$= \tilde{H}_k^{dl} a_{n,k}^{dl} + \tilde{z}_k^{dl} \tag{13}$$

$\tilde{H}_k^{dl} = \Phi_{n,k}^H H_k^{dl} a_{n,k}^{dl}$ is the M×M which is the channel matrix of the down sampled signals and $\tilde{z}_k^{dl} = \Phi_{n,k}^H + G_{n,k}^H z_k^{dl}$ is the noise phase vector.

3.4.1.2 Energy Receiver

As previously stated, each user network has a distinct antenna for energy harvesting, implying all user networks may harvest energies from all subcarriers in each and every down link duration. The influence of noise is considered to be minimal at the EH receiver. As a result, the signal which is received at the user network of k is denoted as

$$y_{n,k}^{EH,dl} = H_k^{dl} x_{n,k}^{dl}$$

$$= H_k^{dl} G_{n,k}^{dl} \Phi a_{n,k}^{dl}$$

$$= \tilde{H}_k^{dl} a_{n,k}^{dl} \tag{14}$$

where, $H_k^{dl} = H_k^{dl} G_{n,k}^{dl}$ denotes the linearity based energy harvesting method and the overall harvested power throughout k user networks is described as

$$P_{n,k}^{lin} = \frac{\alpha \beta_k}{(1-\alpha)\Psi_k} \sum_{j=1}^{k} \sum_{n=1}^{N} \sum_{m=1}^{M} \left|\bar{h}_{m,n,j}^{dl}\right|^2 P_{m,n,j}^{dl}$$

$$= \frac{\alpha \beta_k}{((1-\alpha)\Psi_k} \gamma_{m,n,j}^{dl} \tag{15}$$

Here, β_k signifies the efficiency of converted energy in the linear region based energy harvesting circuit, ψ_k denotes uplink transmitter time of the kth user network, and $h_{m,n,j}^{dl}$ represents the diagonal entries of $= \tilde{H}_k^{dl}$.

3.4.2 Uplink Phase

The many modes of access and the link between the base station and user nodes are established by the uplink process. Because the base stations are compatible with large treating capacity, the modulated sequences are delivered in the uplink during even and odd subcarriers. The nth transmitted symbols based vectors of the kth user network are assumed during time k slot as

$$x_{n,k}^{ul} = G_{n,k}^{ul} = \Phi a_{n,k}^{dl} \tag{16}$$

The received signal at the BS, like the DL phase, is separated by filter, processed using downsampling, and transformed using fast Fourier transform (FFT) into frequency domain signals.

3.5 Proposed NOMA based Genetic Algorithm in Layered PTS FBMC–OQAM Scheme

NOMA is being used in a relatively new area, specifically the power domain. In terms of channel circumstances, it allocates many users to varied power coefficients. Following traditional channel coding and modulation, numerous signals created at the transmitter by many users are immediately overlaid on one another. The same time frequency properties are shared by multiple and different users and recognized in the receivers using SIC based algorithms. As a consequence, as compared to standard OMA, spectral efficiency increases when complexity of the receiver is higher. It is noted that the outdated orthogonal multiple scheme is outperformed by non-orthogonal multiplex schemes. NOMA is being used in a relatively new sector, namely power distribution; specifically, the signal received at the base station during NOMA uplink transmission. It delegates authority to a number of users.

$$Y = \sum_{a=1}^{A} h_a \sqrt{P_a} x_a + n \tag{17}$$

where, p_a is transmit power and x_a is the transmit symbols of ath user and n is additive white Gaussian noise with variance 2 and A is the amount of shared users in the same reserves lab. The transmit power is customized for every user to encourage successive interference cancellations (SIC) at the receiver, ensuring that a user with greater power is recognized accurately. When the receiver invokes SIC, the user with the highest CSI is decoded first. The received signal is then subtracted from the relevant signal element. The signal is processed by the SIC receiver in reducing mandate of strength. It is notable that power level transmissions of different NOMA users vary since they are exposed to diverse channel circumstances. If all of the initial identified symbols are correct, the SINR of NOMA users are provided as

$$SINR_a = \frac{P_a |h_a|^2}{\sum_{b=a+1}^{A} P_b |h_b|^2 \sigma^2} \tag{18}$$

Figure 3.2 displays the NOMA downlink transmission of a double user situation, with various powers distinguishing the users, allocating the identical resources with a total power restriction. The base station provides an overlaid signal consisting of two signals for each user.

The signal of the user with the greater transmit power and lower downlink channel gain is decoded first, while the signal of the other user is considered as noise. It should also be noted that the first identified user has the greatest inter user interference (IUI) and the detection mistake in the first user is passed on to the subsequent users.

3.5.1 Genetic Algorithm in Layered PTS

Figure 3.3 depicts the GA-LPTS scheme block diagram, which is based on the GA-based layered PTS scheme. This approach makes use of OQAM modulated FBMC signal's overlapping structure which is adequate for PAPR reduction in the FBMC/OQAM signal with four up-sampling. The scheme's purpose is to use the prototype filter to minimize the search complexity of the standard PTS technique while not sacrificing system performance. As seen in Figure 3.3, the data block S_m containing M overlapping FBMC/OQAM data is partitioned into V sub-blocks using the neighboring partition technique, that is,

$$s_{m1} = \sum_{V=1}^{V} S_m^v \tag{19}$$

When the candidate vector of phase factors $b_m = [b_{1m}, \bullet \bullet \bullet, b\, v_m, \bullet \bullet \bullet, b_{Vm}]$ is combined with the appropriate V sub blocks in layer 2, the PAPR reduction problem of the FBMC/OQAM signal is stated as $P2$.

Using IoT with NOMA-based GA-LPTS FBMC

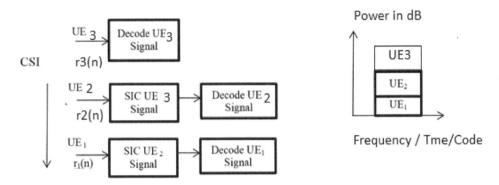

FIGURE 3.2
Downlink power domain NOMA with FBMC.

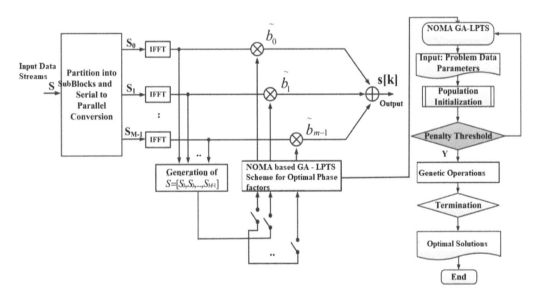

FIGURE 3.3
Proposed NOMA based GA-LPTS scheme.

$$P2: \min_{b_m}\{\max_{k}\left|\overline{[s_m][k]}\right|^2\} \tag{20}$$

$$\tilde{s}_m[k] = \sum_{V=1}^{V} \tilde{b}_m^v s_m^v[k] \tag{21}$$

$$b_m^v = \left\{ e^{\frac{j2\pi l}{w}} \right\} where, \quad l = 0,1,2\ldots W-1; 1 \leq v \leq V; 0 \leq n \leq N-1 \tag{22}$$

3.6 Simulation Results and Analysis

The arithmetical outcomes are offered for a projected NOMA based OQAM modulated FBMC system. The channel bandwidth is assigned as 2.4 GHz and 4-OQAM measured in the simulation. The spectral density of noise is assumed as $N_0 = -175\text{dBm/Hz}$. The user networks are arbitrarily distributed around a 15m radius of the base stations. The Friis channel model is anticipated and the path loss with k = 3, A = 1 and $d\tilde{} = \lambda/4\pi$. Simulation tool is considered as MATLAB R 2018a. Subcarriers assigned are 512 and 1024. Sub carrier index is assigned as 0.5. Phase factors assigned are 1 and -1. Overlapping factor (k) = 4. The numbers of filter co-efficients assigned are 11 and sub-blocks assigned are 4 and 8. Subcarrier frequency for N = 512 is 4.687 MHz. Sampling frequency for N = 512 is assigned as 10 MHz. Subcarrier frequency for N = 1024 is 2.343MHz and sampling frequency for N = 1024 is assigned as 5MHz. Carrier spacing of pilots in the frequency domain is assigned as 4 and raised cosine windowing type is used.

3.6.1 Comparison of PAPR

The proposed power and information transfer using IOT with NOMA based FBMC/OQAM method achieves better reduction of PAPR.

The proposed scheme offers 1.8 dB and 2.5 dB at the complementary cumulative distribution function (CCDF) of 10^{-4} when the phase rotation vectors (U) are assigned as $U = 8$ and $U = 16$ respectively while assigning the subcarrier as N = 512 which is shown in Figure 3.4. Also, Figure 3.5 shows the proposed simulated results of PAPR when the subcarrier N = 1024. The result obtained is a minimum value of PAPR of 2 dB and 2.8 dB when the phase rotation vectors are assigned as $U = 8$ and $U = 16$ respectively at the CCDF of 10^{-4}.

3.6.2 Comparison of Spectral Efficiency

Figures 3.6 and 3.7 show the comparison of spectral efficiency of the system when subcarrier N = 512 and 1024. The proposed scheme offers 42 bits/sec/Hz and 38bits/sec/Hz of high spectral efficiency for phase rotation vectors $U = 8$ and 16 respectively when the subcarrier N = 512, which are compared to the existing systems that show 27 bits/sec/Hz and 21 bits/sec/Hz in [1] and [2] respectively. Also the proposed scheme offers 32 bits/sec/Hz and 28 bits/sec/Hz of high spectral efficiency for phase rotation vectors $U = 8$ and 16 respectively when the subcarrier N = 512. These results are found to be comparatively higher than the existing systems [1] and [2] as shown in Table 3.1.

3.6.3 Computational Complexity

Comparison of relative computational complexity of the proposed scheme for the subcarriers N = 512 and N = 1024 is shown in Figures 3.8 and 3.9 respectively. At first, the results are simulated when the subcarrier N is assigned as 512 and phase rotation vectors U are assigned as 8 and 16. As a result, the proposed system proves to have lower relative complexity of 0.01032 and 0.0209 when N = 512 and 1024 respectively by assigning $U = 8$.

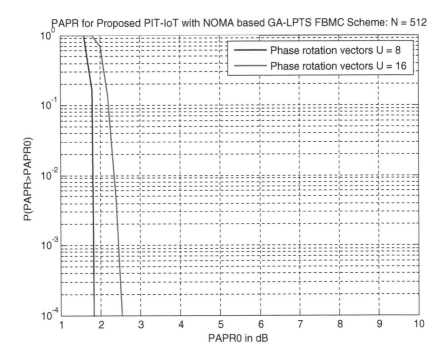

FIGURE 3.4
PAPR for Proposed Scheme when N = 512.

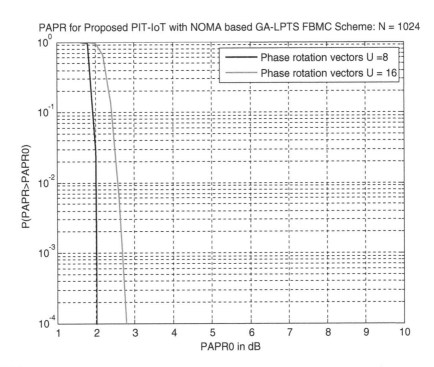

FIGURE 3.5
PAPR for Proposed Scheme when N = 1024.

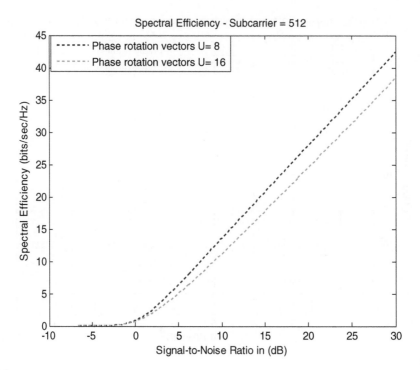

FIGURE 3.6
Spectral efficiency when N= 512.

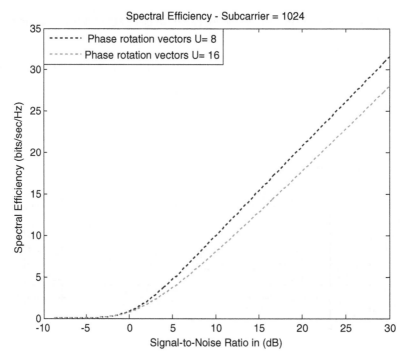

FIGURE 3.7
Spectral efficiency when N = 1024.

TABLE 3.1

Results comparison

Parameters	Conventional Improved DFT spread with GA-LPTS Scheme [2]	Existing Diversity DNA Encryption based LPTS Scheme [1]	Proposed CPIT-IoT with NOMA based GA-LPTS FBMC Scheme
PAPR for 10^{-4} CCDF; N = 512	2.6 dB	2 dB	1.8 dB
PAPR for 10^{-4} CCDF; N = 1024	2.9 dB	2.2 dB	2 dB
Spectral Efficiency N = 512 (at 30 dB SNR)	21 bits/sec/Hz	27 bits/sec/Hz	42 bits/sec/Hz
Spectral Efficiency N = 1024 (at 30 dB SNR)	19 bits/sec/Hz	24 bits/sec/Hz	32 bits/sec/Hz
Computational Complexity N = 512	0.0413	0.0293	0.01032
Computational Complexity N = 1024	0.0461	0.035	0.0209

3.6.4 Golden Section Search Method (Convergence Curve)

The x and z functions in Figure 3.10 show downlink and uplink infrastructures which are equal and have identical optimum values. It also shows base stations only describing x values and determined by the golden section search method. From Figure 3.10, it may be detected that the method of line searching meets after ten repetitions. This suggests that defining the rate of x and the time and weight distributions is not computationally demanding.

3.6.5 Performance of the Optimized Resource Allocation Scheme

The golden section search optimization method is established. This includes listings of schemes such as OOOA, FEOA and FOOA, and EEOA. The various systems are labeled with "O" are characterized by optimum allocation, "F" by fixed allocation, and "E" by equal allocation. In addition, the leading three letters of the structure names reflect the kind of distribution for a certain source. The first letter indicates time allocation, the second letter signifies power allocation, the third letter signifies the type of weight allocation, and the final letter implies the type of allocation, which are "time," "power," "weight," "allocation." For instance, OOOA refers to optimum time, optimal power and optimal weight allocations. Figure 3.11 depicts a scheme of the subjective sum rate for the base station transmitter power. The declining command of routine follows from the plot: OOOA, FOOA, EEOA and FEOA as shown in the graph. The systems with optimal resource allocations fared well, being superior to systems with motionless reserve distributions. It is also shown that the best plan (OOOA) achieves the required level of performance.

3.7 Conclusion and Future Work

Table 3.1 shows the comparison of offered results in terms of the performance parameters. The proposed results are better compared with existing systems such as improved DFT spreading with genetic algorithm based GA-LPTS (layered partial transmit sequence) scheme [1] and diversity DNA encryption based GA-LPTS schemes[2]. From the results

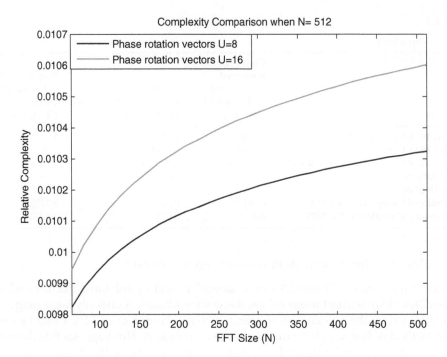

FIGURE 3.8
Computational complexity when N = 512.

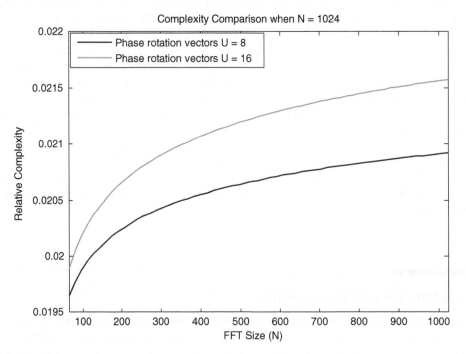

FIGURE 3.9
Computational complexity when N = 1024.

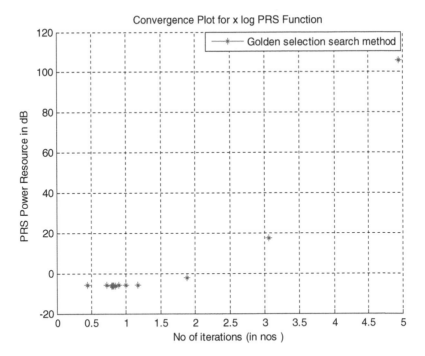

FIGURE 3.10
PRS Function Convergence.

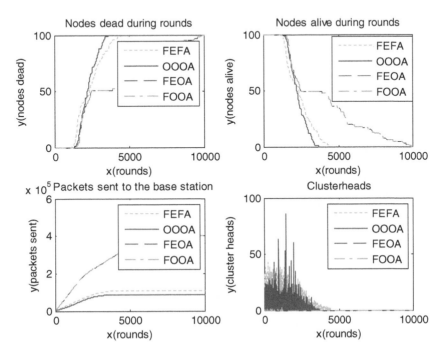

FIGURE 3.11
Performance of Optimized Resource Allocation Schemes.

comparison, the proposed power information transfer–IoT with NOMA based GA-LPTS FBMC scheme offers reduced PAPR of 1.6 dB and 2 dB at the CCDF of 10^{-4} for N = 512 and 1024 respectively. It also provides 42bits/sec/Hz and 32 bits/sec/Hz of high spectral efficiency when N = 512 and 1024 respectively. It also offers 0.01032and 0.0209 of reduced relative computational complexity for N = 512 and 1024 subcarriers respectively . Thus, it is proved that the proposed method offers better performance compared to existing methods.

Security techniques may be developed for FBMC-OQAM systems to increase authenticity, privacy, and protection against intruders. Bio-inspired optimization approaches such as particle swarm optimization, simulated annealing, genetic algorithm, ant colony, wolf optimization algorithm, and others can be used with the presented schemes to achieve superior optimization. Advanced DFT spreading techniques such as complete duplex DFT spreading, broadly linear nyquist criterion DFT spreading, continuous phase modulated DFT spreading, and tone reservation method for DFT spreading may also be employed to increase the overall performance of the FBMC-OQAM system. Hybrid Opto-Electronic FBMC systems are still in the early stages of development. It is planned to operate at greater speeds while deploying them.

Acknowledgements

We thank our parents for their support and motivation.

References

[1] K. Ayappasamy, G. Nagarajan, and R. Senthil Kumaran, "Diversity DNA encryption in layered PTS with FBMC OQAM systems," *Grenze International Journal of Engineering and Technology (GIJET)*, vol. 8, no. 2, pp. 233–239, Jun. 2022.

[2] K. Ayappasamy, G. Nagarajan, and P. Arunagiri, "An improved DFT spread with genetic algorithm in layered PTS for reduced PAPR in FBMC OQAM systems," *IET Digital Library*, Jun. 2022.

[3] T. Kebede and Y. Wondie, "Multi-carrier waveforms and multiple access strategies in wireless networks: performance, applications, and challenges," *IEEE Access*, vol. 10, pp. 21120–21140, Mar. 2022.

[4] L. V. Siying, J. Zhao, and L. Yang, "Genetic algorithm based bilayer PTS scheme for peak-to-average power ratio reduction of FBMC/OQAM signal," *IEEE Access – Special Section on Artificial Intelligence for Physical-Layer Wireless Communications*, vol.8, no. 2, pp. 17945–17955, Jan. 2020.

[5] C. Sexton, Q. Bodinier, and L. A. DaSilva, "Enabling asynchronous machine-type D2D communication using multiple waveforms in 5G," *IEEE Internet of Things*, vol. 5, no. 2, pp. 1307–1322, Apr. 2018.

[6] B. Lim and Y.-C. Ko, "SIR analysis of OFDM and GFDM waveforms with timing offset, CFO, and phase noise," *IEEE Transactions on wireless Communication*, vol. 16, no. 10, pp. 6979–6990, Oct. 2017.

[7] S. Mahama, Y. J. Harbi, A. G. Burr, and D. Grace, "Design and convergence analysis of an IIC-Based BICM-ID receiver for FBMCQAM systems," *IEEE Open Journal of the Communications Society*, vol. 1, pp. 563–577, 2020.

[8] G. Wunder et al., "5GNOW: Non-orthogonal, asynchronous waveforms for future mobile applications," *IEEE Communications Magazine*, vol. 52, no. 2, pp. 97–105, Feb. 2014.

[9] K. Ayappasamy, G. Nagarajan, and P. Elavarasan, "DFT spread C-DSLM for low PAPR FBMC-OQAM systems," in E.S. Gopi (ed.), Machine Learning, Deep Learning and Computational Intelligence for Wireless Communication, pp. 281–302, 2021.

[10] K. Ayappasamy, G. Nagarajan, and P. Elavarasan, "FBMC-OQAM-PTS with Virtual Symbols and DFT Spreading Techniques," *Journal of Advances and Applications in Mathematical Sciences*, vol. 18, no. 9, pp. 817–825, Jul. 2019.

[11] R. Zakaria and D. Le Ruyet, "A novel filter-bank multicarrier scheme to mitigate the intrinsic interference: Application to MIMO systems," *IEEE Transactions on Wireless Communications*, vol. 11, no. 3, pp. 1112–1123, Mar. 2012.

[12] R. Zakaria and D. L. Ruyet, "Intrinsic interference reduction in a filter bank-based multicarrier using QAM modulation," *Phys. Commun.*, vol. 11, pp. 15–24, Jun. 2014.

[13] H. Nam, M. Choi, S. Han, C. Kim, S. Choi, and D. Hong, "A new filter-bank multicarrier system with two prototype filters for QAM symbols transmission and reception," *IEEE Transactions on Wireless Communications*, vol. 15, no. 9, pp. 5998–6009, Sep. 2016.

[14] S. M. A. Oteafy and H. S. Hassanein, "Resilient IoT architectures over dynamic sensor networks with adaptive components," *IEEE Internet of Things Journal*, vol. 4, no. 2, pp. 474–483, Apr. 2017.

[15] S. Mahama, D. K. P. Asiedu, and K. -J. Lee, "Simultaneous wireless information and power transfer for cooperative relay networks with battery," *IEEE Access*, vol. 5, pp. 13171–13178, 2017.

[16] D. K. P. Asiedu, H. Lee, and K. -J. Lee, "Simultaneous wireless information and power transfer for decode-and-forward multihop relay systems in energy-constrained IoT networks," *IEEE Internet of Things Journal*, vol. 6, no. 6, pp. 9413–9426, Dec. 2019.

[17] B. Clerckx, R. Zhang, R. Schober, D. W. K. Ng, D.I. Kim, and H. V. Poor, "Fundamentals of wireless information and power transfer: From RF energy harvester models to signal and system designs," *IEEE J. Sel. Areas Communication*, vol. 37, no. 1, pp. 4–33, Jan. 2019.

[18] F. Mukhlif, K. A. B. Noordin, A. M. Mansoor, and Z. M. Kasirun, "Green transmission for C-RAN based on SWIPT in 5G: A review," *Wireless Networks*, vol. 25, pp. 2621–2649, May 2018.

[19] A. Prathima, D. S. Gurjar, H. H. Nguyen, and A. Bhardwaj, "Performance analysis and optimization of bidirectional overlay cognitive radio networks with hybrid-SWIPT," *IEEE Transactions in Vehicular Technology*, vol. 69, no. 11, pp. 13467–13481, Nov. 2020.

[20] M. R. A. Khandaker, C. Masouros, K. -K. Wong, and S. Timotheou, "Secure SWIPT by exploiting constructive interference and artificial noise," *IEEE Transactions on wireless Communications*, vol. 67, no. 2, pp. 1326–1340, Feb. 2019.

[21] P. Elavarasan and G. Nagarajan, "Optimal phase selection factor for PTS using GPW and RPW in OFDM Systems," *Journal on Computer Science*, vol. 6, pp. 140–147, Apr. 2012.

[22] S. H. Chae, C. Jeong, and S. H. Lim, "Simultaneous wireless information and power transfer for Internet of Things sensor networks," *IEEE Internet of Things Journal*, vol. 5, no. 4, pp. 2829–2843, Aug. 2018.

[23] P. Elavarasan and G. Nagarajan, "A summarization on PAPR techniques for OFDM Systems," *Springer International Journal of the Institution of Engineers* (India): Series B, Volume 96, pp. 381–389, Mar. 2014 .

[24] P. Elavarasan and G. Nagarajan, "Peak-power reduction using improved partial transmit sequence in orthogonal frequency division multiplexing systems," *Elsevier International Journal of Computer and Electrical Engineering*, pp.80–90, 2015.

[25] X. Zhou, R. Zhang, and C. K. Ho, "Wireless information and power transfer in multiuser OFDM systems," *IEEE Transactions on Wireless Communications*, vol. 13, no. 4, pp. 2282–2294, Apr. 2014.

4

5G-NR Wideband MIMO Antenna Design Using Stepped Radiators for Wireless Communication

Mantar Singh Mandloi, Ajay Parmar, Karan Gehlod, Ashish Shakya, Priyanshi Malviya, and Leeladhar Malviya

CONTENTS

4.1 Introduction 39
4.2 MIMO Antenna Design and Results 41
4.3 Conclusion 50
4.4 Future Scope 51
References 53

4.1 Introduction

Nowadays, use of wireless communication is increasing for social as well as for economic activity, and the introduction of 5G new radio (NR) will accelerate it with more applications. The 5G-NR is a new type of radio interface and radio access network which meets the exponentially increasing needs of wireless communication. 5G-NR is the key enabling technology for wireless communication to provide a high data rate as compared to 4G technology. 5G-NR bands are classified as low band, mid band and high band. The frequency range below 1 GHz is considered low band, which are used for massive IoT and mobile broadband. The Sub 6-GHz like 3.4–3.8 GHz, 3.8–4.2 GHz, 4.4–4.9 GHz is considered mid-band, and frequency range above 24 GHz, i.e., mm-Wave, is considered high band 5G NR, which covers the frequency bands of 24.25–27.5 GHz, 27.5–29.5 GHz, 37–40 GHz, and 64–71 GHz [1–7]. The demand for larger bandwidths for the Internet of Things (IoT) network shifted interest from lower frequency UHF (0.3–3 GHz) toward higher frequencies and mm Waves (30–300GHz) [8–9]. Broadband communication has a limited power spectral density at low power and a high data rate ultra-wideband (UWB) signal. Higher data rates are possible by combining UWB with MIMO without increasing the input power from multiple channels, while close spacing affects the compactness [10–13]. UWB and MIMO have better quality for short-range high-speed communications, enhanced capacity, and reliability using multiplexing and diversity gains[14–15].

Varieties of wideband 4G and 5G antenna designs are available in the literature for various frequency bands. A printed slotted patches UWB antenna was designed with series and parallel resistors to control current distribution to widen the impedance band and to improve the efficiency for 0.1–3.0 GHz application[16]. A double layer metamaterial unit had a rectangular patch and rectangular slot introduced on both sides. It is a highly selective dual polarization filter antenna, which covers 3.25–3.85 GHz frequency with a size of 57.2 × 57.2 mm², having maximum gain of 9.0 dBi [17]. For 4G LTE and 5G N78, a 1.66 to 3.95 GHz wideband base station antenna was designed [18]. A dual polarized square-loop with parasitic loops for 3.3–3.6 GHz (5G), and 4.8–5.0 GHz base stations application were also designed [19]. A circularly polarized antenna was designed using cross-slot pairs for 1.95–3.45 GHz bandwidth [20].

To get stable gain pattern using rectangular cells, a multilayer antenna was designed. A bow tie slot was etched in patch in the path of the radiating element to enhance bandwidth. The dimension of Arlon 25N substrate was 40.0 × 40.0 mm² for the 4.98–6.31 GHz frequency band and 7.5 dBi gain was achieved [21]. To make a wideband and isolated antenna, two different types of patches were combined in a design orthogonally. A parasitic element of rod-shape was inserted with some modification at the ends for increasing isolation. The parasitic element diverts the ground current to enhance isolation. The size of the antenna was 40.0 × 42.0 mm² for 7.0 GHz frequency, and the efficiency was 94 percent [22].

A modified rectangular patch with defected ground structure (DGS) was designed to provide wide band performance. The bandwidth of 4.46 GHz was achieved with gain of 7.5 dBi, radiation efficiency of 99 percent, and dimensions of 38.9 × 40.0 mm². The design covered the WLAN, WiMAX, and X-band applications [23]. An inverted π-shaped patch with inverted L-shaped slot was designed to accommodate 4G and 5G applications. The design covered LTE 42/43/46 bands [24]. A 4 elements compact MIMO antenna was designed for 7.2–7.8 GHz X-band satellite applications [25]. For the 5G millimeter wave applications, an end fire dual-polarized phased antenna array was designed to cover 24.5–27.5 GHz [26]. An antenna was designed for gain enhancement in high attenuation cases but the bandwidth becomes low. A metamaterial-inspired antenna was designed for four ports for mm-wave application for 26.0–31.0 GHz frequency band. The designed antenna has dimensions of 31.0 × 48.0 mm², maximum gain of 10.0 dBi, and ECC of 15x10⁻⁴ in the whole band [27]. A hexagonal shaped four element array antenna was designed for 28.0 GHz 5G frequency band on Rogers RT/ Duroid 5880 substrate. The increase in electrical length can be performed by modifying the ground plane, which in turn changes the resonant frequency. The dimension of the array was 45.0 × 20.0 mm² with peak gain of 12.5 dBi, and radiation efficiency of more than 85 percent in band [28]. A dually polarized array with MIMO configuration consisting of eight elements was designed to get directional radiation patterns. The gain in 27.4–28.23 GHz frequency band was 6.6 dBi and the dimension of the Rogers substrate was 150.0 × 100.0 mm² [29].

An antenna featuring different rectangular branches from patch is attached to a finite ground plane for wideband characteristic and covered 23.92–43.8 GHz frequency band. The size of antenna element was 10.0 × 12.0 mm² on plexiglass substrate. The maximum gain of the antenna was 2.08 dBi with 94 percent radiation efficiency. The designed antenna covered 27.5–43.5 GHz frequency bands of 5G [30]. A dual band compact MIMO antenna was designed with 24.10–27.18 GHz and 33–44.13 GHz bands were designed for mm-wave 5G applications [31]. A λ/2 patch based multiband antenna with series of radiating elements resonated at 35.0 GHz frequency, and the λ/3 sized folded stub was used to provide impedance matching at 28.0 GHz. The design covered 28.0–38.0GHz frequency band recommended by the Federal Communications Commission (FCC) for 5G cellular networks. The dimension of 12.6 × 30.0 mm² was used on Rogers RT substrate, and the peak gain of the antenna was 10.23 dBi at 36.0 GHz [32].

Similarly, a 10 element 5G MIMO antenna used monopoles and L-shaped slots as a basic structure. The design covered 3.3–3.6 GHz frequency band having isolation of more than

11.0 dB, efficiency of more than 50 percent, and ECC of less than 0.15 in the whole band. The dimension of the basic unit was 19.0 × 5.0 mm² and the overall dimensions were 14.0 × 75.0 mm² [33]. A T-shaped dual band antenna was designed with polarization diversity. The L-shaped patch and a slot contribute to its T-shape. The design covered 3.4–5.925 GHz frequency bands for LTE 42/43/46 bands. The overall size of the antenna was 150.0 × 80.0 mm² with efficiency of more than 42 percent in each band, and the size of the single element was 16.2 × 3.0 mm² [34]. A 2 x 2 wideband MIMO antenna was designed with dimensions of 15.2 × 7.3 mm² to cover 21.06–29.7 GHz frequency band and isolation was more than 10.52 dB. The gain and ECC were 3.0 dBi and 10⁻⁴ respectively [35]. Another, 2 × 8 MIMO antenna array was designed to cover the frequency band of 27.49–29.42 GHz, with dimensions of 23.61 × 55.18 mm². The design achieved the gain of 11.33 dBi and 88.85 percent of radiation efficiency [36]. A 1 × 2 multi band MIMO antenna was designed to cover the microwave frequency band (2.23–3.70 GHz). The MIMO antenna dimension was 5.1 × 90.2 mm² and the isolation was more than 12.5 dB with radiation efficiency of 73 percent [37]. A hook-shape four-port MIMO antenna with meander line was designed to cover the frequency band of 4.76–10.6 GHz [38].

In this chapter, a four-port coplanar waveguide (CPW) feed wideband MIMO antenna with polarization diversity is presented for 4G/5G frequency range for WLAN, WiMAX, IoT, 5G-NR, and satellite communication. The bandwidth of the proposed prototype is 31.21GHz, and it covers 4.83–36.04 GHz frequency spectrum and provides omnidirectional radiation patterns.

4.2 MIMO Antenna Design and Results

A 2 × 2 MIMO antenna with modified square patch is designed for better diversity performance on Rogers Duroid 5880 dielectric substrate of dimensions 36.0 × 36.0 mm². The height (t) of the substrate is 0.79 mm and the loss tangent is 0.0009. The CST optimized dimensions are given in Table 4.1. The schematic and fabricated views of the proposed MIMO antenna are shown in Figure 4.1(a) and (b).

The size of the modified square patch is 10.0 × 10.0 mm². The design steps are shown in Figure 4.2. The CPW feed is utilized to extend the bandwidth for the various applications of the 4G/5G frequency bands. In step I, a full square patch is used, where the first band extends from 5.36–8.92 GHz (3.56 GHz bandwidth) and the second band extends from 24.24–40.0 GHz (15.76 GHz bandwidth). Design step II is obtained by removing the square of area x^2 from all the four corners of the patch shown in step I. The removal of x^2 area from all four corners extends the bandwidth by a small amount. In step II, the antenna resonates at 5.8 GHz and shows different frequency bands of 4.96–11.328 GHz (6.368

TABLE 4.1

Dimensions of the proposed MIMO

Parameters	ws	ls	pl	pw	wg	fw
Values (mm)	36	36	10	10	7.836	1.90
Parameters	g	lg	fl	x	u	t
Values (mm)	0.214	5.5	5.937	2.5	1.0	1.524

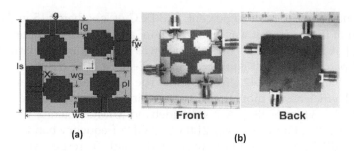

FIGURE 4.1
Proposed MIMO antenna, (a) schematic view, (b) fabricated views.

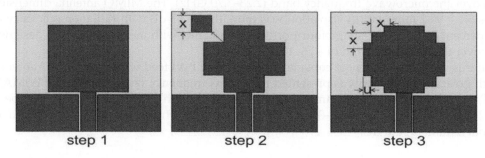

FIGURE 4.2
Design steps for proposed MIMO antenna.

GHz bandwidth), 17.24–18.68 GHz (1.44 GHz bandwidth), 22.88–23.52 GHz (0.64 GHz bandwidth), and 33.52 to more than 40.0 GHz frequency range. Design step III adds area (x-u)^2 to all the four corners of the single element of design step II. In this step II, modified antenna geometry provides very wide bandwidth. Design step III's highest resonant peak is at 6.08 GHz, and has 5.15–34.45 GHz bandwidth. The minimum gain in step III is 2.0 dBi, and minimum efficiency is 39 percent. Design step III has bandwidth of 31.21 GHz for the single element. The proposed radiator is obtained using step III.

Step I shows a narrow band performance due to conventional square patch structure, while with modification at step II, band becomes wide, as well as bands from higher frequencies shifting toward lower frequencies in step III. Step III finally provides a single wideband antenna with 2:1 VSWR (-10 dB impedance band). All the results of S_{11} parameters for step I, step II, and step III are shown and compared in Figure 4.3. Design step III is transformed into the proposed MIMO antenna with polarization diversity. S_{11} parameters of single element (single input single output (SISO)) and MIMO are compared in Figure 4.4. The proposed MIMO exhibits better values of 2:1 VSWR return loss S_{11} in comparison with single element. The multiple peaks of single element are compensated by MIMO and the notch of single element (around 8.0 GHz) is also removed in proposed MIMO.

The S parameters of conventional (square) shaped patch MIMO and proposed MIMO are compared in Figure 4.5. The conventional patch shaped MIMO has one resonant peak at 6.48 GHz frequency in 5.34–7.81 GHz frequency band, and the other band extends from 24.28 to more than 40 GHz with respect to -10 dB impedance bandwidth, and provides isolations S_{12} > 23.47 dB, and S_{13} > 15.39 dB. The conventional patch-based MIMO has multiple bands, whereas the proposed MIMO has a single wideband. The isolation parameters

5G-NR Wideband MIMO Antenna Design

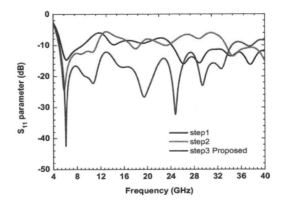

FIGURE 4.3
S parameters of design steps of proposed radiator.

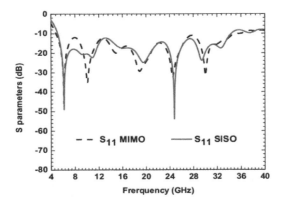

FIGURE 4.4
SISO and MIMO comparison.

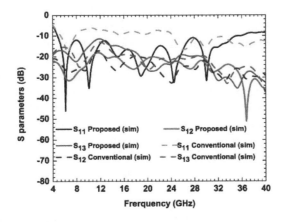

FIGURE 4.5
S parameters of conventional (square) shaped patch MIMO and proposed MIMO.

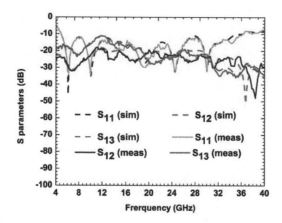

FIGURE 4.6
S parameters of proposed MIMO.wideband MIMO.

S_{12}, S_{13} and S_{14} of proposed MIMO provides more than 10.0 dB isolation. Therefore, the proposed MIMO shows better results than the conventional MIMO for the same substrate size in terms of return loss, bandwidth and isolations.

The return loss and isolation parameters of the four-port proposed MIMO are $S_{11} = S_{22} = S_{33} = S_{44}$, $S_{12} = S_{21} = S_{14} = S_{41} = S_{23} = S_{32} = S_{34} = S_{43}$, and $S_{13} = S_{31} = S_{24} = S_{42}$. Therefore, for simplicity of analysis only S_{11}, S_{12} (or S_{21}) and S_{13} (or S_{31}) parameters are selected throughout the chapter. The results for return loss and isolation parameters are validated using vector network analyzer (VNA), and the far field parameters in anechoic chamber for gain, E-field, and H-field radiation patterns of the proposed MIMO antenna.

The simulated band extends from 4.83–36.04 GHz and provides 31.21 GHz of large bandwidth. The 2:1 VSWR band is selected due to 89 percent power transmission and only 11 percent returned power. The measured band with 2:1 VSWR extends from 4.93–36.01 GHz and provides 31.08 GHz bandwidth. The frequency bands of 0.12 GHz difference are observed between the simulated and measured bands. The difference is due to the fabrication and port coupling losses. The value of isolation parameters (adjacent and diagonal ports) of proposed MIMO for both the simulated and measured bands is more than 10.0 dB. The simulated S_{12} (or S_{14}) > 18 dB and S_{13} > 11.11 dB. The results of return loss and isolation parameters of simulated and measured for the proposed MIMO are shown in Figure 4.6.

To decide the value of isolation, a parameter called surface current distribution (SCD) is used which shows the amount of current distribution on a metallic surface. For port 1 of the proposed MIMO antenna, SCD on peak location is shown in Figure 4.7(a). As the radiation elements are identical at other ports, therefore only port 1 is selected and remaining ports are terminated by 50Ω load. It is seen that most of the current concentrates at port 1 on the feed arm only, when port 1 is excited at 6.11GHz frequency.

The aforesaid proposed MIMO antenna design work is verified by doing parametric sweeps on the critical parameters. The first parametric sweep is done on the fl. The parametric sweep of fl is selected in the range of 5.75 mm to 6.5 mm. It has been observed that for the values of fl < 5.9375 mm, return loss parameter S_{11} represents two frequency bands. For fl > 5.9375 mm, bandwidth in -10 dB impedance band starts shrinking. Therefore, the suitable and optimized value of fl = 5.9375 mm matches with the requirement of the

5G-NR Wideband MIMO Antenna Design

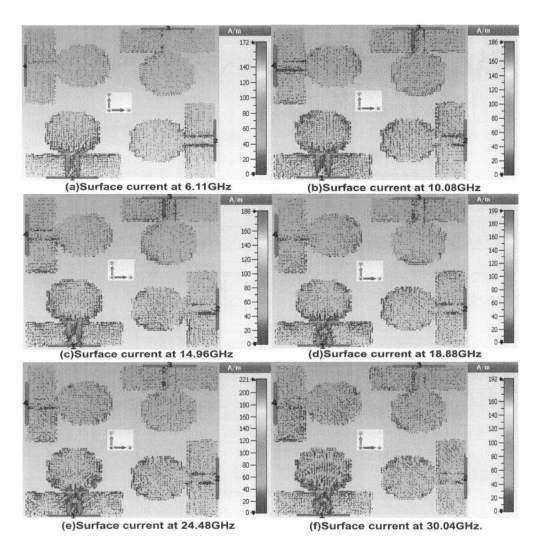

FIGURE 4.7
SCD at peak frequencies of proposed wideband MIMO.

proposed design. The effects of return loss and isolation parameters with fl parametric sweep are shown in Figure 4. 8 (a–c). There are no changes in the values of S_{12} and S_{13} in the whole band for different values of fl. Other than fl = 6.125 mm, shrinkage in bandwidth and notches is observable in return loss S_{11}.

Similarly, the parametric sweep is carried out on parameter u, which represents the difference between 1st cut (x) and introduction of square of size x-u. The parametric sweep on u is carried out for 0.5 mm to 1.5 mm. For u < 1, multiband operation is observable, while for u > 1 bandwidth starts shrinking. Therefore, for the proposed band, the suitable value is u = 1. There are very few changes observable in Figure 4.9 (a–c) for S_{12} and S_{13} isolation parameters. Unnecessary notches are also observable in S_{11} for other than u = 1 values.

The gain of the proposed MIMO antenna after the measurement of the received power in anechoic chamber is given by the FRIIS in equation 1 [39]:

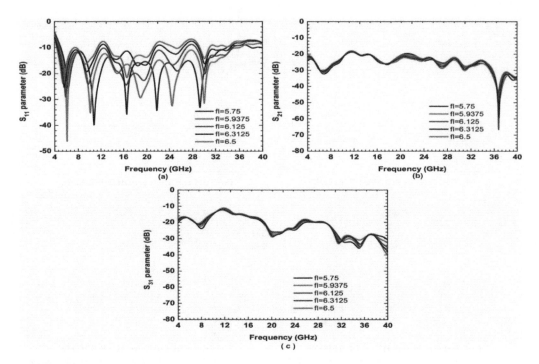

FIGURE 4.8
Effect of parameter fl on S parameters.

$$P_r = \frac{P_t G_t G_r \lambda^2}{(4\pi R)^2}, \tag{1}$$

where, P_t is transmitted power, f_c is carrier frequency, λ is the wavelength, Gt is the transmitter, Gr is the receiver, and R is the distance between the transmitter and receiver.

The simulated and measured MIMO antenna gain in the whole frequency band varies in the range of 2.0–6.0 dBi. At the highest resonant peak, the value of gain is 3.01 dBi. The radiation efficiency in whole frequency band lies between 39 percent and 79 percent, while total efficiency lies between 27 percent and 73.9 percent. The antenna gain variation along with radiation efficiency and total efficiency is shown in Figure 4.10.

The envelope correlation coefficient (ECC) which shows the diversity effect of the proposed MIMO antenna is given by equation 2. For the proposed wideband MIMO, the effect of ECC is evaluated in terms of return loss and isolation parameters, because, the single isolation parameter is unable to express the mutual coupling effect properly among ports. The standard maximum value of ECC is 0.5, which is set by the International Telecommunication Union (ITU) [40].

$$\left|\rho_e(i,j,N)\right| = \frac{\left|\sum_{n=1}^{N} S_{i,n}^* S_{n,j}\right|}{\sqrt{\Pi_{k(=i,j)}\left[1-\sum_{n=1}^{N} S_{i,n}^* S_{n,k}\right]}} \tag{2}$$

5G-NR Wideband MIMO Antenna Design

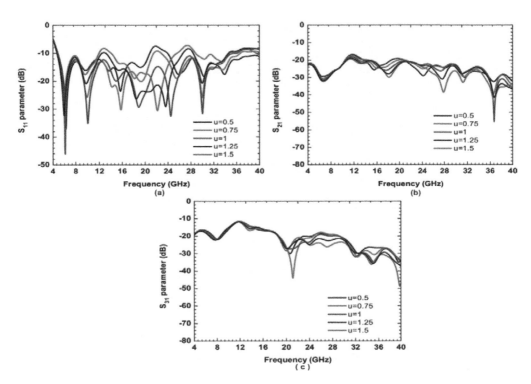

FIGURE 4.9
Effect of parameter u on S parameters.

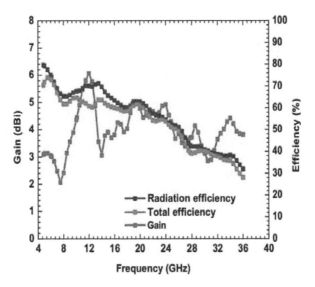

FIGURE 4.10
Antenna gain variation with radiation efficiency and total efficiency.

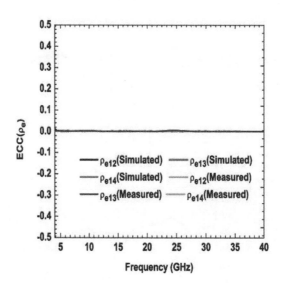

FIGURE 4.11
ECC of the proposed wideband MIMO.

where, i=1 to 4 for four elements, j= 1 to 4 for four elements, and N is the number of radiating elements.

Let ρ_{12} be the ECC between ports 1–2, ρ_{13} the ECC between ports 1–3, and ρ_{14} the ECC between ports 1–4, while including the effects of the other ports also. The results of ρ_{12}, ρ_{13} and $\rho_{14\,are}$ shown in Figure 4.11. The values of ρ_{12}, ρ_{13} and ρ_{14} lie in the range of 1.3×10^{-3}, 9×10^{-3}, and 1.3×10^{-3} in the whole frequency band for both the simulated and measured results. The difference is very small between the simulated and measured values.

The mean effective gain (MEG) is another diversity parameter, and can be obtained using equation 3 [41–44].

$$MEG_j = \frac{1}{2\pi} \int_0^{2\pi} \left[\frac{XPR}{1+XPR} G_{\theta j}\left(\frac{\pi}{2},\theta\right) + \frac{1}{1+XPR} G_{\phi j}\left(\frac{\pi}{2},\phi\right) \right] d\phi \qquad (3)$$

where, XPR is the cross-polarization ratio, $p_{\Theta j}$ and $p_{\Phi i}$ are the probability distribution functions for the azimuthal angle (Θ) and the elevation angle (Φ), $G_{\Theta j}$ and $G_{\Phi j}$ are power gains for azimuthal angle (Θ) and elevation angle (Φ). The value of j=1 to number of radiating elements. The considered values for isotropic and Gaussian environments are: XPR = 6.0 dB (for outdoor case), and XPR = 0 dB (for indoor case).

All the results of the MEG for indoor and outdoor mediums for isotropic and Gaussian environments are shown in Figure 4.12. The values of MEG for XPR=0 dB for isotropic medium is less than -3.0 dB, and for XPR = 6.0 dB MEG is also less than -3.0 dB. Similarly, for the Gaussian medium, MEG for XPR = 0dB is less than -3.0 dB, and for XPR = 6.0 dB, MEG is also less than -3.0 dB. It is clear from the observation that the proposed MIMO antenna is suitable for indoor and outdoor isotropic and Gaussian environments. Therefore, the proposed design can be suitable for IoT applications also.

To describe the diversity performance, total active reflection coefficient (TARC) is also used on the basis of incident and reflected powers. It uses random signals and excitation

5G-NR Wideband MIMO Antenna Design

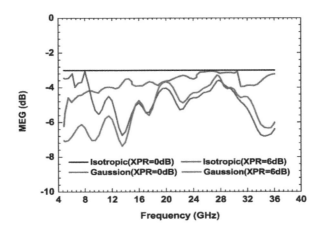

FIGURE 4.12
Mean effective gain of proposed MIMO.

phase angles for determination of active bandwidth of MIMO. For the proposed design, it is calculated between each two ports for their active bandwidth performance. With the help of excitation and scattering vectors, TARC can be computed.

Let a_i and b_i be excitation and scattering vectors respectively. The ratios of b_i with respect to a_i resemble the shape of return loss or isolation parameter in most of the MIMO antenna designs. The TARC can be computed using equation 4 [45–47].

$$\Gamma_a^t = \frac{\sqrt{\sum_{i=1}^{N}|b_i|^2}}{\sqrt{\sum_{i=1}^{N}|a_i|^2}} \qquad (4)$$

The simplified form of TARC is given by equation 5 [48–49] as:

$$\Gamma_a^t = \frac{\sqrt{|S_{ii}+S_{ij}*e^{j\theta}|^2 + |S_{ji}+S_{jj}*e^{j\theta}|^2}}{\sqrt{N}} \qquad (5)$$

where, $\theta = \theta_1, \theta_2$, are excitation phase angles at the ports, $i = 1$ to number of ports, $j = 1$ to number of ports, and N = number of radiating patches.

The results of $TARC_{12}$ (between ports 1 and 2), $TARC_{13}$ (between ports 1 and 3), and $TARC_{14}$ (between ports 1 and 4) are shown in Figure 4.13 (a–c). The best suitable phase combination for $TARC_{12}$ is 0°, 90°; for $TARC_{13}$ it is 60°, 150°, and for $TARC_{14}$ it is 0°, 90°. The active bandwidth for $TARC_{12}$, $TARC_{13}$, and $TARC_{14}$ is approximately 31.0 GHz for -10 dB return loss bandwidth.

The E-field radiation patterns are shown in Figure 4.14 for different peak frequencies. The E-field is calculated at $\Phi=0$ at different Θ at an interval of 10^0. All the E-field radiation patterns are very close to the omnidirectional radiation patterns. Therefore, the proposed design can be utilized for all directions. Similarly, the H-field radiation patterns are shown in Figure 4.15 for different peak frequencies. These values are calculated for H-field at

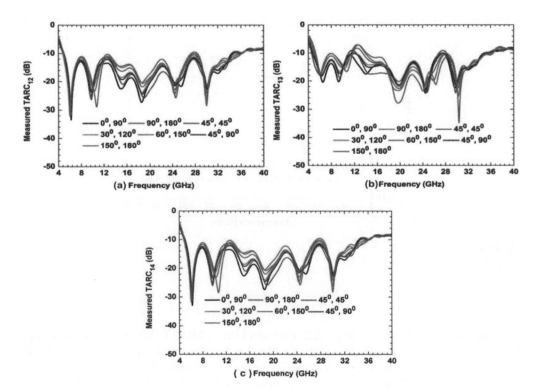

FIGURE 4.13
TARC plots for proposed MIMO.

Φ=90⁰ at different Θ at interval of 10°. These patterns are also omnidirectional and can be utilized for all directions.

The proposed design was fabricated on the Rogers Duroid 5880 dielectric substrate. All the existing designs are compared with the proposed MIMO in Table 4.2. The designs given here use Neltec, and Rogers dielectric substrates. The proposed design has very wide bandwidth in comparison with the references given here. Also, the size is very small in comparison with these references. The effecta of MEG and TARC are extracted for the proposed design, which the other designs are lacking.

4.3 Conclusion

5G-NR is the current research interest given the spectrum shortage and high data rate requirements to accommodate variety of wireless applications. The proposed 5G-NR MIMO covers under and above 6 GHz sub-bands (4.83–36.04 GHz) for 5G-NR, IoT, WLAN, WiMAX, and satellite applications for indoor and outdoor isotropic and Gaussian environments. The proposed design has 31.21 GHz of bandwidth and peak gain of 6.0 dBi. The value of ECC is very much less than the limit set by ITU (ECC <10^{-3}). The omnidirectional radiation patterns and wide TARC bandwidth are the advantages of the proposed MIMO in the present scenario.

5G-NR Wideband MIMO Antenna Design

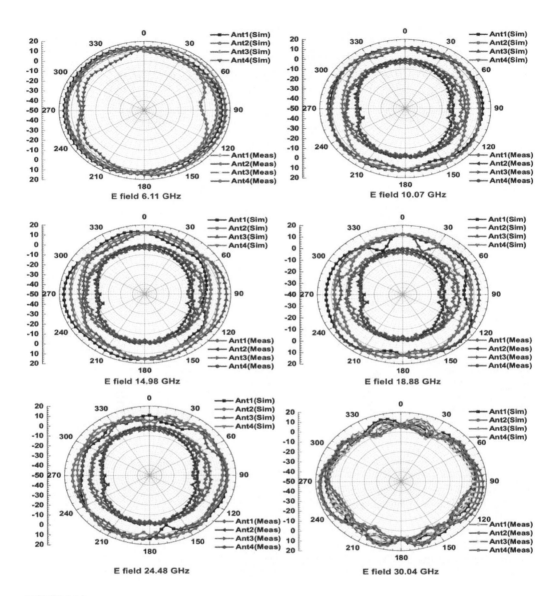

FIGURE 4.14
E-field radiation patterns at different peak frequencies.

4.4 Future Scope

The proposed design maybe extended with SRR, CSRR, DGS, for the next generation of MIMO antennas.

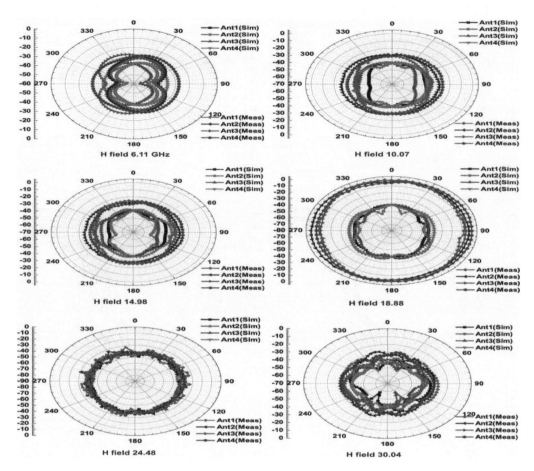

FIGURE 4.15
H-field radiation patterns at different peak frequencies.

TABLE 4.2

Comparison of presented MIMO with existing designs

Reference No.	Dimensions (mm²)	No. of Radiating element	Substrate	Frequency band (GHz)	Bandwidth (GHz)	Gain (dB)	ECC	MEG (dB)	TARC (GHz)	Radiation Efficiency
[7]	38.3 x 38.3	4	Taconic RF-45	3–13.2	10.2	4.1	0.02	-1	26	72%
[9]	26 × 26	2	Taconic RF-35	2.9–11.6	8.7	6.0	0.02	-	-	-
[10]	26 × 26	8	FR-4	3–11	8		0.08	3	-12	-
[11]	110 × 120	2	Rogers RO4350	3–10	7	2.6	9×10^{-5}	-	-	-
[12]	50 × 60	2	F4B	2.98–10	7.02	>2.5	0.003	-	-	80%
[13]	26 x 31	2	FR-4	3.1–11	7.9	5.6	0.001	-	-	85.5%
[25]	40 × 40	4	FR4	7.2–7.8	0.6	9.95	0.02	-	-	-
[27]	48x31	4	Neltec	26–31	5	10	-	-	-	-
[28]	45x20	4	Roger	25.05–34.92	9.871	12.15	-	-	-	85%
[29]	150x100	8	Roger	27.4–28.23	0.83	5.67	-	-	-	-
Proposed work	36x36	4	Roger	4.83–36.04	31.21	2–6	$< 10^{-3}$	<-3	31.0	74%

References

[1] Z. Liu, J. Liu, Y. Zeng, and J. Ma. "Covert wireless communication in IoT network: From AWGN channel to THz band," *IEEE Internet of Things Journal*, vol. 7, pp. 3378–3388, 2020.

[2] S. H. Lee and Y. Sung, "Multiband antenna for wireless UWB dongle applications," *IEEE Antennas and Wireless Propagation Letters*, vol. 10, pp. 25–28, 2011.

[3] G. Bharti, S. Bhatia, and J. S. Sivia, "Analysis and design of triple band compact microstrip patch antenna with fractal elements for wireless applications," *International Conference on Computational Modeling and Security: CMS*, vol. 85, pp. 380–385, 2016.

[4] M. Shafi, A. F. Molisch, P. J. Smith, T. Haustein, P. Zhu, P. D. Silva, F. Tufvesson, A. Benjebbour, and G. Wunder, "5G: A tutorial overview of standards, trials, challenges, deployment and practice," *IEEE Journal on Selected Areas in Communications*, vol. 35, pp. 1201–1221, 2017.

[5] N. Kshetri, "5G in e-commerce activities," *IT Professional*, vol. 20, pp. 73–77, 2018.

[6] J. Nightingale, P. S. Garcia, J. M. A. Calero, and Q. Wang, "5G-QoE: QoE modelling for ultra-HD video streaming in 5G networks," *IEEE Transactions on Broadcasting*, vol. 64, pp. 621–634, 2018.

[7] R. Gomez-Villanueva and H. Jardon-Aguilar, "Compact UWB uniplanar four port MIMO antenna array with rejecting band," *IEEE Antennas and Wireless Propagation Letters*, vol.18, pp. 2543–2547, 2019.

[8] M.S. Khan, A. Iftikhar, R. M. Shubair, A.-D. Capobianco, B. D. Braaten, and D.E. Anagnostou, "Eight-element compact UWB-MIMO/diversity antenna with WLAN band rejection for 3G/4G/5G communications," *IEEE Open Journal of Antennas and Propagation*, vol. 1, pp. 196–206, 2020.

[9] Z. Ali, C. Yin, and X. Zhu, "Compact UWB MIMO Vivaldi antenna with dual band-notched characteristics." *IEEE Access*, vol. 7, pp. 38696–38701, 2019.

[10] S. Shakir, M. Bilal, S. M. Abbas, N. Saleem, Z. Rauf, R. A Wagan, R. Saleem, and M. F. Shafique, "A compact 8-Element 3D UWB diversity antenna system for off device installation," *IEEE Access*, vol. 7, pp. 44117–44127, 2021.

[11] L. Y. Nie, X. Q Lin, Z. Q. Yang, J. Zhang, and B. Wang, "Structure-shared planar UWB MIMO antenna with high isolation for mobile platform," *IEEE Transactions on Antennas and Propagation*, vol. 67, pp. 1–4, 2019.

[12] L. Y. Nie, X. Q. Lin, S. Xiang, B. Wang, L. Xiao, and J. Y. Ye, "High-isolation two-port UWB antenna based on shared structure," *IEEE Transactions on Antennas and Propagation*, vol. 68, pp. 8186–8191, 2020.

[13] A. Khan, S. Bashir, S. Ghafoor, and K. K. Qureshi, "Mutual coupling reduction using ground stub and EBG in a compact wideband MIMO-antenna," *IEEE Access*, vol. 9, pp. 40972–40979, 2021.

[14] K. R. Jha, Z. A. Pandit Jibran, C. Singh, and S.K. Sharma, "4-Port MIMO antenna using common radiator on a flexible substrate for sub-1GHz, Sub-6GHz 5G NR, and Wi-Fi 6 Applications," *IEEE Open Journal of Antennas and Propagation*, vol. 2, pp. 689–701, 2021.

[15] M. Agiwal, H. Kwon, S. Park, and H. Jin, "A survey on 4G-5G dual connectivity: Road to 5G implementation," vol. 9, pp. 16193–16210, 2021.

[16] Y. Qi, B. Yuan, Y. Cao, and G. Wang, "An ultrawideband low-profile high-efficiency indoor antenna," *IEEE Antennas and Wireless Propagation Letters*, vol. 19, pp.346–349, 2020.

[17] S. Ma, H. Zhai, Z. Wei, X. Zhou, L. Zheng, and J. Li, "A high-selectivity dual polarization filtering antenna with metamaterial for 5G application," *Microwave Optical Technology Letters*, vol. 61, pp. 63–67, 2019.

[18] S. Wen and Y. Dong, "A low-profile wideband antenna with monopolelike radiation characteristics for 4G/5G indoor micro base station application," *IEEE Antennas and Wireless Propagation Letters*, vol. 19, pp. 2305–2309, 2020.

[19] G. -N. Zhou, B.-H. Sun, Q. -Y. Liang, S.-T. Wu, Y. -H. Yang, and Y. -M. Cai, "A tri-band dual-polarized shared-aperture antenna for 2G/3G/4G/5G base station application," *IEEE Transactions on Antennas and Propagation*, vol.69, pp. 97–108, 2021.

[20] L. -H. Wen, S. Gao, Q. Luo, Q. Yang, W. Hu, Y. Yin, J. Wu, and X. Ren, "A wideband series-fed circularly polarized differential antenna by using crossed open slot-pairs," *IEEE Transactions on Antennas and Propagation*, vol.68, pp. 2565–2574, 2020.

[21] K. Sun, D. Yang, and S. Liu, "A wideband hybrid feeding circularly polarized magneto-electric dipole antenna for 5G Wi-Fi," *Microwave Optical Technology Letters*, vol. 60, pp. 1837–1842, 2018.

[22] T. M. Guan and S. K. A. Rahim, "Compact monopole MIMO antenna for 5G application," *Microwave Optical Technology Letters*, vol. 59, pp. 1074–1077, 2017.

[23] S. S. Bhatia, A. Sahni, and S. B. Rana, "A novel design of compact monopole antenna with defected ground plane for wideband applications," *Progress in Electromagnetics Research M*. vol. 70, pp. 21–31, 2018.

[24] Y. Li, C. Sim, Y. Luo, and G. Yang, "12-port 5G massive MIMO antenna array in sub-6GHz mobile handset for LTE bands 42/43/46 applications," *IEEE Access*, vol. 6, pp. 344–354, 2018.

[25] A. Dkiouak, A. Zakriti, M. E. Ouahabi, N. A. Touhami, and A. Mchbal, "Design of a four-element MIMO antenna with low mutual coupling in a small size for satellite applications," *Progress in Electromagnetics Research M*, vol. 85, pp. 95–104, 2019.

[26] H. Li, Y. Li, L. Chang, W. Sun, X. Qin, and H. Wang, "Wideband dual-polarized end-fire antenna array with overlapped apertures and small clearance for 5G millimeter wave applications," *IEEE Transactions on Antennas and Propagation*, vol. 69, pp. 815–824, 2021.

[27] Z. Wani, M. P. Abegaonkar, and S. K. Koul, "A 28-GHz antenna for 5G MIMO applications," *Progress In Electromagnetics Research Letters*, vol. 78, pp. 73–79, 2018.

[28] H. Ullah and F. A. Tahir, "A broadband wire hexagon antenna array for future 5G communications in 28 GHz band," *Microwave Optical Technology Letters*, vol. 61, pp. 696–701, 2018.

[29] M. Ikram, M. S. Sharawi, K. Klionovski, and A. Shamim, "A switched-beam millimeter-wave array with MIMO configuration for 5G applications," *Microwave Optical Technology Letters*, vol. 60, pp. 915–920, 2018.

[30] A. Desai, T. Upadhyaya, and R. Patel, "Compact wideband transparent antenna for 5G communication systems," *Microwave Optical Technology Letters*, vol. 51, pp. 781–786, 2019.

[31] A. Desai, C. D. Bui, J. Patel, T. Upadhyaya, G. Byun, and T. K. Nguyen, "Compact wideband four element optically transparent MIMO antenna for mm-wave 5G applications," *IEEE Access*, vol. 8, pp.194206–194217, 2020.

[32] S. F. Jilani, Q. H. Abbasi, Z. U. Khan, T. H. Loh, and A. Alomainy, "A Ka-band antenna based on an enhanced Franklin model for 5G cellular networks," *Microwave Optical Technology Letters*, vol. 60, pp. 1562–1566, 2018.

[33] J. Y. Deng, J. Yao, D.Q. Sun, and L.X. Guo, "Ten-element MIMO antenna for 5G terminals," *Microwave Optical Technology Letters*, vol. 60, pp. 3045–3049, 2018.

[34] V. Li, C. Y. D. Sim, Y. Luo, and G. Yang, "Multiband 10-antenna array for sub-6 GHz MIMO applications in 5-G smartphones," *IEEE Access*, vol. 6, pp. 28041–28053, 2018.

[35] P. Waghmare, P. Gupta, K. Gehlod, A. Shakya, and L. Malviya, "2X2 wideband array MIMO antenna for 5G spectral band," *IEEE 5th International Conference for Convergence in Technology (I2CT)*, pp. 1–4, 2019.

[36] L. Malviya, A. Parmar, D. Solanki, P. Gupta, and P. Malviya, "Highly isolated inset-feed 28 GHz MIMO-antenna array for 5G wireless application," *Elsevier Third International Conference on Computing and Network Communications (CoCoNet '19)*, pp. 1–4, 2020.

[37] L. Malviya, M. V. Kartikeyan, and R. K. Panigrahi, "Multi-standard, multi-band planar multiple input multiple output antenna with diversity effects for wireless communication," *Int. Journal of RF Microwave & Computer Aided Eng.*, pp.1–8, 2018.

[38] L. Malviya, K. Gehlod, and A. Shakya, "Wide-band meander line MIMO antenna for wireless applications," *IEEE Access*, pp. 1–4, 2018.

[39] K. Okada and J. Pang, "Millimeter-wave CMOS phased-array transceiver supporting dual-polarized MIMO for 5G NR," *IEEE Custom Integrated Circuits Conference (CICC)*, pp. 1–8, 2020.

[40] L. Malviya and S. Chauhan, "Multi-cut four-port shared radiator with stepped ground and diversity effects for WLAN application," *Int. Journal of Microwave and Wireless Technologies*, pp. 1–10, 2019.

[41] P. Gupta, L. Malviya, and S. V. Charhate, "5G multi-element/port antenna design for wireless applications: A review," *Int. Journal of Microwave and Wireless Technologies*, pp. 1–21, 2019.

[42] L. Malviya, R. K. Panigrahi, and M.V. Kartikeyan, "MIMO antennas with diversity and mutual coupling reduction techniques: A review," *Int. Journal of Microwave and Wireless Technologies*, vol. 9, pp. 1763–1780, 2017.

[43] Z. Niu, H. Zhang, Q. Chen, and T. Zhong, "A novel defect ground structure for decoupling closely spaced E-plane microstrip antenna array," *Int. Journal of Microwave and Wireless Technologies*, pp. 1–6, 2019.

[44] X. Li, L. Yang, J. W. Zhao, Y. Liu, and F. Z. Sun, "A bottom-feed omni-directional circularly polarized antenna array," *Int. Journal of Microwave and Wireless Technologies*, pp. 1–6, 2019.

[45] Z. Xiu, Q. Zhang, and L. Guo, "A printed multiband MIMO antenna with decoupling element," *Int. Journal of Microwave and Wireless Technologies*, vol. 11, pp. 413–419, 2019.

[46] A.K. Awasthi and A.R. Harish, "Wideband tightly-coupled compact array of dipole antennas arranged in triangular lattice," *Int. Journal of Microwave and Wireless Technologies*, vol. 11, pp. 382–389, 2019.

[47] Z. Jiakai, Z. Qi, L. Haixiong, D. Jun, and G. Chenjiang, "Wideband radar cross section reduction of a microstrip antenna with square slots," *Int. Journal of Microwave and Wireless Technologies*, vol. 11, pp. 341–350, 2019.

[48] L. Malviya, R.K. Panigrahi, and M.V. Kartikeyan, "A 2×2 dualband MIMO antenna with polarization diversity for wireless applications," *Progress in Electromagnetics Research C*, vol. 61, pp. 91–103, 2016.

[49] L. Malviya, R.K. Panigrahi, and M.V. Kartikeyan, "Four element planar MIMO antenna design for long-term evolution operation," *IETE Journal of Research*, vol. 64, pp. 367–373, 2018.

5

Advanced Wireless Communication and Sensor Networks: Applications and Simulations

Ciro Rodríguez and Isabel Moscol

CONTENTS
5.1 Introduction 57
5.2 Problems in Wireless Sensor Networks 58
    5.2.1 Lack of Electrical Energy 58
    5.2.2 Lack of Technological Support to Help Patients and Doctors 58
    5.2.3 Poor Use of Resources in Agriculture 58
5.3 WSN Applications 58
    5.3.1 Energy 58
    5.3.2 E-Health 58
    5.3.3 Buildings 59
    5.3.4 Military 59
    5.3.5 Industrial Detection 59
    5.3.6 Agriculture 59
5.4 Simulators for Analysis Prior to Actual Deployment of a WSN 61
5.5 Conclusions 62
References 63

5.1 Introduction

Wireless communication and the impact of telecommunications have made an astonishing evolution in recent years. The use of wireless networks is nowadays an alternative for organizations of all kinds to be competitive.

Wireless communication is communication in which the communication (sender/receiver) is not linked by a physical propagation medium but uses the modulation of electromagnetic waves through space. In this sense, physical devices are only present in the transmitters and receivers of the signal, which include antennas, laptops, PDAs, mobile phones, etc.

Wireless sensors, as well as information and communication technologies, in recent years, have managed to open doors to changes that will improve the quality of life of people, changes that are reflected in the areas of energy, health, in buildings, making the management of activities practically automatic and less risky. In agriculture, they have made it possible to improve productivity and the optimal management of resources. In this chapter, we intend to give an overview of the main uses of wireless sensor networks in areas such as energy, e-health, buildings, and agriculture.

5.2 Problems in Wireless Sensor Networks

5.2.1 Lack of Electrical Energy

The lack of electricity in remote or difficult to access areas is a problem for the resident population or companies located in such a place, therefore a solution is proposed using wireless sensor networks that allow the collection or harvesting of electricity using renewable energy sources, the most common and profitable way in the long term being the collection and storage of energy by means of solar panels.

5.2.2 Lack of Technological Support to Help Patients and Doctors

WSNs assist in patient monitoring and reduce medical errors. The large amount of data generated by the sensors is further processed and converted into valuable information that can help doctors to visualize more detail or perceive aspects hidden from human perception to increase the chance of success in a diagnosis or clinical treatment.

5.2.3 Poor Use of Resources in Agriculture

The inappropriate use of agricultural resources decreases the quality of the product and generates unnecessary expenses; however, with the information provided by a wireless sensor system it would be possible to know, for example, the level of soil moisture, the temperature of the environment, as well as to optimize the irrigation system. This would reduce water consumption on a large scale, saving resources and optimizing the final production indicators.

5.3 WSN Applications

5.3.1 Energy

There is a different way of providing the power supply and that is by harvesting energy from the environment surrounding the sensor device with the implementation of energy harvesting transducers (sensors and actuators). Energy harvesting provides the power and energy needed to operate the sensor node, which will not require battery maintenance throughout its operational lifetime. In effect, energy harvesting enables perpetual sensors.

5.3.2 E-Health

The work developed by [1] proposes a working prototype of an electrocardiograph implemented through wireless sensor networks (WSN) to measure the cardiac signals of athletes.

On the other hand, in [2] an e-Health architecture based on wireless sensor networks for healthcare is implemented. The proposed structures of the system to be carried by the patient to be monitored are shown. The physiological variables implemented under the proposed architecture and the results so far fall under the concept of e-health, an emerging field at the intersection of medical informatics, public health and business [3]. A case study has been proposed [4] for the detection of cardiovascular arrhythmias in non-critical patients during rehabilitation sessions in closed spaces.

5.3.3 Buildings

The TinyDomus project proposes software for managing the deployment of a wireless network with two autonomous devices called motes. Each module has a USB interface to connect them to a desktop computer; in addition to this, three LEDs are used with which they can interact, as well as a series of sensors to monitor certain environmental and physical aspects. The functionality that the TinyDomus project aims to achieve is to simulate a home automation system. According to the temperature and luminosity data, two sensors are activated, simulating the behavior of an air conditioning system or regulating the switching on of the lighting systems in the room where we are. The Hall effect sensor, on the other hand, would be responsible for managing the security system.

5.3.4 Military

WSNs are involved in military intelligence gathering such as tracking an enemy, monitoring the battlefield, or tracking and classifying moving targets [5]. They are also used for property protection, surveillance, and border control. An example is a fence or boundary surveillance system comprising a robot, a camera and two types of sensor nodes, for the ground and the fences. The network reports the acquired data to the base station which sends control messages to the camera to focus on the location where the detected event has occurred. Additionally, commands can be sent to mobile robots that extend the communication distance of the system.

5.3.5 Industrial Detection

Industries are increasingly implementing WSNs to improve efficiency of machine performance, most importantly, continuous monitoring in the maintenance area. They measure vibrations and lubricant levels. The National Institute of Standards and Technology (NIST) developed the P1451 intelligent transducer interface standard to enable complete plug-and-play networks and sensors in industrial environments. Factories continue to automate assembly lines and implement advanced in-line quality control checks triggered by the sensors, for instance, pH probes for a suitable spectral sensor that can serve as miniature spectrometers, or optical sensors that can replace existing instruments and perform composition and structural property measurements. Sensor inputs feed databases to display real-time information on a large or small scale [6].

5.3.6 Agriculture

[7] Presents the design of a wireless sensor network for irrigation in citrus crops in Brazil. The pilot system was developed on 6 hectares to evaluate it under field conditions. The sensor nodes are 50 meters apart. Nine sensor nodes were used, consisting of an LM35

temperature sensor and a capacitive sensor to measure soil moisture, all processed by a Microchip PIC16F88 microcontroller to send data every 15 minutes. The actuator node has similar characteristics to the previous one, but with a valve and flow meters. A field station receives the data coming from the sensor nodes at hourly intervals and has a weather station connected to it. The communication range is estimated at 500 meters, to be able to transmit to the base station (server) in charge of processing the information.

Precision farming saves costs and time in agricultural production and optimizes resources through the incorporation of various technologies, such as the Nanoenvi Smart Agro System. The integration of precision farming techniques can help farmers to optimize their management systems. Critical environmental parameters such as air and soil temperature and humidity, leaf wetness, conductivity and soil water content can, with proper monitoring, be controlled to ensure resource optimization and thus environmental and economic sustainability.

In one case, an agricultural control system consists of two network nodes, also called motes, in charge of collecting information from the environment inside a greenhouse and adjusting to the climate and environment required by a certain crop. The computational load of the actions to be performed falls on the PC, which will be connected to one of the motes in order to be part of the total system.

The Polytechnic University of Cartagena proposes the design of a wireless prototype to adjust the position of the dendrometers of an agronomic farm [8]. For this purpose, they carried out several studies of controlled deficit irrigation in the facilities of the "Tomás Ferro" Agroalimentary Experimental Station during the last few years. To do this, they initially deployed wired sensor networks with a centralized architecture, where they used a Campbell Scientific datalogger as a storage and data acquisition element. They used sensors to monitor the state of the environment, soil, and vegetation; some of the types used were tensiometers, FDR probes, thermadiometers, and dendrometers.

Design of a wireless sensor network for precision agriculture: The design of the solution is presented [9], in which the topology of the network was considered and also the fact that the protocol corresponding to the design must provide reliable data and at the same time consume as little energy as possible. On the other hand, the fact that a network of this type should be able to grow without affecting its quality was also considered, as well as the fact that the information should be presented in a user-friendly software that is easy to analyze.

Wireless sensor networks for precision agriculture in smallholder regions: This article [10] presents a proposal for a sensor network architecture in the aspects related to node typology, sensor integration, WSN technology, topology, and network routing. The fundamental line of technical development of the sensor network focuses on the use of wireless sensor technologies better known as WSN (Wireless Sensor Networks / IEEE 802.15.4 standard) for the design of a distributed system to capture information on multiple critical parameters in the evolution of crops in smallholder regions.

New trends in wireless sensor technologies for the productive improvement of agricultural crops: A case study of the use of a WSN in an agricultural application is presented in [11], where a disease known as mildew is present in grapevines, which considerably affects the production of grapes. By means of a system integrated by WSN for the collection of information, as has been developed in other cases, such as for example in [12], all the variables that favor the growth and spread of the disease can be kept under control and the system will give a warning when they are fulfilled and there is a possibility that the disease has developed in a localized area. In this way it is possible to determine if the disease is really attacking and to proceed to use a treatment to solve it.

Applications and Simulations

TABLE 5.1

Protocol optimization technologies

Source	Protocol	Optimization technology
[16]	Geographical and opportunistic routes	Greedy strategy and flexible redirections with depth adjustment
[17]	Agent-based approach (ABA)	Establish a route based on node energy and number of crossings
[13]	Energy-aware and avoidable routing protocol in vacuum (EAVARP)	A flexible redirection strategy based on the avoidance of hollow zones
[14]	Energy efficient grid based geographic routing (EMGGR)	Gateway selection algorithm with packet forwarding mechanisms
[15]	Energy location based geographic routing (EEL)	Localization of NADV (normalized advance rate) and TOA (time of arrival)
[18]	Enhanced vector routing forwarding (VBF)	Geographic routing strategy with a tube-route radius

5.4 Simulators for Analysis Prior to Actual Deployment of a WSN

In the case of WSN, testing any methodology/algorithm can be costly and difficult depending on the environment/medium in which it is required to be applied. Therefore, it is important to test before proceeding to the actual deployment of WSN. In the selected studies it has been possible to identify tools/simulators used for analysis prior to actual deployment, as shown in Table 5.1. As can be seen from this data, the most used tool in WSN is the network simulator (NS). Specifically, NS2 is the old version of the network simulator, which has now been replaced by the new, more commonly used version NS3.

Simulators play an important role in a complex WSN environment, as they allow verification of the network design prior to actual deployment. Four types of WSN communication architecture have been identified: one-dimensional, two-dimensional, three-dimensional, and four-dimensional. Table 5.1 summarizes the three communication technologies—acoustic, optical, and electromagnetic—which are used in WSNs [13]. Each communication technology is classified according to different attributes, mainly by communication range and data transfer rate. There are also other attributes that are used as a measure in various studies, such as latency which refers to the delay in WSN data transmission, transmission power which refers to energy consumption during data transmission, cost which refers to the cost of implementation and operation of each technology, directionality which can refer to omnidirectional transmissions, where a signal is transmitted and received from any of the possible directions, or unidirectional transmissions, where a signal is transmitted or received in one direction only, the loss of the signal propagation path, also called attenuation, and power consumption [14–15].

Proposed WSN network architecture: The network architecture is based on a coordinator node (or Gateway), which is responsible for collecting data from each sensor node (see Figure 5.1) and transmits the data via GPRS to a central server. The sensor nodes are those that process and transmit the information from the sensors to the coordinator node. It is proposed to use a star topology, because initially the prototype network has only three sensor nodes.

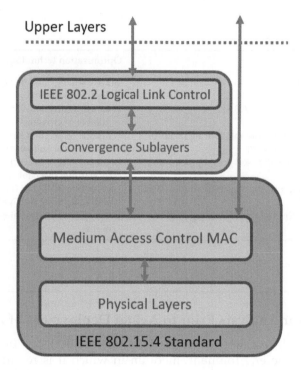

FIGURE 5.1
Protocol stack for the 802.15.4 standard.

We can infer that the tools "network simulator (NS2 and NS3 versions)" are constantly being used to test WSNs (Figure 5.2). In addition, there is a certain preference for three-dimensional (3D) communication architectures. A detailed and reliable verification of the WSN before actual deployment is essential. Therefore, a more detailed analysis of the tools is necessary to identify specific advantages and limitations.

5.5 Conclusions

Wireless sensor networks are on the rise: we can say that it is and will be a trend in the coming years and its use will become more and more widespread, and one of the reasons is the low cost of installation compared to wired sensors. In addition to this we can add the advances in sensor technology. In the case of energy, which is often scarce but ceases to be a problem thanks to energy harvesting, we no longer worry about this issue, as well as the cost of batteries, and so we are talking about saving money and increasing energy efficiency. In the area of healthcare, improving the quality of life of patients is being achieved by using tiny wireless sensors that can monitor vital signs, and environmental and location sensors, that can be integrated into a WWBAN (Wearable Wireless Body Area Network). They can monitor patients with previously detected diseases. The area of biomedicine is providing new solutions with the use of wireless sensor technology and the use of these sensors in health has many advantages such as reducing medical

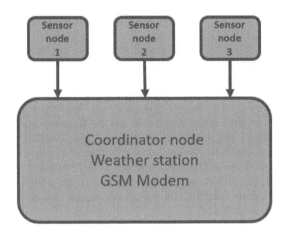

FIGURE 5.2
WSN architecture.

errors, increasing the quality of medical care, improving the efficiency of caregivers and the comfort of patients.

In the area of building maintenance, WSNs also play an important role, as they help in the monitoring and proper functioning of automated systems, and once detected, possible current and future problems can be addressed with automated maintenance. In the area of agriculture, wireless sensors improve agriculture by making it more accurate, thus saving time and cost in production.

References

[1] J. Medina, "Functional prototype of an electrocardiograph implemented by means of wireless sensor networks (WSN)," Thesis, Universidad de San Buenaventura, Colombia, 2011.

[2] H. Cervantes de Ávila, J.I. Nieto Hipólito, J. de D. Sánchez López, M.E. Martínez Rosas, and A. Hawa Calvo, "An e-Health architecture based on wireless sensors network," *Difu100ci@, Revista de difusión científica, ingeniería y tecnologías*, vol. 6, no. 2, Art. no. 2, Dec. 2012.

[3] D. Malan, T. Fulford-Jones, M. Welsh, and S. Moulton, "CodeBlue: An ad hoc sensor network infrastructure for emergency medical care," International workshop on wearable and implantable body sensor networks, London, UK, vol. 6, p. 5, Apr. 2004.

[4] D. Martínez, F. Blanes, J. Simo, and A. Crespo, "Wireless sensor and actuator networks: A characterisation and case study for medical applications in enclosed spaces," International Multiconference on Computer Science and Information Technology, p. 8, 2008.

[5] G. Oreku and T. Pazynyuk, *Security in Wireless Sensor Networks*, vol. 1. London: Springer, 2016.

[6] H. Xu, Y. Tu, W. Xiao, Y. Mao, and T. Shen, "An Archimedes curve-based mobile anchor node localization algorithm in wireless sensor networks," in *8th World Congress on Intelligent Control and Automation*, Jinan, China, Jul. 2010, pp. 6993–6997.

[7] A. Torre-Neto, R.A. Ferrarezi, D.E. Razera, E. Speranza, and C. Lopes, "Wireless sensor network for variable rate irrigation in citrus," *Precision Agriculture*, vol. 6, p. 9, 2005.

[8] R. Narváez, "Diseño de un prototipo inalámbrico para ajustar la posición de los dendrometers de una explotación agronómica," Tesis de grado, Universidad Politécnica de Cartagena,

Colombia, 2013. Available at https://repositorio.upct.es/bitstream/handle/10317/3224/pfc5117.pdf?sequence=2&isAllowed=y
[9] D. Valdiviezo, "Design of a wireless sensor network for precision agriculture," Thesis, Pontificia Universidad Católica del Perú, 2009.
[10] A. López Fidalgo, "Wireless sensor networks for precision agriculture in smallholder regions," presented at the Congreso Nacional del Medio Ambiente, Spain, 2009.
[11] R. Fernández Martínez, A.V. Pernía Espinoza, A.S. Sanz García, J. Las Heras Casas, and J. Cendón, "Nuevas tendencias de tecnologías de sensores inalámbricos para la mejora productiva de cultivos agrícolas," Universidad de León, Madrid, 2010, p. 11.
[12] D. Park, S. Cho, and J. Park, The realization of greenhouse monitoring and auto control system using wireless sensor network for fungus propagation prevention in leaf of crop. In D. Ślęzak, Th. Kim, A. Stoica, and B.H. Kang (eds.), *Control and Automation*. CA 2009. Communications in Computer and Information Science, vol 65. Berlin and Heidelberg: Springer. https://doi.org/10.1007/978-3-642-10741-2_4.
[13] M.F. Ali, D.N.K. Jayakody, Y.A. Chursin, S. Affes, and S. Dmitry, "Recent advances and future directions on underwater wireless communications," *Arch Computat. Methods Eng.*, vol. 27, no. 5, pp. 1379–1412, Nov. 2020.
[14] Z. Xi *et al.*, "Research on underwater wireless sensor network and MAC protocol and location algorithm," *IEEE Access*, vol. 7, pp. 56606–56616, 2019.
[15] I. Ullah, J. Chen, X. Su, C. Esposito, and C. Choi, "Localization and detection of targets in underwater wireless sensor using distance and angle based algorithms," *IEEE Access*, vol. 7, pp. 45693–45704, 2019.
[16] M. Ayaz, A. Abdullah, I. Faye, and Y. Batira, "An efficient dynamic addressing based routing protocol for underwater wireless sensor networks," *Computer Communications*, vol. 35, no. 4, pp. 475–486, Feb. 2012.
[17] R.W.L. Coutinho and A. Boukerche, "Data collection in underwater wireless sensor networks: Research challenges and potential approaches," in *Proceedings of the 20th ACM International Conference on Modelling, Analysis and Simulation of Wireless and Mobile Systems*, Miami, Florida, USA, Nov. 2017, pp. 5–8.
[18] M. Ghaleb, E. Felemban, S. Subramaniam, A.A.A. Sheikh, and S.B. Qaisar, "A performance simulation tool for the analysis of data gathering in both terrestrial and underwater sensor networks," *IEEE Access*, vol. 5, pp. 4190–4208, 2017.

6

Advanced Wireless Communication: Technology Overview, Challenges, and Security Issues

Amit Kumar, Adesh Kumar, Geetam Singh Tomar, and Aakanksha Devrari

CONTENTS

6.1 Introduction to Wireless Communication Networks 65
6.2 Overview of Generations in Wireless Communication 66
6.3 OFDM 69
6.4 Motivation and Need 70
6.5 Challenges in Wireless Communications..74
      6.5.1 Privacy, Secrecy, and Security 75
      6.5.2 Communication Infrastructure...75
      6.5.3 Wireless Energy and Power Transfer 75
      6.5.4 Spectrum Utilization 76
      6.5.5 Modulation and Coding 76
6.6 Conclusion and Future Work...77
Acknowledgments 78
References 78

6.1 Introduction to Wireless Communication Networks

The growth of wireless multimedia communication systems is the current and future trend in the era of technology. In the recent past, a lot of stress has been placed on the use of mobile wireless systems over wired communication networks. The current scenario shows the low data rate services available to users, but the demand for a broadband communication system [1] is becoming the major requirement in both private as well as public sectors. To create communication through multimedia mobile systems, the transmission rate should be as high as possible, in the range of megabits per second. If the transmission rate of data is megabits, then the value of time delay will be greater than the one-time symbol. This delay can be improved using the adaptive equalization technique but there are practical complications in terms of cost and hardware complexity [2]. One solution to overcome the problems of multipath fading and achieve low complexity is OFDM. It reduces the multipath fading effect with the help of a parallel data transmission approach. OFDM [3] is a multicarrier transmission method that transmits a single stream of data over several subcarriers at a lower rate. It can be understood as a multiplexing method or a modulation process. The main reason for using OFDM is its robustness to interference and

DOI: 10.1201/9781003326205-7

frequency selective fading. A single carrier system if subjected to single fading can cause complete link failure whereas a multicarrier system is less affected due to small portion exposure.

The most common scenario in the field of high data transmission is the acceptance of multicarrier interfacing systems such as OFDM. Broadcasting systems of Europe and Australia such as DAB and DVB also utilize the OFDM method. It offers multipath tolerance [4] and has high spectral efficiency, which makes it a suitable choice for broadcasting. The data transmission in OFDM occurs with the help of a parallel set of carriers with low bandwidth. The frequency between the individual carriers is the reciprocal of the symbol period. A proper time window slot is used for the synchronization at the receiver. OFDM uses IFFT technique [5] in the transmitter end and the FFT technique in the receiver end. The data rate of the OFDM system increases by increasing the number of carriers.

OFDM divides the existing bandwidth into several subcarriers, which are orthogonal to each other. In OFDM, the symbol duration is generally increased to enhance the strength of the system in terms of delay. The problem of inter symbol interference is eliminated using cyclic prefix [6] until the interval of the prefix is greater than the actual delay value. A cyclic prefix means the repetition of the last data samples at the beginning of the data sequence. It is also responsible for the low complexity equalization in the frequency domain.

The rapid development of wireless communication networks has begun and fifth generation mobile systems are being launched in the early 2020s. The 4G cellular systems focus on existing technologies, including WWAN, WLAN, and Bluetooth [7]. The 3G system was merely focused on new hardware and new standards. The recent developments in mobile radio coverage and the internet have enhanced the demand for "internet in the pocket" which will increase the data throughput [8] and at the same time reduce the cost of system hardware. Several generations have evolved in mobile telecommunications technology [9]. Every generation of cellular telecommunications has its own needs and improvements. This is the same for all generations of cellular technology.

6.2 Overview of Generations in Wireless Communication

The era of the mobile industry has shown tremendous growth as we move with technological advancement. The wireless technology improvement is defined by the changes in various parameters from time to time along with the band of frequency. These variations are classified as the generations or 'G'. The first generation is represented by 1G, the second generation is represented by 2G, the third generation is represented by 3G, the fourth generation is denoted by 4G, and the fifth generation is denoted by 5G. Each generation has its distinct features, standards, technology, bandwidth, etc. differentiating them from each other. Table 6.1 presents the generations of cellular technologies.

First generation (1G): It was introduced in the early 1980s when the first cellular system was brought into existence by Nippon Telephone and Telegraph (Japan) and was based on analog systems. Later this cellular concept reached European countries. The popular communication system of this era was based on frequency modulation known as TACS and NMT systems. The system was based on analog technology and launched in Finland, Denmark, Sweden, Norway, and Iceland in 1981. These systems have a channel capacity of 30 kHz and frequency ranging from 824 to 894 MHz. The first cellular mobile phone belongs to 1G. Continuous revolution in this field and continuous improvement in time

TABLE 6.1
Generations of cellular technologies [10, 11, 12]

Technology	1G	2G	3G	4G	5G	6G
Requirements	No authorized requirements and it uses analog technology.	No authorized necessities and it uses digital technology.	ITU's IMT-2000 is essential with 144 Kbps mobile data rate, 384 Kbps pedestrian, and 2 Mbps data rate inside.	ITU's IMT innovative essentials comprise the facility to operate up to 40 MHz frequency with the radio channels.	In order to properly care for virtual reality and ultra-high-definition video, data rates of at least 1 GB/s or higher are utilized.	The competency of handling immense data volume and high-data-rate connectivity per device. eMBB, URLLC, mMTC, AI communication, Tangible internet, optimum throughput, maximum network capability, maximum energy competence, and greater data security
Data bandwidth requirements	1.9 Kbps data rate	14.4 Kbps–384 Kbps data rate	1 Mbps Data rate	2 Mbps – 1 Gbps Data rate	1 Gbps and larger than (based on request)	1 terabyte per second
Main network	PSTN	PSTN-based delivery network for packets.	Packet delivery network	IP-based networks.	5G communication and interfacing, Flatter IP system	Cyber twin-based network for 6G core networks is cognitive service architecture.
Service	Analog voice-based services	Digital technology based voice has higher capacity and packetized information.	Integration of high-quality audio, video, and information.	Access to dynamic data, high definition streaming, wearable devices, and a universal roaming facility.	Dynamic information processing, wearable computing devices, HD flowing, universal roaming facility.	The 6G system is projected to provide 1,000 times closer wireless communication than 5G, with far higher data rates and lower latency, and artificial intelligence-based wireless communication tools will be driven.
Standards	NMT, AMPS, Hicap, CAPD, TACS & ETACS	GPRS, GSM, and EDGE based	CDMA and WCDMA20-00 technology	All-access divergence comprising MC-CDMA, OFDMA, and Network LMPS	CDMA & BDMA	FDD, TDD, MWT, Massive MIMO, Dense networks, Future PHY / MAC & Duplex methods
Multiple access method	FDMA	TDMA & CDMA	CDMA	CDMA	CDMA & BDMA	DOMA, NOMA, SCMA & MDAS
System	Analog	Digital	Broad bandwidth/ CDMA/IP	Broadband LAN/ MAN/WAN/PAN and WLAN	Broadband LAN/MAN/ WAN/PAN/ WLAN and cutting-edge technologies grounded using OFDM modulation applied in 5G	5G and satellite networks will be integrated into 6G systems. The following is a list of the satellite systems that have been created in different countries: • Galileo (European Union) • GLONASS (Russia) • GPS (United States of America) • COMPASS (China)

could not be fulfilled by 1G to meet the security and efficiency requirements of the band spectrum.

Second generation (2G): 2G systems were launched around 1991. These systems rely on the use of multiple base stations uniformly scattered all over the world. These stations utilize several multiple access techniques including CDMA, FDMA, and TDMA to establish communication between users. Later on, advanced versions of 2G arrived, such as 2.5G GPRS and the upgraded version 2.75G Enhanced GPRS or EDGE GPRS (EGPRS). The system of 2G technology depended on digital systems providing good security and high spectrum efficiency.

Third generation (3G): The main objective of the 3G system is to provide high-speed data. In the third generation systems, the data rate increases up to 14 Mbps and can be greater than this. 3G is based on International Mobile Telecommunications (IMT-2000) technical standards. These standards support a data speed of at least 200 Kbps. Apart from mobile telephony, higher speed is required for gaming and portable devices like tablets. It also provides security standards by supporting platforms like user authentication. 3G was comprised of three technologies, i.e. CDMA 2000, W-CDMA, and TD-SCDMA.

Fourth generation (4G): Fourth generation systems tend to be the successor of second and third generation systems [13, 14]. This system establishes the perception of anytime, anywhere, and anyhow. The best feature of 4G is its unified access and quality of service. 4G technology is defined as "MAGIC" meaning "Mobile Multimedia, Anywhere, Global Mobility Solutions Over Integrated Wireless and Customized Services." It provides very high data rates, up to 1 Gbps. OFDM technology is utilized in the 4G wireless system. The formats used in OFDM are FBMC, UFMC, and GFDM. Every format has its limitations and advantages. Therefore, adaptive methods can be used in which more flexible communication can be provided for fifth generation mobile systems.

Fifth generation (5G): The standards are not defined and properly set by the governing bodies. The network capacity of the 5G network [15, 16] is expected to be 10,000 times greater when compared to the current data network capacity. 5G assumes a peak data rate of 10 Gbps along with a cell edge data rate of 100 Mbps. The latency is expected to be less than 1ms. Beyond speed, the technology is predicted to unleash a vast 5G Internet of Things network, in which systems can meet the communication requirements of billions of connected devices while preserving a stable balance of latency, speed, and cost.

Sixth generation (6G): Wireless communications technologies that support cellular data networks are now being developed under the 6G standard in telecommunications. It is the expected successor to 5G and will most likely be even faster. The future generation of 6G wireless communication systems is targeted at bringing communication facilities for future requirements in the 2030s, while retaining wireless communication system sustainability and competitiveness, because 6G communication systems will employ revolutionary technologies providing high reliability, low energy consumption, high throughput, network densification, and huge connections, to name a few [17]. The 6G system will also carry on the previous generational pattern with the addition of new technology. Artificial intelligence, implants, smart wearables, sensing, driverless cars, computing reality devices, and 3D mapping are among the innovative facilities. The capacity to switch large amounts of data and actual data rate connection per device is the greatest significant need for 6G wireless communication networks.

The 6G system will improve upon 5G in terms of performance and user QoS while also introducing some novel characteristics. This will keep the system and user data safe. It will deliver useful services. The 6G communication system is envisioned as a worldwide communication infrastructure. In many circumstances, the per-user data rate in 6G is estimated to be around 1 Tb/s. The 6G technology is expected to provide 1,000 times the number of simultaneous wireless connections as the 5G technology. In addition, ultra-long-distance communication with a latency of less than one millisecond is expected. The most intriguing aspect of 6G is the incorporation of fully maintained AI for driving self-governing systems. Video traffic is expected to increase in 6G communications to outnumber other types of data traffic. The terahertz (THz) spectrum, AI, and optical wireless are the most important technologies that will propel 6G forward.

6.3 OFDM

OFDM is used for the transmission of multicarriers [18] where the available band spectrum is divided into multiple subcarriers and then modulation of each carrier is performed using a data stream of low data rate. The available spectrum is efficiently utilized with the help of subcarriers being orthogonal to each other. This leads to the cancellation of inter-carrier interference (ICI) between the carriers.

OFDM converts the selective frequency channels to independent subchannels and reduces the issue of multipath fading in wireless systems. The high-speed wireless LAN standards, LTE and WiMAX, use OFDM and MIMO wireless technologies. The requirement of the high data rate of 100 Mbps in 4G systems is supported using MIMO OFDM systems [19]. Applications that require high throughputs such as interactive gaming, broadband systems, HDTV, and multicast video also utilize OFDM technology. The technology advancement demands the development of MIMO OFDM systems in various respects, and these are proving to be the future wireless technology, along with IEEE 802.16 and IEEE 802.11n wireless standards.

In a traditional parallel data system, frequency band is split into 'N' non-overlapping frequency channels. Each channel is modulated with the help of a distinct symbol. To eliminate interference within the channel it is better to avoid spectral overlapping. To cope with the inefficiency caused by multipath fading and impulsive noise, it was proposed to use FDM along with a parallel data stream.

The difference between the non-overlapping multicarrier approach and the overlapping multicarrier method is seen in Figure 6.1. The bandwidth utilization is lowered by 50 percent when an overlapping approach is used. For the realization of this technique, the crosstalk between the carriers must be reduced to maintain orthogonality.

The term "orthogonal" refers to a relationship between carrier frequencies in a certain system. The carriers are placed in such a manner that signals are received with the help of filters and demodulators. In this kind of receiver, guard bands are used between carriers which reduces the spectrum efficiency of the system. However, in an OFDM system, these carriers can be rearranged in a special manner making the sidebands overlap with each other. Here the receiver is a bank of de-modulators that lowers down carriers to a particular dc level and thus the resultant data can be recovered after integration over a period.

If the distance between the carriers is a multiple of the factor $1/T$, it means carriers are linearly independent with respect to each other, i.e. orthogonal. Figure 6.2 shows the OFDM subchannel spectra and OFDM signal. The OFDM technology was used in a variety of military applications in the early 1960s, including KATHRYN and CINEPLEX. In the 1980s, OFDM was studied for high-density recording and modems with high speed. These techniques were also used in quadrature amplitude modulation. In the late 1990s,

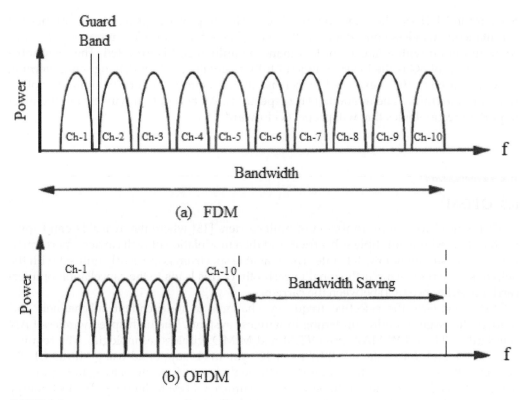

FIGURE 6.1
OFDM signal concept: (a) conventional multicarrier method, (b) orthogonal multicarrier modulation method [18, 19]. Figure 6.1 (a) is a graph showing frequency on x axis and power on y axis with different channels from 1 to 10 showing guard band between individual channels.

OFDM came into use for wideband data systems in digital subscribers ADSL, HDSL, DAB, and VDSL.

The key advantages of the OFDM technique are as follows:

- It is one of the most successful methods for overcoming the effects of multipath distortions; also its hardware complexity is less compared to an equalizer-based single carrier system.
- The capacity of OFDM can be significantly increased for time-varying channels by using the data rate following the signal-to-noise ratio of subcarriers.
- It is powerfully resistant to narrowband interference as only small portions of subcarriers are affected.

6.4 Motivation and Need

Currently, the OFDM systems are restricted to a single user for data transmission onto subcarriers at a given point in time. For supporting multiple users, OFDM uses multiple

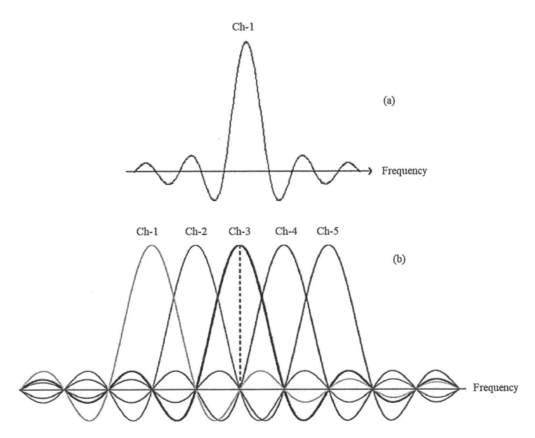

FIGURE 6.2
(a) Spectra of OFDM subchannel, (b) Spectra OFDM [20].

access schemes like time division and frequency division. Generally, in a two-way communication system, the major requirement is that the system should support multiple users. The OFDM system can be made functional for multiuser applications. The implementation of multiuser OFDM systems is a hot area of research in the field of wireless communication systems. It was first suggested by Wahlqvist that these systems are reliable and can be implemented for various applications. Strong interest has been shown by researchers in making OFDM work for multiuser systems. One notable example following this trend is OFDMA, a combination of OFDM and FDMA.

OFDMA is also identified as a multiuser OFDM system modulation and multiple access methods for 4G wireless systems. It can be seen as an extension of OFDM and a popular choice for high data speed systems like IEEE 802.11a/g local area networks (LANs) and fixed broadband systems such as IEEE 802.16a. Nowadays OFDMA has gained significant interest and is therefore being adopted amongst the modes of physical layer in the IEEE standard 802.16-2004. In OFDMA, the subcarriers are divided into subchannels, which are then assigned to different users for making concurrent communications. To preserve orthogonality between the subcarriers' signals all the users must be synchronized with base stations.

OFDMA is the multiuser variety of the OFDM modulation technique. Multiple access through OFDMA can be achieved by allotting subsets of each subcarrier to individual

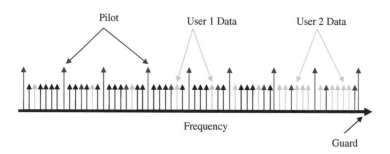

FIGURE 6.3
OFDMA carriers [21].

users, as shown in Figure 6.3. It allows simultaneous data transmission at a lower rate from several users.

The several advantages of OFDMA over OFDM using time-domain statistical multiplexing are as follows:

- It allows simultaneous transmission of data at a low data rate from different users.
- In this method, it is possible to avoid the carrier pulse.
- Transmission power is lowered for the user with a low data rate.
- It allows constant and shorter delay.
- It simplifies the multiple access problem by avoiding collision which improves the robustness of OFDM to interference and fading effects.

OFDMA [22] technique is usually carried in the digital environment and there are so many methods available for its implementation. One such method for the implementation of OFDMA is FPGA. FPGA is user programmable and therefore a designer can fully control the design without the requirement of manual fabrication technique. This module in itself combines speed, power, and other important attributes (such as density of an ASI) along with the processor to give multiple benefits to the OFDM system. This allows the user to reprogram the FPGA to meet future requirements at the time of design fabrication. It will prove to be the optimal choice for the implementation of OFDMA due to its flexibility in design and low cost as compared to other techniques.

The key difference between an OFDM and an OFDMA system is that in the former, users are distributed based on time, while in the latter, users are allocated based on both time and frequency, as depicted in Figure 6.4. This is beneficial for LTE because it allows frequency-dependent scheduling. It could conceivably, for example, make it more likely that user one has a better radio link quality in a specified bandwidth area of the available bandwidth. The primary bandwidth is split up into multiple subcarriers in OFDM systems. Each of these subcarriers can be tuned separately. Hundreds of subcarriers with a substantive separation between them are frequent in the OFDM system. This increases the technology's robustness in mobile propagation situations.

When transmitting the OFDM signal, the symbols that need to be delivered after these symbols have been modulated are suppressed. These symbols are then employed as input bands in an inverse fast Fourier transform. Performing this procedure results in the generation of OFDM symbols, which are then broadcast. In the process of using IFFT, a transformation from the frequency domain to the time domain is carried out. However, in order

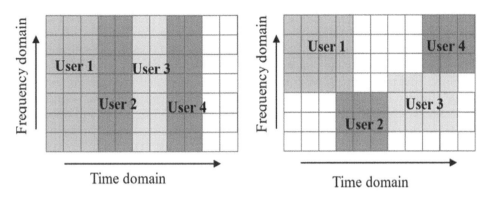

FIGURE 6.4
OFDM vs. OFDMA [23, 24].

to prevent interference between the OFDM symbols during transmission, a cyclic prefix is added to each symbol before transmission begins. LTE's cyclic prefix has an initial symbol that is 5.2 microseconds, the remainder of the symbols are 4.7 microseconds, and there is an extended cyclic prefix for cells that are larger.

At the OFDMA system's receiving end, an FFT operation transforms the symbol back to the frequency domain. Signals coming from several sources are transmitted over multiple frequencies simultaneously in an FDM system. A guard band space is provided between subcarriers in order to eliminate the overlapping of signals, and each frequency range is modulated individually by a different data stream. OFDM is also known as discrete multi-tone modulation because it uses some kind of QAM [25] to modulate a large number of evenly spaced subcarriers rather than a single carrier because there are more carriers to modulate. This spread-spectrum technology enhances data communications productivity by improving the throughput.

In addition, issues with multi-path signal cancellation and interference in the spectrum can be considerably mitigated by selectively modulating the "clean" carriers or avoiding carriers with severe bit-rate defects. Both of these strategies can be implemented simultaneously. OFDM, much like FDM, uses multiple subcarriers, but the subcarriers in OFDM are closely separated to prevent interference, and guard bands are not used between subsequent subcarriers because the frequencies (subcarriers) are orthogonal to one another. A data stream moving at an exceptionally high rate is broken up into a great number of data streams moving at a low rate simultaneously in an OFDM system. After that, each of the smaller data streams is moved to its own data subcarrier and modulated using techniques such as QAM or PSK [26]. In order to transmit the same amount of data, OFDM requires less bandwidth than FDM does, which ultimately results in improved spectral efficiency. Because equalization is performed on a subset of subcarriers rather than a single broad carrier, an OFDM system, such as WiMAX, is more resistant to interference and frequency-selective fading in an NLOS environment. This is because OFDM systems, like WiMAX, perform equalization on a subset of subcarriers.

The effect of ISI is lessened as a result of the parallel OFDM subcarriers having a longer symbol period than a single carrier system with a cyclic prefix would have. The OFDM spread-spectrum technique is utilized in the transmission of digital television in Australia, Japan, and Europe; digital audio broadcasting in Europe; ADSL modems; and global wireless networking (IEEE 802.11g) [27].

OFDMA is very similar to OFDM in that it uses a number of closely spaced subcarriers, but the subcarriers in OFDMA are grouped together. Every group has a specific subchannel that serves as its name. It is not necessary for the individual subcarriers that make up a subchannel to be physically next to one another. There may be several receivers connected to a single subchannel in the downlink at any given time. When transmitting in the uplink direction, a transmitter might be given one or more subchannels to use. The term "subchannelization" refers to the process of allocating different channels to subscriber stations (SSs) based on the circumstances of the channels they use and the data they require. A mobile WiMAX base station (BS) has the ability to allocate more transmit power to devices that have a lower SNR and less power to devices that have a higher SNR within the same time slot. The current demand for ubiquitous telecommunication technologies should be met by the capacities of 4G and 5G standards to enable services. OFDM technology is in great demand as the whole frequency spectra are distributed into several narrowband subchannels.

The hardware implementation of multicarrier-based broadband communication technologies is a challenging task. The main focus of the research area is to explore the clustered OFDMA on the FPGA platform. The design of OFDMA with different FFT and IFFT will explore the solution relating to optimal hardware requirements for the OFDMA transreceiver chip. The main aim of the research is to provide low-demand hardware, memory, timing, delay, and power consumption with FPGA for high-speed computation methods. It is possible to check the complexity of hardware chips with different FFT algorithms to estimate which FFT provides the optimal hardware utilization on FPGA. Different consumers see different channel quality. A channel with a deep fade may be favorable for another user. It also enables multiuser diversity and the ability for various users to broadcast across different portions of the broadband spectrum in the traffic channel.

The CORDIC [28] algorithm has been utilized for several signal-processing applications. The CORDIC integrated FFT will solve the trigonometric and mathematical problems in using FFT and IFFT computations. The scalable design and implementation of the OFDMA chip on FPGA will give faster switching and frequency support, which is the basic need for 5G communication. The state of the art work will be that the designed chip can be configured for MIMO systems and provides the hardware solution for machine-to-machine or device-to-device communication. The scalable OFDMA chip has the advantage of supporting multiuser activity and provides low latency, maximum frequency, and best efficiency.

The FPGA implementation of OFDMA [29] will provide optimal hardware, maximum frequency support, minimum hardware delay, low power consumption, faster switching, larger throughput, and high mobility. The realization of the OFDMA hardware chip on the latest FPGA will be a great achievement for the researchers looking for the domain that satisfies device-to-device or machine-to-machine communication. The synthesized code will be processed for STA, floor planning, and focused layout design [30]. It will enable the industries for the new revolution in machine-to-machine communication.

6.5 Challenges in Wireless Communications

Advanced wireless communications have security as the major concern. Apart from that, communication infrastructure, spectrum allocation, energy efficiency, wireless power transfer dynamic allocation, coding, and modulation are the concerns of future communication systems [31, 32].

6.5.1 Privacy, Secrecy, and Security

Wireless communication networks are transferring a large volume of sensitive data. Wireless transmissions, on the other hand, are broadcast, making the information sent vulnerable to eavesdropping. Novel communication technologies should be applied to enhance information privacy to proficiently support miscellaneous high secrecy complex uses. Exploring wireless propagation characteristics could mitigate the safety concerns about the performance of wireless communication. The difficulty is finding out how to get the most out of such an investigation while yet conserving the secrecy of real customers. The wireless community remains interested in advanced communication techniques that expand privacy and secrecy.

A wireless LAN is an alternative to a wired LAN. Wireless LANs are attractive and a more common standard choice for an extensive spectrum of commercial consumers as a common drive connectivity solution. However, one of the most significant disadvantages is that wireless LANs are insecure, and data processed over them can be readily broken and manipulated. Because data exchanged through a wireless network is broadcast for the entire neighborhood to hear, wireless network security is significantly more crucial and mandatory than wired network security. When sensitive data is delivered over the airways, wireless structures should not be used unless exact precautions are included. A defined and specific level of security is required for all wireless networks. When sensitive data, such as that from financial institutions, banks, military networks, or information on terrorists, is transmitted through a wireless network, special safeguards should be taken to preserve privacy and secrecy. Otherwise, it is easy to see how something useful might turn destructive. Cyber security [33, 34] is a major concern nowadays and different types of attacks occur. Penetration, denial of service attacks, and sabotage are examples of wireless attacks against wireless system information or wireless networks. The security [35, 36] can be handled by integrating the cryptography decryption and encryption algorithms by embedding the hardware chip [37, 38] in the network or the communication system.

6.5.2 Communication Infrastructure

Communication infrastructure [39] is a very important aspect of meeting the increased demand for wireless services and involves using a unique system structure to improve resource consumption. Numerous base stations will collaborate to serve multiple users utilizing separate antenna arrays, contributing to a high spectrum utilization efficiency and dynamically adapting to changes in service needs and propagation conditions in cloud-assisted joint signal processing. Understanding the potential everyday benefits of real-time channel evaluation, management in the time and frequency domains, and integrated handling at fully connected base stations, while minimizing system complexity and implementation costs, is the objective. Wireless infrastructure, whether cellular or satellite, is a business-to-business rather than a consumer-to-business market segment. As a result, performance demand has always driven the market rather than cost. Discrete compound technology components make up the majority of an RF card in the actual base station or with a low-noise slab down converter module in the satellite receiver.

6.5.3 Wireless Energy and Power Transfer

As a result of the huge number of connected nodes predicted in 5G and 6G networks, a variety of potential answers have been proposed for the construction of self-sufficient communication systems. Integration of WIPT in particular allows active energy replacement for wireless devices and is an advantageous approach for powering energy-constrained wireless networks. This is especially critical for upcoming wireless networks, as the next generation of IoT devices will necessitate more power due to smart processing's huge computing requirements. In contrast, WIPT presents various technological challenges due to the limited range of wireless power transfer. In the majority of instances, the highest power efficiency that may be achieved in the far field cannot exceed 50 percent. Therefore, further research must be conducted on far field to improve energy transmission efficiency and directivity in a variety of communication requirements and technologies. SWIPT is a revolutionary concept for wireless communication systems [40] that enables wireless nodes to recharge their rechargeable batteries using RF signals (instead of static lines or conventional energy sources) while decoding data. Forthcoming developments in WPT efficiency will expose a plethora of potentials for the equipment's indispensability in different types of embedded systems, such as wireless cameras and sensors. The following are just a few of WPT's many practical advantages. Due to its broadcast nature, it is able to supply energy to a number of different receivers at the same time.

- Low level of complexity
- The ER hardware size and expense
- Mobile device's long-term viability
- Small-power sensor devices, such as RFID tags, can benefit from this technology

6.5.4 Spectrum Utilization

Spectrum sensing [41] is the process of gathering information about spectrum usage and the presence of a primary user (PU) in a specific geographical region. In addition to time, frequency, and spatial dimensions, polarization and valuable information can be incorporated to deliver more specific data regarding spectrum prospects. Energy detection is the most often utilized method due to its low computational and execution complexity. This method does not require any previous knowledge of the PU signal, and it just compares the received signal power to a predefined threshold to detect the signal. Matching filter and feature detection are more advanced algorithms that are more difficult for the instrument but provide improved sensing accuracy, particularly with low SNR values. The feature detection takes advantage of the principal signal's periodicity or statistical features to discriminate between different types of PUs. New spectrum bands, such as millimeter-wave and terahertz, are being researched due to the scarcity and desirability of wireless-communications-appropriate bandwidth (THz). The RF band possesses more bandwidth and the coverage region is often less due to the losses during propagation. The application of guiding beamforming communication technology is the greatest common mitigating method. The issue is determining how to swiftly update beamforming recommendations when the environment changes, particularly in multiuser environments. The development of cost-effective THz wireless transceivers is another application that requires attention.

6.5.5 Modulation and Coding

Everything is intelligently possible because of the IoT [42]. People will have more convenience and higher quality existence in the IoT era when products like smart cities and smart homes are implemented. 5G serves as a precursor and basis for accomplishing this goal. The base station in a 5G mobile communication system provides adaptive coding and modulation depending on channel status information provided by the user and enhances spectrum efficiency by combining multiple modulation types and code rates. As a result, the receiver's estimated channel state must be mapped to a value that represents the overall channel state and transmitted back to the base station. Furthermore, a good technique for achieving considerably higher spectral efficiency in forthcoming wireless networks is to use improved modulation and network coding algorithms, as well as novel modulation resources. Individual wireless generation is controlled using the different coding schemes as the technology is progressed. For example, in 2G, convolutional codes were explored, in 3G and 4G, turbo codes and improved turbo codes were used, and in 5G, LDPC and polar codes were used. The majority of the investigated codes were created with limited coding rate capabilities, implying that their performance was optimized for specified coding rates. For channel coding in 5G, we'll need improved error correction algorithms. LDPC and polar codes are used for error correction to meet 5G communication requirements. LDPC codes are utilized for the information channel while polar codes are used for the control channel in 5G new radio (NR). This code has a significant coding gain. It has several advantages, including low computation power, high throughput, and low latency. The LDPC and Polar Codes encoding and decoding [43] methods are used for efficient and error-free communication in 5G.

Techniques known as NOMA [44] are attracting a lot of attention as potential components of the 5G communication system. The primary motivation behind NOMA's incorporation into 5G networks is that it enables multiple users to make efficient use of the same set of time and frequency resources. NOMA approaches fall into one of two categories: either the code domain or the power domain [45]. In the power domain, NOMA is able to accomplish multiplexing, while in the code domain, NOMA is also able to accomplish multiplexing.

6.6 Conclusion and Future Work

Fifth generation (5G) methods are approaching standardization in the wireless domain. Several results for considerably successful system performance have been developed because of investigation. Industry can now relate these solutions because of the growth of enabling technologies comprising modulation, adaptive coding, iterative decoding systems, and space-time coding. The chapter presented an introduction to advanced wireless communication and an overview of the cellular technologies starting from 1G to 6G for wireless communication. IoT is the noble technology of 5G and 6G that uses the features of configurability and adaptability that will be the focus of future system requirements and features in 5G system and performance. Advancements have been seen in modulation techniques for different generations. Adaptive wideband CDMA, multicarrier OFDMA, adaptive TDMA, and UWB receiver elements are among the systems that make up 4G and 5G. OFDM is the new wireless communication modulation system that includes 5G. OFDM is a high-data-rate communication technique that combines the benefits of QAM and FDM. OFDM is an effective method for achieving high data rates without sacrificing spectral efficiency. OFDMA is an improvement to OFDM that allows several devices to share a wireless channel, allowing for more efficient use of the channel. In 4G and 5G networks, OFDMA is already in use. In the wireless LAN context, the current Wi-Fi standard (IEEE 802.11ax) uses OFDMA for both uplink and downlink data transfer. It has been pointed out that communication infrastructure, cyber security, energy efficiency, spectrum allocation, coding, wireless power transfer dynamic allocation, and modulation are the concerns of future communication systems. LDPC and polar codes are extensively used in current communication systems and will be in the future. In the future, research will focus on the hardware chip solution and effective communication for device-to-device and machine-to-machine with scalable configuration and design. On the other hand, 5G specifications firmly establish the necessity of developing new technologies. The data rate, latency, capacity, resource sharing, reliability, and energy bit are some of these. To achieve these challenging requirements, researchers are focusing their efforts on 6G wireless communications, which will empower a change with new hardware technologies, security, and applications.

Acknowledgments

We thank our parents for their support and motivation.

References

[1] Berezdivin, R., Breinig, R., & Topp, R. (2002). Next-generation wireless communications concepts and technologies. *IEEE communications magazine*, 40(3), 108–116.

[2] Zhuang, W., & Ismail, M. (2012). Cooperation in wireless communication networks. *IEEE Wireless Communications*, 19(2), 10–20.

[3] Chhaya, L., Sharma, P., Bhagwatikar, G., & Kumar, A. (2017). Wireless sensor network based smart grid communications: Cyber-attacks, intrusion detection system and topology control. *Electronics*, 6(1), 5.

[4] Chhaya, L., Sharma, P., Kumar, A., & Bhagwatikar, G. (2017). Communication theories and protocols for smart grid hierarchical network. *Journal of Electrical and Electronics Engineering*, 10(1), 43.

[5] Sood, S. Singh, A., and Kumar, A. (2013). VHDL design of OFDM transreceiver chip using variable FFT. *Journal of Selected Areas in Microelectronics (JSAM), Singaporean Journal of Scientific Research (SJSR)*, 5, 47–58.

[6] Baliyan, N., Verma, M., & Kumar, A. (2017). Channel capacity in MIMO OFDM system. In *Proceeding of International Conference on Intelligent Communication, Control and Devices* (pp. 1113–1120). Singapore: Springer.

[7] Kumar, A., Kumar, A., & Tomar, G. S. (2022). Hardware chip performance of CORDIC based OFDM transceiver for wireless communication. *Computer Systems Science and Engineering*, 40(2), 645–659.

[8] Kumar, A., Kumar, A., & Devrari, A. (2021). Hardware chip performance analysis of different FFT architecture. *International Journal of Electronics*, 108(7), 1124–1140.

[9] Armstrong, J. (2009). OFDM for optical communications. *Journal of Lightwave Technology*, 27(3), 189–204.

[10] Kumar, B. A., & Rao, P. T. (2015, July). Overview of advances in communication technologies. In *2015 13th International Conference on Electromagnetic Interference and Compatibility (INCEMIC)* (pp. 102–106). IEEE.

[11] Ohmori, S., Yamao, Y., & Nakajima, N. (2000). The future generations of mobile communications based on broadband access technologies. *IEEE Communications Magazine*, 38(12), 134–142.

[12] Lakshmanan, M. K., & Nikookar, H. (2006). A review of wavelets for digital wireless communication. *Wireless Personal Communications*, 37(3), 387–420.
[13] Nakajima, N., & Yamao, Y. (2001). Development for 4th generation mobile communications. *Wireless Communications and Mobile Computing*, 1(1), 3–12.
[14] Akyildiz, I.F., Gutierrez-Estevez, D. M., Balakrishnan, R., & Chavarria-Reyes, E. (2014). LTE-advanced and the evolution to beyond 4G (B4G) systems. *Physical Communication*, 10, 31–60.
[15] Akyildiz, I. F., Gutierrez-Estevez, D. M., & Chavarria-Reyes, E. (2010). The evolution to 4G cellular systems: LTE-advanced. *Physical communication*, 3(4), 217–244.
[16] Wang, C. X., Haider, F., Gao, X., et al. (2014). Cellular architecture and key technologies for 5G wireless communication networks. *IEEE Communications Magazine*, 52(2), 122–130.
[17] Bagwari, A., & Singh, B. (2012, November). Comparative performance evaluation of spectrum sensing techniques for cognitive radio networks. In *2012 Fourth International Conference on Computational Intelligence and Communication Networks* (pp. 98–105). IEEE.
[18] Fadhil, T. A., Fadhil, H. A., & Aljunid, S. A. (2008). The application of OFDM in digital multimedia broadcasting. *2008 International Conference on Electronic Design*. IEEE.
[19] Hussein, E. A. (2019). Enhancing the BER and ACLR for the HPA using pre-distortion technique. *International Journal of Electrical and Computer Engineering*, 9(4), 2725.
[20] Sampath, H., Talwar, S., Tellado, J., Erceg, V., & Paulraj, A. (2002). A fourth-generation MIMO-OFDM broadband wireless system: Design, performance, and field trial results. *IEEE Communications Magazine*, 40(9), 143–149.
[21] Mach, P., & Bešták, R. (2006). *Implementation of OFDM into broadband wireless networks*. Czech Technical University in Prague.
[22] Yin, H., & Alamouti, S. (2006, March). OFDMA: A broadband wireless access technology. In *2006 IEEE Sarnoff Symposium* (pp. 1–4). IEEE.
[23] Difference between OFDM and OFDMA. DifferenceBetween.com. www.differencebetween.com/difference-between-ofdm-and-vs-ofdma
[24] OFDMA. 4G-LTE/LTE-A: Coursework for Computer Networks II. www.gta.ufrj.br/ensino/eel879/trabalhos_vf_2014_2/rafaelreis/ofdma_scfdma.html
[25] Hong, S., Sagong, M., Lim, C., Cho, S., Cheun, K., & Yang, K. (2014). Frequency and quadrature-amplitude modulation for downlink cellular OFDMA networks. *IEEE Journal on Selected Areas in Communications*, 32(6), 1256–1267.
[26] Wei, Z., Zhang, L., Wang, L., et al. (2020). Multi-user high-speed QAM-OFDMA visible light communication system using a 75-μm single layer quantum dot micro-LED. *Optics Express*, 28(12), 18332–18342.
[27] Mohammed, M. L., & Hasan, F. S. (2022). Design and performance analysis of frequency hopping OFDM based noise reduction DCSK system. *Bulletin of Electrical Engineering and Informatics*, 11(3).
[28] Angarita, F., Canet, M. J., Sansaloni, T., Perez-Pascual, A., & Valls, J. (2008). Efficient mapping of CORDIC algorithm for OFDM-based WLAN. *Journal of Signal Processing Systems*, 52(2), 181–191.
[29] Jallouli, K., Mazouzi, M., Diguet, J. P., Monemi, A., & Hasnaoui, S. (2022). MIMO-OFDM LTE system based on a parallel IFFT/FFT on NoC-based FPGA. *Annals of Telecommunications*, 1–14.
[30] Shajin, F. H., & Rajesh, P. (2022). FPGA realization of a reversible data hiding scheme for 5G MIMO-OFDM system by chaotic key generation-based paillier cryptography along with LDPC and its side channel estimation using machine learning technique. *Journal of Circuits, Systems and Computers*, 31(5), 2250093.
[31] Koyuncu, H., Bagwari, A., & Tomar, G. S. (2020). Simulation of a smart sensor detection scheme for wireless communication based on modeling. *Electronics*, 9(9), 1506.
[32] Arya, G., Bagwari, A., & Chauhan, D. S. (2022). Performance analysis of deep learning based routing protocol for an efficient data transmission in 5G WSN communication. *IEEE Access*, 22(24), 9731.

[33] Yadav, M., Singh, K., Pandey, A. S., Kumar, A., & Kumar, R. (2022). Smart communication and security by key distribution in multicast environment. *Wireless Communications and Mobile Computing*, 2022.

[34] Kumar, N., Mishra, V.M., & Kumar, A. (2021). Smart grid and nuclear power plant security by integrating cryptographic hardware chip. *Nuclear Engineering and Technology*, 53(10), 3327–3334.

[35] Parreño, I. F., & Avila, D. F. (2022). Analysis of the cybersecurity in wireless sensor networks (WSN): A review literature. In A. Rocha, C. H. Fajardo-Toro, & J. M. Riola (eds.), *Developments and Advances in Defense and Security* (pp. 83–102). Singapore: Springer.

[36] Mishra, V. M., & Kumar, A. (2022). FPGA integrated IEEE 802.15. 4 ZigBee wireless sensor nodes performance for industrial plant monitoring and automation. *Nuclear Engineering and Technology*, 54(7), 2444–2452.

[37] Kumar, A., Kuchhal, P., & Singhal, S. (2015). Secured network on chip (NoC) architecture and routing with modified tacit cryptographic technique. *Procedia Computer Science*, 48, 158–165.

[38] Kumar, N., Mishra, V. M., & Kumar, A. (2020). Smart grid security with AES hardware chip. *International Journal of Information Technology*, 12(1), 49–55.

[39] Revathi, S., Shrivastava, A., Yussupova, A., Hidayatulloh, A. N., Saltanat, D., & Mishra, A. (2022, April). Role of wireless communications in digital economy in the present context. In *2022 6th International Conference on Trends in Electronics and Informatics (ICOEI)* (pp. 703–709). IEEE.

[40] Clerckx, B., Popović, Z., & Murch, R. (2022). Future networks with wireless power transfer and energy harvesting. *Proceedings of the IEEE*, 110(1), 3–7.

[41] Sood, E. V., & Singh, E. M. (2025). Spectrum utilization by using cognitive radio technology. *Communications*, 14, 2110is.

[42] Malik, P. K., Sharma, R., Singh, R., et al. (2021). Industrial Internet of Things and its applications in industry 4.0: State of the art. *Computer Communications*, 166, 125–139.

[43] Fang, J., Che, Z., Jiang, Z. L., et al. (2017). An efficient flicker-free FEC coding scheme for dimmable visible light communication based on polar codes. *IEEE Photonics Journal*, 9(3), 1–10.

[44] Trivedi, V. K., Ramadan, K., Kumar, P., Dessouky, M. I., & Abd El-Samie, F. E. (2019). Enhanced OFDM-NOMA for next generation wireless communication: A study of PAPR reduction and sensitivity to CFO and estimation errors. *AEU – International Journal of Electronics and Communications*, 102, 9–24.

[45] Huang, Y., Zhang, C., Wang, J., Jing, Y., Yang, L., & You, X. (2018). Signal processing for MIMO-NOMA: Present and future challenges. *IEEE Wireless Communications*, 25(2), 32–38.

Part II

Advanced Wireless Sensor Networks: Architecture, Consensus, and Future Trends

7
Advanced Wireless Sensor Networks: Introduction and Challenges

Pooja Joshi, Somil Kumar Gupta, Kapil Joshi, and Jyotshana Bagwari

CONTENTS
7.1 Introduction 83
7.2 Flashback to Sensor Network Childhood 84
    7.2.1 Transitions in WSN Technology 84
7.3 Characteristics of WSN 86
7.4 Fundamental Architecture 87
    7.4.1 The Sensing Sub-System 89
        7.4.1.1 Data-centric Architectures 89
        7.4.1.2 Hierarchical Architectures 90
        7.4.1.3 Location-based Architectures 90
        7.4.1.4 Mobility-based Architectures 90
        7.4.1.5 QoS-based Architectures 90
        7.4.1.6 Network Flow Architecture 90
        7.4.1.7 Multipath-based Architectures 91
        7.4.1.8 Heterogeneity-based Architectures 91
7.5 Applications where Advanced WSN is Applied 91
    7.5.1 Structural Health Monitoring 91
    7.5.2 Traffic Control 92
    7.5.3 Telemedicine 92
    7.5.4 5G Communication (Radio Nodes) 92
    7.5.5 Pipeline Monitoring 94
    7.5.6 Precision Agriculture 94
7.6 Conclusion 95
References 95

7.1 Introduction

Typically, a wireless sensor network (WSN) is a technology that monitors physical or environmental properties like temperature, humidity, sound, luminosity, or pollutants and transfers these parameters via a network to another point for further analysis and visualization. The major advantage of the WSN is that it requires no or fewer infrastructures, and it can be configured itself. Due to these features of the WSN, it is becoming the preferred tool to capture the environmental parameters in those locations which are out of the human ambit. In the past couple of decades, wireless sensor networks are becoming popular due to the development of smart sensors with the help of growing micro-electro-mechanical system (MEMS) based technology [1]. As compared to the traditional sensors, these smart sensors, because of limited processing power and computational resources, are compact and cheaper. A wireless sensor node is a device that combines sensing and computing capabilities. These sensor nodes are smart enough to detect, evaluate, and acquire data from the environment, and then relay that information to the user depending on a local decision-making process [2]. A WSN often has tens of thousands of sensor nodes that communicate via radio transmissions.

7.2 Flashback to Sensor Network Childhood

To understand the popularity of modern WSN, it becomes essential to track its inception. As we see, the development of many modern technologies happened generally in the military and industrial domains, and the underlying framework of WSN was also developed by the military, but the WSN applications that we see today are far different to their original versions developed many decades ago. The first-of-its-kind system, the Sound Surveillance System (SOSUS), was designed by the US military in the 1950s to identify and track Soviet submarines [3]. The newly developed acoustic sensors named hydrophones were used in that network. Those sensors were placed all over the Atlantic and Pacific oceans. This system is still in use, not for tracking submarines but for monitoring aquatic life.

The very first attempt to implement distributed networks was made by the United States Defense Advanced Research Projects Agency (DARPA) in the 1980s by initiating the Distributed Sensor Network (DSN) program [4]. The prime objective of this program was to develop the hardware for a globally distributed network, the Internet of today. Because of this program, many academic institutions like Carnegie Mellon University and the Massachusetts Institute of Technology (MIT) have shown interest in researching WSN technology. As a result, WSNs were utilized to monitor air quality, detect forest fire, prevent natural disaster, forecast the weather, and many more purposes.

Later on, big businesses like IBM and Bell Labs etc. started developing large-scale industry-driven applications using WSN features in various domains such as power distribution, wastewater management, and specially curated automatic assembly lines for manufacturing systems [5]. Priority was given to performance and operability in the development of earlier applications in the fields of science, technology, heavy industry and the military, over various crucial parameters such as hardware, standards, power utility, and scalability.

7.2.1 Transitions in WSN Technology

In the twentieth century, due to limited resources (hardware and software), there was no application based on WSN that could be used for large-scale industry. Many academic organizations and industries made joint ventures to develop such kinds of systems by removing obstacles to development. These gradual developments can be seen in Figure 7.1.

By reducing the cost and energy used per sensor and streamlining development and maintenance duties, several of these projects and standard organizations want to make it simpler to deploy WSNs in large numbers in minor industrial and commercial applications. Figure 7.2 provides a concise summary of the history of sensor networks.

Advanced Wireless Sensor Networks

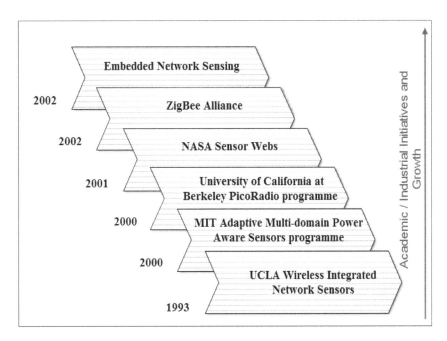

FIGURE 7.1
Development of WSN-based industrial applications.

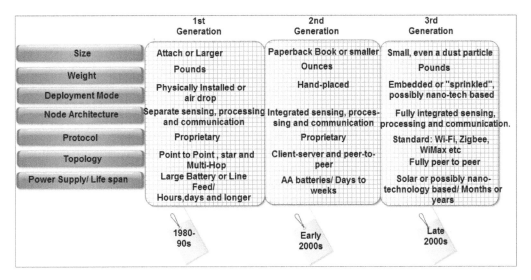

FIGURE 7.2
History of commercial sensor networks.

The major objective of such types of development by different organizations was to position WSN in non-heavy-industrial applications by means of reducing the overall cost of each sensor.

7.3 Characteristics of WSN

Following are some major characteristics of a WSN:

- Power utilization of a WSN is restricted.
- Ability to handle node failures.
- Node homogeneity is a term used to describe the consistency of nodes.
- Large-scale deployment capability.
- Capacity to withstand adverse environmental conditions.
- Easy to utilize.

These are a handful of WSN's most important and common traits. However, the properties of WSN for diverse uses may differ significantly. They can also share similar traits. Most of those features are likely to vary depending on the application. It may be appropriate in some instances, but it may not be considered in others. Some common characteristics are shown in Figure 7.3.

- As we know, the restricted power resources of sensor networks pose several issues. The size of the nodes sets a limit on the battery's capacity. The software and hardware architecture must take energy efficiency into serious consideration. For WSN nodes that are powered by non-rechargeable batteries, specific power-management algorithms are required [6]. The power-devouring qualities of the node must be noted for three components: the sensing circuitry, the digital processing unit, and the radio transceiver unit.
- Depending on the demand, the nodes and their components may be employed in several power setting modes, such as "always-on," "standby," or "hibernation." Because the energy policy varies by application, it may be permissible to turn off a subset of nodes to save energy in some cases, but in others, all nodes must be operational at the same time. This aids in the operation of WSN in low-power mode [7].
- To work even in adverse settings and maintain their responsiveness, the sensor nodes must become self-contained.
- The amount of energy consumed for radio transmission must be reduced while constructing the WSN; however, nodes can utilize more energy for computing and/or filtering.
- Because of the large number of nodes and comparatively high node compactness, the scalability of routing methods employed in WSNs is a hot topic. A good routing protocol must be scalable and easily adaptable to network topology changes [8].
- Mobility is a component that determines the performance of wireless sensor networks. Indeed, in an IP-based architecture, a mobile node must retain connectivity with other sensors while also allowing access to the Internet. Each of these features has its own significance [9].

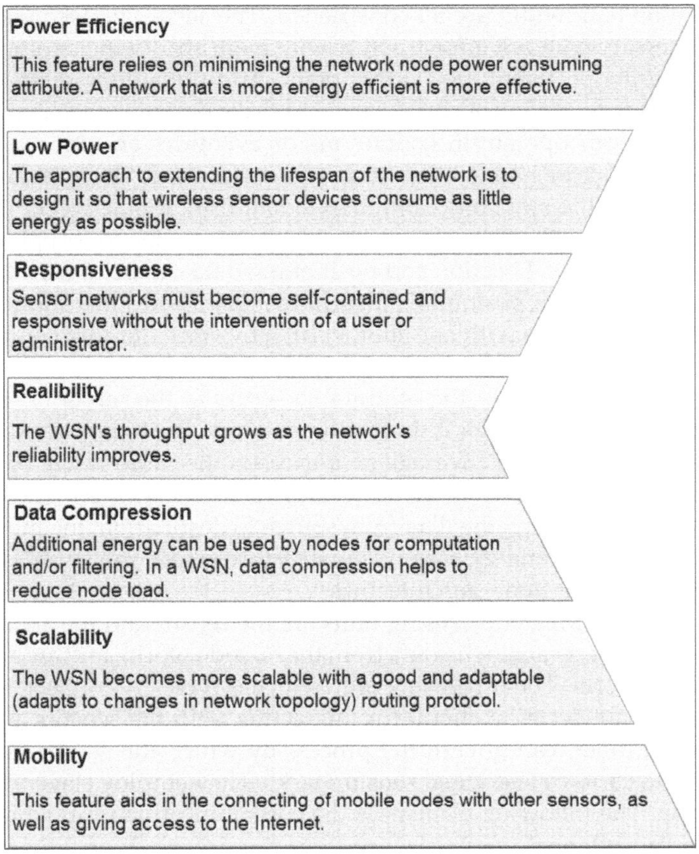

Power Efficiency
This feature relies on minimising the network node power consuming attribute. A network that is more energy efficient is more effective.

Low Power
The approach to extending the lifespan of the network is to design it so that wireless sensor devices consume as little energy as possible.

Responsiveness
Sensor networks must become self-contained and responsive without the intervention of a user or administrator.

Realibility
The WSN's throughput grows as the network's reliability improves.

Data Compression
Additional energy can be used by nodes for computation and/or filtering. In a WSN, data compression helps to reduce node load.

Scalability
The WSN becomes more scalable with a good and adaptable (adapts to changes in network topology) routing protocol.

Mobility
This feature aids in the connecting of mobile nodes with other sensors, as well as giving access to the Internet.

FIGURE 7.3
Major characteristics of a WSN.

7.4 Fundamental Architecture

A wireless sensor network is a mobile system composed of autonomous, geographically distributed devices that use sensors to keep track of physical or environmental parameters. The WSN system includes a gateway that offers distributed nodes and wireless communication to the wired world. It can also be described as a self-configured and infrastructure-free wireless network that collects data from various physical or environmental variables, such as temperature, sound, vibration, pressure, motion, or pollutants, and cooperatively transmits it to a central location or sink for viewing and analysis. A wireless sensor network typically consists of sensor nodes. Radio signals can be used by a sensor node to connect with other nodes [10].

This sensor network is made up of thousands of small sensors that connect to one another and then to a base station, also known as a sink. The Internet or another backbone network is used to collect this information. After that, the user can access the entire system. Information is wirelessly transmitted from each sensor to a base station. The information is relayed to the base with the assistance of other sensors. Their processing power, storage, and communication bandwidth are all constrained. The sensor nodes must self-organize and bring the necessary network infrastructure with them after being deployed. The inbuilt sensors start gathering pertinent data at that point. Additionally, wireless sensor devices respond to requests from a control site to carry out tasks or deliver sensing samples [11].

The sensor nodes can operate in continuous or event-driven modes. Between users and the network, a sink or base station serves as an interface. By injecting queries and retrieving results from the sink, one can retrieve data from the network. Wireless sensor networks can have any architecture, from a basic star network to a sophisticated multi-hop wireless mesh network. Location and positioning data can be obtained via the Global Positioning System and local positioning algorithms [12]. Nodes in a star topology or star network cannot communicate with one another directly since they are connected only to a centralized communication hub.

Each node acts as a client while the central hub serves as the server or sink because all communication must pass through it. In a radio network, multi-hop routing is a form of communication when the radio coverage area exceeds the radio range of a single node. Therefore, a node can use other nodes as relays to travel to a particular location. By employing fewer nodes to relay the data from source to destination, the multi-hop meshes architecture aims to reduce energy usage. This structure is seen in Figure 7.4 and serves as the foundation for a sensor node's architecture.

The two subunits that make up sensing units are the sensor and the analogue to digital converter. A sensor collects analogue data from the outside world, and an ADC transforms that data into digital form. The processing unit, which is typically connected to the storage unit and consists of an electrical circuit for interacting with the sensors and a microprocessor or microcontroller, can govern the process by which the sensor nodes cooperate with other nodes to carry out assigned sensing tasks. They employ clever data processing and manipulation. The following element is the communication unit, which comprises a radio system and an antenna.

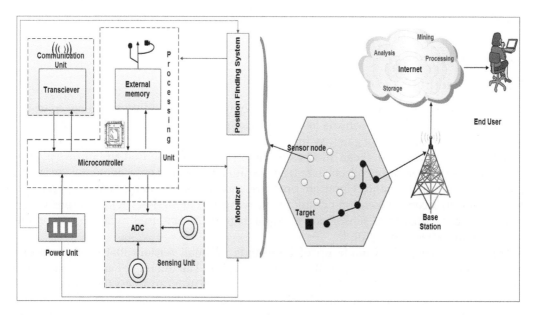

FIGURE 7.4
Block diagram of a WSN sensor node.

Data is transmitted and received using radio transceivers, which have an internal antenna or a connection to an external antenna. Signals are transmitted and received using an antenna. The node is connected to the network by the transceiver unit. The power source, which is typically a battery or an embedded energy harvesting system, is stable. The other node components depend on the application [13]. Additionally, it has a mobilizer, a power generator, and a location detecting system. This is how a sensor network communicates.

There are many suggested routing protocols for the commuter routing path from the source node to the base station. Through a gateway, the wireless sensor network communicates with the local or wide area network. The gateway serves as a link between the other network and the wireless sensor network. The most widely used wireless standard for sensor node connectivity is ZigBee technology. There are other wireless specifications and services as well. The IEEE 802.15.4 working group offers a standard for low-power device communication, and often sensors and smart meters employ one of these standards for connectivity. It can link sensors running at 2.4 GHz with a data range of 250 kbit/s.

7.4.1 The Sensing Sub-System

To study the behaviour of a physical object in the environment with a computer system, sensors play a vital role as an interface between the system and that object. Most of the physical objects in our surroundings manifest themselves by releasing a certain amount of energy (either heat or mechanical).This energy can be identified by force or pressure applied by/to the object if the object is at rest. If the object is in motion then properties such as angular velocity and linear acceleration etc. are useful to detect the energy. These sensors are capable of sensing this energy by generating analogue signals. The prime responsibility of a physical sensor is to transform these different types of energies into the form of electrical energy without actually changing or modifying the essential aspect of the physical signal or the physical energy.

The analogue signal which is produced by the sensor is continuous both in time and in amplitude. There is an accelerometer to sense linear acceleration. This linear acceleration represents some physical movement and this physical movement is related to physical activities. For example, in Active Volcano Monetary we measure seismic waves which produce some physical vibration on the surface of the earth. In structural health monitoring, we measure different types of acceleration generated by physical structures such as bridges and complex buildings. In healthcare, we monitor different types of human movements such as movements of joints and muscles. In transportation we monitor the drive quality of trains and measure the vibration generated by the train. We have different types of acoustic sensors to measure the distribution of pressure in pipelines, and the quality and type of fluid transported in pipelines; we emit sound signals into the pipeline and based on the amount of signal received on the other side and the change in phase or frequency of this signal, it is possible to know the quality of pressure-flow inside the pipeline [14].

Due to multiple limitations and difficulties, WSN does not have a standard design that could be used for all of its numerous taxonomic classifications. As a result, eight different WSN approaches have been suggested based on application requirements, as follows.

7.4.1.1 Data-centric Architectures

Massive sensor nodes are randomly distributed across this design and interact with one another without the use of global network IDs, making it difficult to identify the precise node for a query in the absence of ID [15]. As a result, data may be sent to and received from any point inside the zone, which uses a lot of power. In order to mitigate the issue of duplicate data and energy drain, the routing mechanism has been detailed in Sensor Protocols for Information through Negotiation (SPIN). This design aims to send data via the most effective path rather than attaining scalability.

7.4.1.2 Hierarchical Architectures

Hierarchical architecture is advantageous when taking scalability in a broad area network into consideration. In order to ease communication between cluster heads and inside the same cluster, the sensor node establishes the cluster with a cluster head that forms an intercluster and an intracluster. Cluster heads employ a multi-hop routing approach to provide data to the base station after receiving it from the sensor nodes. To fit the design, a variety of methods—including low-energy adaptive clustering hierarchy (LEACH)—have been developed. This architecture aims to minimize the power consumption of a large network [16].

7.4.1.3 Location-based Architectures

The position of the sensors is used by the routing algorithms in a location-based architecture to transmit and receive packets. Numerous methods, such as GPS receivers, triangulation, and positioning the sensors in well-known locations, can help sensors determine their location. To decrease the amount of electricity used for data routing, several methods have been suggested. Examples of algorithms include anchor location services (ALS), trajectory-based forwarding (TBF), and others [17].

7.4.1.4 Mobility-based Architectures

In a mobility-based design, the sensors (sink, intermediate, and source) may change their positions dynamically across period. Routing and interconnection are difficult in a mobility-based system. Compared to a static network, sensors utilize energy to gain access to the network. Because of this, a number of theories have been put out, including source encrypted authenticated data (SEAD), data MULEs (mobile data collector), and others.

7.4.1.5 QoS-based Architectures

One may define the QoS design in terms of reliability, latency, energy use, and other factors. To transmit and receive data in accordance with preset requirements, this architecture is required. The complexity of the QoS design is a result of the maintenance costs.

7.4.1.6 Network Flow Architecture

The method of improving the entire sensor link is called network flow based architecture. The links can be described using a variety of attributes, such as data transmission latency, power usage, and the time difference between two sensor nodes. This kind of architecture is used to choose the optimal course given a collection of requirements. Examples include maximum lifespan routing, minimum cost forwarding, and maximum lifetime data collection and aggregation. Different pathways can be used in this design to connect the drain and origin nodes.

7.4.1.7 Multipath-based Architectures

The multipath-based architectural solution is effective for distributing data payloads across channels and real-time data streaming applications that require greater reliability. Power efficiency and dependability are the goals of such architecture.

7.4.1.8 Heterogeneity-based Architectures

A heterogeneity-based architecture may be created using a variety of tasks and sensor kinds, such as sensing, communication, battery-operated and power-consuming nodes. The energy efficient heterogeneous cluster (EEHC) and other designs are included in this category. This architecture's goal is to utilize resources as effectively as possible.

7.5 Applications where Advanced WSN is Applied

7.5.1 Structural Health Monitoring

A well-established discipline of civil engineering, structural health monitoring, is one of the applications of WSN. This discipline inspects the integrity of complex architectures such as bridges, large buildings, industrial complexes, big aeroplanes, cargo ships, etc.

Normally, there are four types of inspections that can be done at regular intervals to check the status of the structures. Assimilation of WSN technology into this discipline has made the inspection easy, smooth, and effective. With the introduction of sensor nodes into the process, large structures can be inspected at much higher accuracy and more detailing can be obtained. Following are conventional methods through which inspection can be made:

- Visual inspection, where the municipality or the company which is responsible for the inspection should send people on a regular basis probably once a day to see whether the Pig components of the structure are working properly and check that there is no visible damage to the structure [18]. This inspection is labor-intensive, tedious, inconsistent, and subjective.
- The second type of inspection is called basic inspection, which takes place at least once a year. Observation is done by experts to inspect the integrity of the deck, the cables, the suspension cables, the towers, and so on. During basic inspection, force excitation is applied to the structure, for example using impact hammer vibrators, so that they can inspect the response of the bridge to this type of excitation. They also see how the structure reacts to ambient excitation, for example, to wind, cars driving on the structure. If the oscillation of the structure or bridge or building not as it should be then they recommend a detailed inspection.
- Detailed inspection takes place at least once every five years. It is quite costly and involves bulky devices as the inspection is assisted by instruments, for example X-rays, acoustic signals, and infrared signals, to monitor both microscopy and the global health of the structure.
- According to technical needs, if there are some microscopic fractures then they recommend a special inspection; this special inspection does not really focus on the entire structure but is localized so in this case a highly sensitive, highly accurate instrument will be used.

7.5.2 Traffic Control

Here we focus on ground transportation in many countries. It is not just a matter of driving a car from home to the office—ground transportation links a variety of supply chain systems and other types of delivery systems. One of the main problems of traffic control or ground transportation is congestion. In many countries, congestion is a big problem and unfortunately nobody profits from this. For example, cars consume large amounts of fuel, releasing unhealthy polluting gasses into the environment. In 2009, the American urban mobility report indicated that 4.2 billion hours were wasted because of congestion and 2.8 billion gallons of electrical power were used as a result. Overall congestion cost the country about 87 billion US dollars, an increase of 50 percent over the previous decade, so it is reasonable to harness the power of technology to predict congestion and provide drivers with alternative paths so that this problem can be alleviated [19]. Therefore, to overcome this problem we use sensors to estimate the number of cars on the street. Of course, there are radar systems and camera systems, because when we talk about wireless sensor networks or any other wired sensor network, these are additional technologies to enrich our knowledge. One proposed solution is inductive loops, every 10 or 50 meters in the street. An inductive loop is just a simple conductor or coils.

7.5.3 Telemedicine

Another application of wireless sensor networks is telemedicine. Here we consider one example. If you have a problem you go to a doctor; you have to sit in a controlled environment while the doctor takes some measurements, maybe an electrocardiogram, an electromyogram, blood pressure, pH tests to determine the concentration of acid in the body, and so on [20]. Wireless sensor networks simplify this task by deploying sensors on patients' bodies. These sensors are wireless; they are also unobtrusive so they can be hidden under a jacket or the surface of the skin. The sensor network can sense useful biomedical data and either save the data locally or communicate with one another to transfer the data to a remote station where a doctor can monitor the patient's health condition. One of the applications is gastroparesis. Normally the stomach and intestine are responsible for driving out digested food from the body, and the vagus nerve especially is responsible for contacting the stomach muscle so that food can be driven into the large intestine. But if for some reason the vagus nerve is not working properly or the patient has some stomach or intestinal problem, one of the symptoms is constipation. Scientists produced a very small wireless mobility sensor and this node consists of a pressure sensor and temperature sensor. A recent iteration also contains a micro-camera inside it. It is capable of communicating data wirelessly and it can be ingested into the body without any harm. It gives more accurate information regarding the whereabouts of the modulated sensor. We don't merely sense the presence of these wireless sensor nodes, as is the case with nuclear medicine, but they are also very active as they consistently sample pressure, temperature, and pH, while being completely harmless. The patient can go to the office and work undisturbed without feeling scrutinized at all, and it enables great flexibility not only in diagnosis but also in monitoring patients with gastroparesis.

7.5.4 5G Communication (Radio Nodes)

The potential to sense RF frequency around and on-site makes cognitive radio important since it allows users to change radio operational parameters such as energy, bandwidth, and modulation to improve the system, the speed, avoid collisions, and ensure compatibility between the primary users (PU) and secondary users (SU) [21]. The prime user has immediate access to the network infrastructure, and their channel access rights are treated as property rights. SU, who have a low priority, are always seeking to make adaptive use of the same airwaves while avoiding interference with PU transmissions:

CR's main goals are as follows:

(i) highly dependable communications whenever and wherever they are needed, and
(ii) efficient use of the radio spectrum.

CR provides two significant characteristics for unplanned spectrum usage, one of which is cognitive aptitude. The radio senses the surroundings in order to locate a vacant radio spectrum at a certain location at a set period. The identified regions are then picked for good communication without interfering with other users. The cognition cycle is a phase cycle in which the CR detects its current state and responds to its operational environment, as reflected in Figure 7.5. Spectrum sensing is the process where cognitive radio detects the accessible spectrum bands, collects their data, and then looks for gaps in the band.

- Spectrum analysis: entails estimating the unused spectrum discovered by energy detection.
- Spectrum mobility: maintaining the standards for flawless communication during the move to the best spectrum.
- Spectrum sharing: providing a good technique for coexisting xG users to schedule their spectrum usage.

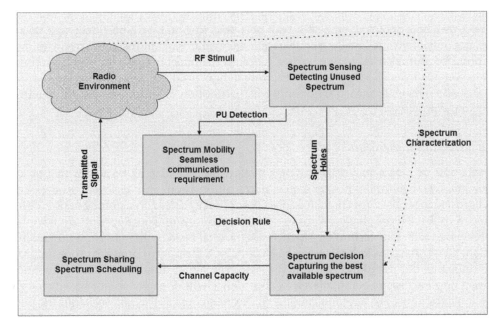

FIGURE 7.5
Cognitive cycle of sensor node.

- Spectrum decision: cognitive radio analyzes the data rate and bandwidth of the data transmission before making a spectrum decision.

Then, based on the spectrum characteristics and user requirements, the appropriate spectrum band is chosen. Once the spectrum band has been identified, the communication can begin. Therefore, it can be stated that CR technology helps in many ways to resolve the above-mentioned problem:

- keeping the spectrum clear
- providing priority service to higher-priority users during an emergency's peak communications period
- improved spectrum efficiency through dynamic spectrum access
- getting legacy and new devices and system

7.5.5 Pipeline Monitoring

Another application for wireless sensor networks is pipeline monitoring. We may not see it every day but pipelines surround us, the vital infrastructure for civilized life that transports water, oil, and gas and also transports our wastes. The integrity of this infrastructure is so vital because of its length and complexity. It is very difficult to diagnose when leakage happens on a pipeline; fortunately, we can use simple sensors. We can embed the sensor in pipelines to measure the pressure, temperature, and acoustic, all of which are quite vital to determine the flow of the fluid inside the pipeline. A Pluto pipeline generates a hotspot at the location of the leak whereas gas pipelines generate a folded path due to the gas pressure relaxation [14]. It travels at a higher propagation velocity in metal pipeline than in PVC, therefore a large number of commercially available sensors are available to detect and localize thermal anomalies, like fibre optic sensors, temperature sensors, and acoustic sensors [14]. One example of pipeline monitoring is the pipeline project in the US where researchers deploy a wireless sensor networks to monitor domestic sewage. In some cases, not only domestic sewage but also rainwater runoff and industry wastewater can be intermingled and released together. This has a great impact not only on the environment but also on our lives so in this project the researchers used pH sensors, pressure sensors, and acoustic sensors to measure not only the distribution of water but also the quality of water in the pipeline.

7.5.6 Precision Agriculture

With the rise of sensor based technologies, many sectors such as heavy industries, consumer markets, health, the power sector, home automation, and agriculture have seen a rapid growth in terms of production and accuracy. Every sector is becoming smart with the inclusion of these technologies. Conventional methods adopted in agriculture have been amalgamated with new technology to increase the total yield, with minimum human intervention. Sensor based wireless networks have helped farmers to produce crops according to the need of the population within the available agricultural land. The conventional agrarian approaches are being altered in accordance with the new sensor based concepts. Precision agriculture is a hot topic among WSN researchers as this sector carries the potential for tremendous reforms that can be brought out as technology advances day by day. Crops can be checked for any disease at the early stages of production thus removing any possibility of wastage [22].

7.6 Conclusion

The objective of this chapter is to provide a broad overview of wireless sensor networks. It starts with the basics of wireless networks and gradually advances toward modifications of the technology. It discusses the major features of WSN such as architecture, sensor node capabilities, their topologies, power utilization, etc. A brief introduction to the generations of modern wireless senor networks is also provided. After that, a generalized architecture of WSN is discussed with all critical components. Some prominent applications which are using WSN in different sectors are also described. Future work in the field of WSN is briefly noted in some sections of this chapter. WSN is playing a major role in developing the smart applications of almost every domain. Conventional procedures are being changed with the inclusion of sensor-based technologies such as WSN. Researchers are making progress by suggesting new approaches to make it more convenient and applicable to different sectors.

References

[1] Matin, M.A., & Islam, M.M. (2012). Overview of wireless sensor network. In *Wireless Sensor Networks – Technology and Protocols*. InTech. https://doi.org/10.5772/49376

[2] Senouci, M.R., & Mellouk, A. (2016). Wireless sensor networks. In *Deploying Wireless Sensor Networks* (pp. 1–19). Elsevier. https://doi.org/10.1016/B978-1-78548-099-7.50001-5

[3] SourceTech 411. (n.d.). The evolution of wireless sensor networks. https://sourcetech411.com/evolution-wireless-sensor-networks

[4] Chong, C.-Y., & Kumar, S.P. (2003). Sensor networks: Evolution, opportunities, and challenges. *Proceedings of the IEEE*, 91(8), 1247–1256. https://doi.org/10.1109/JPROC.2003.814918

[5] Gogolák, L., & Fürstner, I. (2020). Wireless sensor network aided assembly line monitoring according to expectations of industry 4.0. *Applied Sciences*, 11(1), 25. https://doi.org/10.3390/app11010025

[6] Silva, A., Liu, M., & Moghaddam, M. (2012). Power-management techniques for wireless sensor networks and similar low-power communication devices based on nonrechargeable batteries. *Journal of Computer Networks and Communications*, 2012, 1–10. https://doi.org/10.1155/2012/757291

[7] Dimitrievski, A., Filiposka, S., Melero, F.J., Zdravevski, E., Lameski, P., Pires, I.M., Garcia, N.M., Lousado, J.P., & Trajkovik, V. (2021). Rural healthcare IoT architecture based on low-energy LoRa. *International Journal of Environmental Research and Public Health*, 18(14), 7660. https://doi.org/10.3390/ijerph18147660

[8] Kim, J., Lee, C., & Rhee, J.-K. K. (2011). Traffic off-balancing algorithm for energy efficient networks. In L. Wosinska, K. Sato, J. Wu, & J. Zhang (Eds.), *2011 Asia Communications and Photonics Conference and Exhibition* (p. 83100C). IEEE. https://doi.org/10.1117/12.904448

[9] Al-Rahayfeh, A., Razaque, A., Jararweh, Y., & Almiani, M. (2018). Location-based lattice mobility model for wireless sensor networks. *Sensors*, 18(12), 4096. https://doi.org/10.3390/s18124096

[10] Chavan, S.V., Ladgaonkar, B.P., & Tilekar, S.K. (2018). An overview of sensor nodes for wireless sensor network applications: A review. *Jetir*, 5(1).

[11] Kumar, S., Vijay, S., & Srivastava, P. (2021). Design and energy consumption issues in wireless sensor network. *IJCRT*, 9(8).

[12] McGrath, M.J., & Scanaill, C.N. (2013). Sensor network topologies and design considerations. In *Sensor Technologies* (pp. 79–95). Apress. https://doi.org/10.1007/978-1-4302-6014-1_4

[13] Janhunen, J., Mikhaylov, K., Petäjäjärvi, J., & Sonkki, M. (2018). Wireless energy transfer powered wireless sensor node for green IoT: Design, implementation and evaluation. *Sensors, 19*(1), 90. https://doi.org/10.3390/s19010090

[14] Adegboye, M.A., Fung, W.-K., & Karnik, A. (2019). Recent advances in pipeline monitoring and oil leakage detection technologies: Principles and approaches. *Sensors, 19*(11), 2548. https://doi.org/10.3390/s19112548

[15] Biswal, A.K., & Samantaray, M. (2020). A novel approach for localization in sensor network. *International Journal of Engineering Research and Technology, 9*(1). https://doi.org/10.17577/IJERTV9IS010195

[16] Razaque, A., Mudigulam, S., Gavini, K., Amsaad, F., Abdulgader, M., & Krishna, G.S. (2016). H-LEACH: Hybrid-low energy adaptive clustering hierarchy for wireless sensor networks. *2016 IEEE Long Island Systems, Applications and Technology Conference (LISAT)*, 1–4. https://doi.org/10.1109/LISAT.2016.7494136

[17] Zhang, R., Zhao, H., & Labrador, M.A. (2006). The Anchor location service (ALS) protocol for large-scale wireless sensor networks. *Proceedings of the First International Conference on Integrated Internet Ad Hoc and Sensor Networks – InterSense '06*, 18. https://doi.org/10.1145/1142680.1142704

[18] Riquarts, Kurt. (1987). *Science and Technology Education and the Quality of Life: Papers Submitted to the 4th International Symposium on World Trends in Science and Technology Education, Kiel, 4–12 August 1987*. Institute for Science Education at Kiel University.

[19] Nasir, M.K., Md Noor, R., Kalam, M.A., & Masum, B.M. (2014). Reduction of fuel consumption and exhaust pollutant using intelligent transport systems. *Scientific World Journal, 2014*, 1–13. https://doi.org/10.1155/2014/836375

[20] Darwish, A., & Hassanien, A.E. (2011). Wearable and implantable wireless sensor network solutions for healthcare monitoring. *Sensors, 11*(6), 5561–5595. https://doi.org/10.3390/s110605561

[21] Nguyen, V.T., Villain, F., & le Guillou, Y. (2012). Cognitive radio RF: Overview and challenges. *VLSI Design, 2012*, 1–13. https://doi.org/10.1155/2012/716476

[22] Akhter, R., & Sofi, S.A. (2021). Precision agriculture using IoT data analytics and machine learning. *Journal of King Saud University – Computer and Information Sciences, 34*(8), 5602–5618. https://doi.org/10.1016/j.jksuci.2021.05.013

8

Wireless Sensor Networks: Routing Protocols and Cross-Layer Solutions

N. Thirupathi Rao, Eali Stephen Neal Joshua, and Debnath Bhattacharyya

CONTENTS
8.1 Introduction 97
    8.1.1 Technical Challenges in Cross-Layer Design 98
    8.1.2 Problem Statement 99
    8.1.3 Network Terminology 99
8.2 Literature Review 100
8.3 Materials and Methods 102
8.4 Proposed Scheme 103
    8.4.1 Cross-Layer Designing-Routing Scheme 103
        8.4.1.1 Architecture and Building Blocks 104
        8.4.1.2 Energy Efficiency in Cross-Layer Techniques 105
8.5 Simulation Results 107
    8.5.1 Multiple Access Schemes 108
    8.5.2 Pooling in Multiple Access Schemes 108
    8.5.3 Routing, Energy Efficiency, and Network Lifetime 110
        8.5.3.1 Congestion Control in Wireless Networks 111
        8.5.3.2 Cross-Layer Design and Optimization 112
        8.5.3.3 The Need for a General Framework for Cross-layer Design in Wireless Systems 114
8.6 Conclusion 115
References 115

8.1 Introduction

WSNs usually use a tiered protocol architecture, but our cross-layer protocol takes the place of that architecture. The idea behind the design is called "completely unified cross-layering." In this method, the data and abilities of different communication layers are combined into a single protocol. A cross-layer protocol does this with initiative determination, received-based contention, local congestion management, and distributed duty cycle operation. Researchers have been looking at better ways of sending data and voice via communication networks. Their focus has been on developing more efficient systems. Within the framework of this research project's communication network design, fork-packet communication was used between the nodes. Massive volumes of packets are seen to enter the network when a non-homogeneous bulk packet flow model is used. It was presumed that each and every datum and speech packet that entered the first buffer would successfully traverse the gap between it and the second and third buffers at the first node. This particular network architecture is suitable for use with a wide variety of communication channels, including radio and satellite.

All incoming messages are broken up into a variable number of packets and stored in buffers until it is time for them to be sent. Some networks put a significant emphasis on the prompt transmission of messages. To say that the non-homogeneous compound Poisson process is adequately described by the fact that packets arrive in buffers is an understatement. It has come to our notice that academics who do literature reviews don't pay much attention to parallel and serial communication networks that include non-homogeneous compound Poisson bulk arrivals and dynamic bandwidth allocation. In the case of the heterogeneous Poisson process, there is no such thing as a stationary arrival. This kind of non-homogeneous Poisson process is shown by the Poisson processes known as compound Poisson processes. In the course of this study, the researchers developed and analyzed a model of a communication network consisting of three nodes, complete with non-homogeneous Poisson compound bulk arrivals and dynamic bandwidth allocation. This model performs an outstanding job of anticipating network performance indicators and keeping an eye on traffic for sites that have MAN, WAN, and LAN networks and where traffic happens in bursts and is not consistent. Measures of network performance include the average buffer delay, the throughput of transmitters, the utilization of nodes, and the average number of packets that are stored in each buffer on average. One method for assessing the effectiveness of communication networks is to use numerical examples. The effect of the model's sensitivity to changes in input parameters on performance metrics is investigated by using a batch size distribution that is consistent throughout. When batch sizes are distributed binomially, it is possible to investigate the behavior of the network. The Poisson distribution is used to simulate arrivals in this model, and it is compared to a communication network. Using this strategy, it is possible to set up and monitor MAN, WAN, and LAN networks with more ease.

As a consequence of this, we devised a mathematical model and put it through its paces in order to evaluate the efficiency of this sort of communication network. It is a widely held opinion that Poisson processes combined with dynamic bandwidth allocations have an effect on the amount of time it takes for data to be sent in each node.

8.1.1 Technical Challenges in Cross-Layer Design

Wireless sensor networks (WSNs) have been the subject of many studies. The capacity of WSNs to collaborate and correlate is used by most of the proposed communication protocols to increase energy efficiency.

Because they all employ the same standard layered protocol design, these protocols are all unique. No effort has been made to fine-tune these protocols to increase overall network performance while using less energy. Due to WSNs' limited energy and processing capability, the best alternative to tiered protocols is cross-layer design, or collaborative networking layer optimization and design. As a result, an increasing amount of research has concentrated in recent years on creating cross-layer wireless sensor network protocols. It has recently been shown that cross-layer integration and design solutions save much more energy.

Three main factors may be used to explain this shift. Because of the restricted processing, storage, and power of wireless sensor nodes, this technique is required. Much additional effort is involved in layering protocols, making them inefficient. Protocols also need to consider the characteristics of low-power radio transceivers and the wireless channel circumstances. Even though much recent work has been done on cross-layer design and protocol building for WSNs, there is no systematic approach to describe cross-layer interactions and leverage them to your advantage. Ad hoc and sensor networks with several hops may benefit from decentralized resource allocation techniques developed for multihop wireless ad hoc networks. By tackling resource allocation optimization issues at several layers simultaneously, current research has attempted to create effective cross-layer design methodologies. As a result of this approach, most extant studies look at each layer of resource allocation independently.

Cross-layer design strategies based on the collaborative solution of optimization problems for resource allocation at distinct levels have been examined in recent studies. Many outstanding research concerns arise when WSN protocols are designed using systematic cross-layer design methodologies. Using a cross-layer method has benefits and drawbacks, which we will discuss in this chapter. Cross-layer design is implemented by looking at relevant literature and identifying topics for future investigation. Cross-layer design architectural concerns and considerations are also raised. Design and development cannot always be kept separate when using a cross-layer methodology. As a result, subsequent iterations will be more challenging to improve upon. Unintended dependencies between functions are more likely to occur in systems without layers because of the lack of layers.

8.1.2 Problem Statement

A cross-layer design seeks to find the optimal way between these four variables by solving a 4-tuple that is given by route (or next hop), operating channel, transmission power, and throughput. The goal of this endeavor is to find the most efficient path. This is done in order to satisfy the performance needs of the application while also preserving the available energy resources. The data obtained from a sensor are sent to the central processing unit in this manner. The process of cross-layer optimization includes a number of steps, including the assignment of transmit power, the selection of a route at the network layer, and the selection of a channel at the link layer. When selecting how to distribute resources for a certain transmission, the 4-tuple is used to ascertain the network's operational point so that appropriate actions may be taken.

When it comes to utility functions, the 4-tuple is used as a basis for presenting how each node wants to employ various PHY layer resources in order to increase throughput. This is accomplished by displaying the 4-tuple in a certain sequence. The fact that the resources of the PHY layer also have an effect on the longevity of the network is a cause for worry. Utility functions will analyze the benefits in performance and the power costs associated with a particular allocation of resources before settling on the optimum operating point. To put it another way, the significance of the application plays an essential role in establishing the nature of the concessions that a sensor must make.

8.1.3 Network Terminology

In order to solve the problem of intelligent routing, the sensor nodes and mesh nodes that make up our distributed wireless sensor network (WSN) will be named V_i and M_i

respectively. The sensor nodes and mesh nodes are clustered into Cl(Vi,Mi) structures and connected to one another via the use of directed link set El in order to produce a two-tier network. The architecture that can be seen in space Sin is used to connect all of the nodes together. Figure 8.2 depicts the relationship between G (Cl (Vi, Mi), El. T=tvx|vxZ, Eavail=evx|vxZ, and W|Vi|XW|Vi|Vi|, the total number of residual energies of all network nodes, should all be specified. SN, the total number of scattered WSN sub-networks (sensor and mesh networks), should also be described.

When K' is set to a low value, there is a limit placed on the number of nodes that may be used to direct traffic. Because there is such a wide array of transmission powers and operating ranges that may be employed by communication protocols, it is quite possible that route lengths would shift as a result. Because of the different ways in which transmitters and receivers use energy, it is still ideal for the network to limit the number of hops that may occur inside a channel, even if doing so requires it to send out more power. As a direct consequence of this, the length of our routes is constrained in accordance with the configuration of our network. It is possible for this to change based on the current state of the network.

In order to carry out this research, we are going to assume the following:

All of the clusters provide reliable data when an exponential distribution is applied to the arrival timings of events and when a parameter is used.

By adjusting their frequency, which they are aware of, sensor nodes have access to a number of different operating channels from which to choose.

Sensor nodes and other mesh nodes are able to interact with one another through software radio as well as other methods.

All of the nodes that make up a cluster get omnidirectional antennas as well as gains that are the same.

In order to facilitate general design analysis, sensor nodes and mesh nodes do not have power supplies attached to them.

As was discussed in Section 2, mesh nodes are organized in a hexagonal pattern to provide the best possible network design and coverage.

The Global Positioning System (GPS) is what sensor nodes turn to in order to figure out where exactly they are located inside the network. In the not-too-distant future, we are going to investigate sensor localization. The central control station, which has a mesh node as part of its infrastructure, serves as the broadcast sink for the network. Because each hop chooses its own operational channel, there is no one channel that is used over the whole of the trip.

The M/M/3/3 queuing technique that is employed in each sensor node only makes use of three channels and not any lines. We are able to make the assumption that the cluster head gives us real-time information on the state of the channel since our sensor nodes are permanently installed. Because of this, we are able to infer that the fading will occur in a gradual manner.

8.2 Literature Review

In their research publications, a number of academics focused on queuing models and how the Poisson process might be used in queueing model-based communication networks. In this part, references are made to a number of publications and pieces of research that are relevant to the current model.

The authors looked at several communication network models and various different models with a variety of inputs in [1]. It was very necessary to evaluate the functionality of the network model once packets started arriving. Within the scope of this inquiry, the authors looked at the network model inputs as well as the performance of the model.

The authors provide a comprehensive explanation of queuing models that include random transmission phases in [2–4]. Random transmission will take place whenever packets are either added to the network or leased from the network model under consideration. The authors of [5] devoted a significant amount of space to a discussion on the use of the Poisson process in line models. The authors investigate the usage of the Poisson process, including how to use it, what kinds of problems it may be applied to solve, and how it can be applied to solve those problems. Consideration should also be given to the conditions in which the Poisson process is rendered ineffective.

When it comes to designing communication networks, the authors of [6] spent a significant amount of time discussing the benefits of a variety of distributions and queuing models. The authors of [7] focused their attention on network traffic, more specifically traffic on mobile networks. In this research, a wide variety of models are dissected in great detail in order to get an idea of how busy mobile network are. In addition to that, it offers the results of these simulations. In [8], the authors investigated potential solutions to the problem of network congestion by using load-dependent queuing models. The results provide us with a more distinct notion of which models need to be promoted and manufactured in the foreseeable future. In [9], an original model of a network was suggested, and it was based on the idea of waiting in line. As an innovative strategy or piece of technology, they looked at time-shared systems that allowed for the delivery of bulk packets to network nodes.

In [10–12], the design of communication networks may make use of queuing models with bulk arrivals at each node, which can be implemented with two or three nodes. In order to ascertain how data packets reach each node in the network, the network model makes use of the bulk arrival model in addition to the non-homogeneous Poisson arrivals. [13] examines a unique three-node network model of individuals waiting in line. Utilizing the Duane process and dynamic bandwidth allocation at the model's receiver end were two of the new tactics or approaches that were employed in the investigation of how packets were delivered to the network model. The authors of [14] developed the queuing model by creating and evaluating communication networks. This expansion included the addition of additional features and constraints to the model. The authors looked at many models of interdependent queuing and analyzed them. In these models, the operation or processing of data packets at one node is impacted by the data or packets at the other node.

In a recent work [15], the arrival process was modelled using a representation based on a non-homogeneous Poisson process. They are of the opinion that the immigrants do not have somebody in their lives to whom they are committed. On the other hand, a message that is delivered to a source in an in-store or forward communication network is fragmented into an unpredictable number of packets that is based on the overall size of the message. When examining the buffers, it is necessary to take into consideration the fact that the arrival of packets is a non-homogeneous compound Poisson process, which combines the Poisson process with the compound Poisson process.

The researchers describe work in [16–17] in which they constructed a completely new model for a three-node network that is based on models of waiting. They explored how packets arrive in the network model based on the Poisson process, with dynamic bandwidth allocation at the receiver end of the network model. This was done as a new technique or method. In [18–19], an innovative application model that included a phase-type output transmission was investigated.

The purpose of this study was to investigate a forked communication network model that possessed a non-homogeneous Poisson binomial bulk mode of arrival for bulk packets as well as a dynamic bandwidth allocation transmission mechanism. The focus of the investigation was on the forked communication network model. It has never been tried before to combine parallel and series integrated communication networks with non-homogeneous compound Poisson processes, binomial bulk arrivals, and dynamic bandwidth allocation. The performance of several network models is investigated in this study.

8.3 Materials and Methods

We may be able to make smarter and more effective use of the resources at our disposal with the assistance of a mathematical model that describes the construction of communication networks. Queuing models will be used to assist in the development of a mathematical model that will be used to support and establish a communication network. In these models, the queues are represented as several kinds of linked and interlinked communication networks. In queuing models, incoming messages are treated as customers, the buffers at nodes are treated like waiting lines, and all other network events are treated like the model's service. The modeling of communication networks based on queuing theory is a logical approach that may be used to explain how different network system functions work. In certain networks, the time of day is used to calculate when a message will arrive at its destination. The most accurate way to characterize the process of data packets arriving in buffers is as a non-homogeneous compound Poisson process. The research pays relatively little attention to parallel and serial mixed communication networks with non-homogeneous Poisson bulk arrivals and dynamic bandwidth allocation. This is because these types of networks combine both types of communications. One method that may be used to demonstrate the unpredictability of the network's bulk arrivals is a non-homogeneous Poisson process.

The purpose of this study was to investigate a communication network consisting of three branching nodes. In this network simulation, these packets came in a non-homogeneous compound Poisson process packet flow, and a dynamic bandwidth allocation strategy was used to assign them to their respective destinations. The proposed model will be of significant assistance in modeling the traffic on communication networks under a wide range of different scenarios. This architecture works well for local area networks, metropolitan area networks, and wide area networks. It is conceivable that the traffic on these networks may sometimes behave in an unpredictable and burst manner.

Different difference-differential equations are used in order to generate the probability-generating functions of three nodes that are located inside a network. The performance of the network could be evaluated by observing the frequency with which the buffers were empty, the rate at which the transmitter was sending data, the degree to which each node was being used, and the length of time that it took each buffer, on average, to send a packet. By analyzing the numerical example that was provided, it was able to get an idea of how effective the communication system was. Both the Poisson arrival and the Duane processes models were investigated to find out how this most recent model stacked up against them.

Studies that compare two different things indicated that a recently constructed model was accurate in its assessment of the performance of communication networks.

8.4 Proposed Scheme

8.4.1 Cross-Layer Designing-Routing Scheme

As we've seen, the many features of wireless networks have an effect on the development and functioning of each protocol layer. These effects might be positive or negative. Although we still make use of a tiered argument structure, the traits and characteristics of a number of different layers are still brought up in the course of our conversations. When it comes to packet scheduling, the most important aspects to take into consideration are the media access control, the channel dynamics, and the power management at the physical layer. TCP, for example, has a poor performance when used over wireless networks. Using information at lower levels, it has been shown that TCP is able to differentiate between the many different causes of packet loss. Information is shared across multiple layers of a system that has been designed using cross-layer architecture. This is done in order to make the most of the resources provided by the network and to provide the system with a high degree of flexibility. In a cross-layer design, the formation of each layer is accomplished by using a limited set of key factors and control knobs. Other levels consult the parameters to decide how their control knobs should be adjusted in response to the present condition of the network. Parameters are utilized in this way by other levels. It is usual practice to consider cross-layer design to be an optimization problem consisting of a large number of optimization variables and constraints that have an effect on many levels. After the problem with optimization has been fixed, the control knobs of a layer may have their settings adjusted to their ideal levels. Figure 8.1 provides a visual representation of one possible use of this strategy.

The following are some of the reasons why cross-layer design is beneficial to wireless network architecture. It is feasible to employ the traditional architectural design method when designing wired networks; however, this results in a restricted search area for the

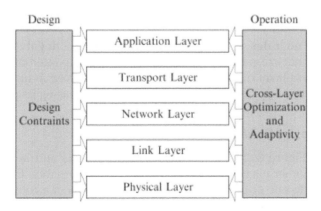

FIGURE 8.1
Cross-layer design view implementation in wireless sensor networks.

best potential adaption. Because wireless networks, in contrast to landline networks, have a limited number of resources, they are required to explore a more expansive optimization space that is comprised of several layers. Second, the protocol stack that is now in use was developed for networks that are physically connected to one another via the use of cables. It is possible that wireless networks that are substantially dissimilar to this one will not be able to use it. For instance, the concept of "connection" has undergone tremendous change throughout the course of history. In addition to this, the physical distance between two nodes may have a considerable effect on the connection between them. Research must be conducted on both wired and wireless networks concurrently, despite the fact that the two types of networks couldn't be more unlike. Take the Transmission Control Protocol (TCP) over a mobile network as an example. Thirdly, in contrast to wireline networks, wireless networks have design characteristics that are more tightly intertwined among their many tiers. Establishing a distinct "separation idea" for wireless networks would be challenging due to their decentralized nature. More and more people are using wireless networking protocols and algorithms, which make it easier for devices on different levels to connect with one another. This trend is expected to continue. It has been shown throughout the research that there has been a significant rise in performance. Figuring out how to squeeze every last bit of performance out of your program is a significant element of the cross-layer design process. When picking parameters and weighing the potential performance benefits against the additional complexity introduced by the combination of new information gleaned from several layers [20].

Interactions and unanticipated repercussions of the law. Traditional architectural design enables designers to focus on one particular issue at a time without being required to take into account the other components of the protocol stack. Everything is affected by the cross-layer design. It is imperative that, at all costs, unintended effects be avoided.

A graphical representation of the connections between the many parts that make up a system. In cross-layer designs, it is standard practice to include an adaptation loop as a component of a separate protocol. It is possible that a dependency graph with a node for each important parameter and a directed edge illustrating how the parameters are dependent on one another would be of great use to us.

It is essential to keep time and stability separate from one another. When two different adaptation loops regulate and use the same parameter, averaging and temporal separation are two strategies that may be used to maintain stability in the system. In the dependent network, every loop has to be backed up by proof of stability for interactions that take about the same length of time. To do this task successfully will need a significant amount of work.

When there are no rules to follow, cross-layer design devolves into chaos. The implementations of our cross-layer architecture need to be updated on a consistent basis. The lifespan of a cross-layer design as well as the costs involved in updating it need to be carefully evaluated in comparison to the possibilities of increasing its performance.

8.4.1.1 Architecture and Building Blocks

As can be seen in Figure 8.2, a network is formed by hosts and network nodes being connected to one another via communication links. The information that is created by the applications is sent to one or more remote devices by means of a communication device. The nodes that make up a network collaborate with one another and share control information in order to allow data to be sent from one device to another. The nodes and devices in a network are able to interact with one another because of the connections that link

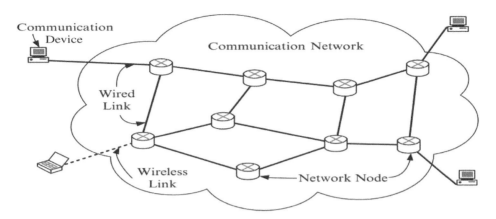

FIGURE 8.2
The communication network of the proposed model.

them to one another. Metcalfe's law, which holds that the value of a network is equal to the square of the number of users, is often used to determine how much a communications network is worth. A vast communication network, on the other hand, calls for a significant number of intricate processes to be carried out in order to ensure the efficient movement of data from one device to another. Even if it were technically feasible to combine all of these functions into a single module, there is no need to do so. It would be more effective to use the tried-and-true strategy of "divide and conquer." The duties associated with the communication task are segmented into more manageable chunks and categorized according to the degree to which they are dependent upon one another. Each level of the subtasks is in charge of a different facet of the communication. To get anything in return, you have to first make use of the services provided by people who are lower on the food chain.

As a consequence of this, these services are presented in a way that makes differences and complexity more difficult to understand. One way that things in different systems may more readily interact with one another is via the use of a protocol. In most cases, it describes how a subtask that is contained inside a layer ought to be carried out. As long as the common interfaces are kept in mind, using this technique makes it much easier to construct a communication network. This is because each individual subtask can be planned and optimized independently, which allows for this strategy's full potential. Computer networks are able to interact with one another thanks to a set of protocols known as TCP/IP, which is often referred to as the Internet protocol suite. The application layer, the transport layer, the network layer, and the data link layer are the several tiers that make up this structure. It is constructed from a number of different protocols. In addition to that, there is a third layer that serves as the glue that holds everything together.

8.4.1.2 Energy Efficiency in Cross-Layer Techniques

In traditional networks, optimization is typically carried out by analyzing the objectives of each layer based solely on the data collected at that level. This process does not take into account any design considerations or information obtained from higher levels, which is not the case in contemporary networks. As a consequence of this, the approach is most successful in the immediate vicinity, but it is less successful on a global scale. As the performance of the wireless system has increased, researchers have been hard at work developing cross-layer optimization methods to enhance the system's flexibility, scheduling, and resource allocation. Within these approaches, the PHY layer, the MAC layer, and the higher layers are all interconnected and interdependent. Approaches that are effective across levels share data and information with one another and communicate with one another in order to establish a distinctive plan. When these two levels interact with one another, for instance, the Application layer modifies QoS and source coding in response to PHY's provision of channel status information (CSI). However, the PHY's rate and power might change based on the parameters of the application as well as the traffic. Congestion management at the transport layer may be modified depending on the manner in which the PHY layer and the transport layer work together and make use of CSI that is PHY-based. The most qualified users might be selected by the MAC layer scheduling controller, but this is contingent on the manner in which the network layer protocols interact with the physical layer and get information from it (PHY). In the body of research that has been done, a great number of strategies that place an emphasis on communication and information sharing across different levels have been investigated. In this chapter, we will investigate an adaptive cross-layer packet transmission method while still adhering to other quality of service (QoS) criteria.

It is not a newly discovered phenomenon that cross-layer energy reduction occurs when the system properties are random. It has been the subject of debate in a number of articles and books in the past. The major emphasis of research into wireless technology has been on developing protocols and methods that use the least amount of power feasible. In the body of published research, a multitude of potential solutions have been offered with the goal of extending the battery life of handheld mobile devices and lowering the amount of energy required by base stations, all while keeping the same level of overall power constraint. Cross-layer design for energy-efficient wireless communications places an emphasis on system-based techniques for optimizing transmission energy; resource management across time, frequency, and spatial domains; and energy-efficient hardware implementations. In this chapter, we look at cooperative communications, cross-layer self-organized networks, and the challenges of future radio access networks that go beyond the Long-Term Evolution Advanced (LTE-Advanced) standard.

Current research in the field of physics has shown and investigated a number of different strategies that can cut the amount of energy that multiantenna systems use. The various options include, but are not limited to the following: using the cooperative beamforming framework to save energy, provides a cross-layer optimum weight design for single-beam beamforming and a sub-optimal weight design for multi-beam beamforming. Both of these designs are great for single-beam beamforming. The authors presented a cross-layer approach as a means of transitioning between a MIMO system with two transmit antennas and a single-input multiple-output (SIMO) system in adaptive MIMO systems. This transition would take place in adaptive MIMO systems. It is possible that this may result in fewer people using mobile devices. The authors of [21] investigated the efficiency with which Rayleigh fading networks consume energy. They found that upper-layer design features may be matched to the results of their research. In the following research, an effort is made to improve the PHY's energy efficiency by using strategies that adapt and optimize rate and power, making use of the information provided by the data-link layer. Utilizing cross-layer design strategies and a discrete-time queuing model over Rayleigh fading channels are two examples of ways that sensor networks may be improved for increased energy efficiency. Coordinated adjustment of PHY and data-link layer parameters in wireless networks that have a finite amount of energy might potentially save a significant amount of energy. These parameters include modulation order, packet size, retransmission limit, and others. When the device was in sleep mode, the authors investigated a cross-layer architecture that used adaptive modulation and coding to conserve energy. If there aren't enough packets to fill the buffer, there are a few options for cross-layer design to think about or even channels to add when making a decision on how to transmit data in order to save energy. This does not have any effect on the transmission technique that was used in these tests since it is believed that the buffer size is unlimited.

The authors of [22] have made use of game theory in order to design energy-efficient algorithms that take cross-layer concerns into account. As part of a game-theoretic and cross-layer optimization technique, each node determines, based on its own power usage, how large the contention window ought to be. This decision is made in conjunction with the cross-layer optimization. A problem with energy-efficient optimization in wireless code division multiple access (CDMA) data networks that incorporate power management and multiuser detection was addressed using a noncooperative game-theoretic technique. [23] investigated a noncooperative power management game for optimal energy efficiency with a fairness limit on the maximum received powers in cognitive CDMA wireless networks with primary and secondary users on the same frequency band. In this, a game-theoretic energy-efficient model is used to investigate the topic of cross-layer power and rate control for wireless networks with quality of service (QoS) restrictions.

[24] discusses cross-layer design, which helps cut down on the amount of energy that is used. It presents a cross-layer window control method in the transport layer to increase throughput and a power management technique in the PHY layer to reduce the amount of power that a wireless multihop network consumes. Both of these strategies are intended to be implemented at the transport layer. The researchers' investigation focused on the total power consumption of multihop wireless networks. We present a distributed technique that is both low power and minimal complexity. Cross-layer design in wireless standards is the topic of inquiry in [25], illustrating the cross-layer design that will be used for mobile WiMAX.

As a consequence of the design, there is a reduction in the number of lost packets as well as the amount of power used, while the throughput is raised. Wireless local area networks (WLANs) based on the IEEE 802.11 standard have the potential to be used in a cross-layer and energy-efficient way to carry VoIP packets (WLANs). This method improves a station's energy economy and WLAN use, without having any negative impact on the quality of the voice that is sent. MPEG-4 video transmission over a slow-fading channel provides a two-phase energy-efficient resource allocation technique for numerous users to manage their sleeping time against scalability trade-off across the PHY, communications, and link layers. This technique can be used to improve the overall quality of the video transmission. [26] provides a model for a WLAN system that significantly lessens the amount of power required by the system via a dynamic frequency selection approach and an energy-efficient channel allocation mechanism. Green base stations in 2G and HSPA networks can save energy by shutting off some resources; nevertheless, this does not impact users' quality of service (QoS). Both a real-time dynamic approach of activating and deactivating resources and a semi-static method in which resources stay constant for extended periods of time are proposed by [27].

8.5 Simulation Results

In the field of wireless research, the idea of random waypoints is used rather often. In spite of the fact that the stationary speed distribution was not recognized in the past, recent research has shown that it is unique from the uniform distribution. In addition, the authors assert that the average speed of the nodes will slow down as vmin becomes closer and closer to zero. This period, which often continues for a considerable amount of time, is referred to as a "transient phase." Throw out all of the simulated data collected during this time period in order to prevent the simulations from being skewed. The simulation of this phase may take up to a thousand seconds when the minimum speed is low, despite the fact that the majority of simulations described in the available research require less than 900 seconds to complete. Even if the "mobility" paradigm is applied and the minimum speed is set to zero, eventually all nodes will stop moving altogether. In order to ensure that the results of a simulation are reliable, it is necessary to exclude the simulation of the transient phase of a mobility model. The standard method for accomplishing what is known as a "perfect simulation" involves ensuring that fixed conditions are maintained.

8.5.1 Multiple Access Schemes

In a wireless network, mobile users are responsible for sharing the wireless channel. MAC, which is a key function of network control, informs users on how to share a channel effectively (for example, with high throughput or spectrum efficiency), and it is responsible for ensuring that this occurs (i.e., users should have equal chance of accessing the channel). The many kinds of access methods may be categorized in a variety of different ways, as shown in Figure 8.3. In wireless networks, channelization is a standard operational procedure. During the process of channelization, wireless channels are broken down into subchannels, which may be frequency bands, time slots, or spreading codes. After some time has passed, users are given access to subchannels. This section covers some of the most important random-access algorithms for wireless data networks, including ALOHA, ALOHA/slotted ALOHA, carrier sense multiple access, and carrier sense multiple access with collision avoidance.

8.5.2 Pooling in Multiple Access Schemes

Pooling systems contain just one server for every m station, in contrast to queues, which have multiple servers and many stations (users) (stations). Each station, or user, is dealt

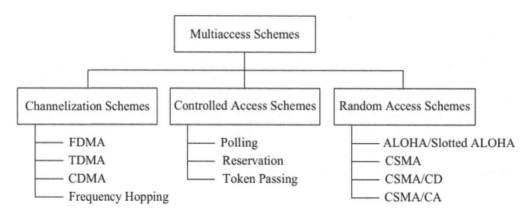

FIGURE 8.3
The various schemes in multiple access.

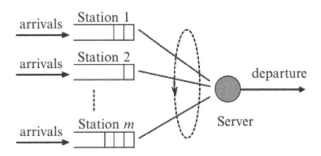

FIGURE 8.4
The pooling scheme from one to many.

with on an individual basis. A pooling system that is currently operational is shown in Figure 8.4. As depicted in the illustration, a line is drawn across each station in the picture, and this line temporarily keeps customers at that station. The patrons are free to arrive at any of the stations in whatever order they like. After using the system and requesting a service from the server, each user is required to log out of the system. In addition, the operating hours are established in a number of different ways. If you have a pooling system that can handle the queue for a single client as well as buffers of any size, you have the ability to do all of these things. The name of this kind of buffer gives away the fact that each station in a system with a single-client buffer can only accommodate one consumer at a time. Those customers who come after the buffer is sufficiently full are not accommodated.

The server will eventually be able to give service to each individual user as long as all of the users connect to the system and remain buffered. A server has the option of proceeding to the subsequent station either in accordance with a present sequence (such as cyclic) or at random. The time that elapses between the conclusion of one service and the commencement of another is referred to as the switchover time. There are three main types of service policies that are included in a voting system. The term "limited-service polling systems" refers to election administration methods that limit the total number of voting locations that may be attended to at one time. When a server visits a station that uses the limited-1 method, that station can only accommodate one client at a time to receive service. The system is deemed to be a gated service system so long as the server continues to service a station until either it is empty or all of the clients who came before the station being pooled have been served. Even if the buffer at a station is full, the server will continue to serve all users until the buffer at that station is drained again, even if the buffer at the station is full. In the next section, we will investigate a voting system using a system with only one voting station. Figure 8.5 presents an example of a gated service system with a m value of 1.

Depending on the requirements of the customer, different coloured blocks represent the amount of time that will be required to meet those requirements. Take, for instance, the Poisson arrival technique with a rate as an example. Let's say that the length of time it takes to serve a client is denoted by X, and we'll set E[X] to be equal to 1/. Therefore, the use of the system is =/. The length of time that the jth transitional phase lasts is denoted by the variable Vj. This is what takes place whenever the system is extended to accommodate a new customer. It is necessary to wait until all of the people who came before it have been served before it can be attended to. The total amount of time that a customer must wait in a line is determined by combining a number of factors, including the amount of time that a customer spends being serviced, the amount of time that a customer waits in

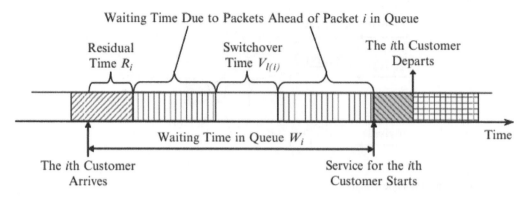

FIGURE 8.5
The gated scheme with one pooling station.

a queue, and the amount of (i). Both current and potential customers get their moment in the spotlight at certain points in the sales cycle. The only customers who have a reasonable expectation of being serviced are those who came before the current service hour; the rest of Ni's customers, including myself, will have to wait until after Vl (i).

8.5.3 Routing, Energy Efficiency, and Network Lifetime

Reduced power consumption is one of the most important factors to think about when designing wireless networks. Because mobile nodes are powered by rechargeable batteries, effective energy management has the potential to lengthen the battery life of a mobile node, which in turn increases the durability of the network. In wireless sensor networks, where the batteries of the sensor nodes may or may not be able to be recharged, these issues take on an especially essential role. Because it is such a tough problem to overcome, it is imperative that every layer of a wireless protocol stack be constructed with energy efficiency in mind. Routing is a part of managing and maintaining a network, but these activities also have an effect on how efficiently a wireless network uses its resources. The sending and receiving of data in wireless networks require a significant amount of energy. It is feasible to govern the flow of data through the network in such a way that the batteries of all nodes discharge at the same rate. The network's lifetime is one of the most significant performance criteria to consider while designing wireless networking technologies that are efficient in terms of energy consumption. Depending on the kind of network and the type of application being used, the failure of a certain number of nodes or the need to break the network apart might mean any one of a number of different things. We will talk about a routing design that demonstrates how energy efficiency affects routing in wireless networks. Take, for example, the multihop wireless network that is shown in Figure 8.6.

The majority of the limited amount of battery power that is available to each node is used for the transmission and reception of data. Each node should utilize a route with many hops so that data may be sent to gateways. Adjust the level of the transmission power so that it is as low as is practically possible for the receiver. This is a straightforward approach to the idea of a wireless sensor network. The network may be visually depicted using a directed plot with the notation G (N, L), where N refers to the total number of nodes and L refers to the total number of connections that are directed. In order to establish a connection, I need to be aware of the set of nodes that fall inside the "Si" category, which

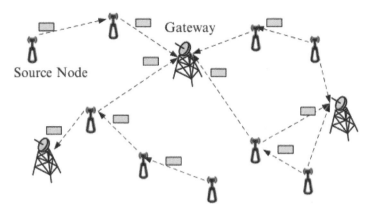

FIGURE 8.6
The traffic generation from the source to destination.

stands for "full transmit power." Initial energy, denoted by Ei, is shown for each node. The following model of energy use will be used, as it has been decided: Node I use energy et ij when it sends a data unit to its neighbor, Node j. In order for Node I to get a data unit from Node j, it is necessary for it to spend an amount of energy that is equal to er j. Assume there are C unique commodities, each of which has its own set of nodes that define its starting point and its endpoint. Each commodity c C has O(c) origin nodes, with data being created at Node I at a rate Q(c)I, and D(c) destination nodes, with each node functioning as a data sink. Let the value q(c) ij reflect the projected rate that the routing algorithm will achieve while sending commodity c from Node I to Node j. An iterative flow augmentation approach is shown to be a potential solution to the problem of achieving maximum lifespan routing in [28]. Through the use of the loop, a commodity c in O(c) is able to determine the least expensive path to each of its target nodes in D(c). After that point, the flow will grow by an amount that is equal to Q(c) I, where Q(c) I is the increment size for the route that is the least expensive. Following that, the power expenses and connection prices of every node are computed. The most recent connection costs are used to determine which paths have the lowest fees, and this procedure is repeated as many times as necessary until all of the energy in a node has been used.

8.5.3.1 Congestion Control in Wireless Networks

The primary responsibility of the transport layer is to maintain an acceptable level of traffic at all times. In a communication network that uses store-and-forward, the buffer of a router is shared by multiple different sessions at the same time. When the rate at which things are leaving is lower than the rate at which new things are entering, a buffer overflow will occur. When the buffer reaches its capacity, packets are discarded, and all sessions that are dependent on the bottleneck router experience significant lag time. Therefore, the primary objective of congestion management is to identify (possible) bottlenecks and reduce the rate at which sources provide data.

In order to make the most of the capacity of the network, service providers might consider increasing their prices when the level of network congestion has lessened. A technique known as "additive-increment-multiplicative-decrease" is used in order to make adjustments to the transmission rate at both ends of a TCP connection. TCP has to use timeouts or repeated acknowledgments to determine if the network is busy. This is because routers are not actively participating in the communication process (so that they may be stateless and manage huge traffic volumes).

Congestion is the root cause of any and all packet loss, and acknowledging this fact is a precondition for considering the validity of this argument. When a timeout occurs, the Transmission Control Protocol (TCP) assumes that the network is at capacity, and as a result, it slows down the pace at which data is sent. This assumption is correct for landlines, as opposed to wireless networks, which are more likely to suffer from transmission problems or route failures, both of which may result in lost or missed packets. The throughput of the wireless connection is denoted by the equation $S = G$. This assumes a transmit rate of G and a packet loss rate of p packets lost per second (1 p). If there is no congestion, the Transmission Control Protocol (TCP) should increase its transmission rate G rather than decrease it in order to achieve a high throughput. Since TCP is unable to differentiate between loss caused by transmission issues and loss caused by congestion, a wireless connection in the end-to-end path presents a challenge for the protocol (s). TCP does not operate efficiently in multihop wireless networks due to the high number of route failures and problems at the MAC layer. Mobility is the key factor that contributes to route failure; topological changes may also have an impact on routes.

If there is congestion, the Media Access Control (MAC) layer will lose a frame, which will lead the routing layer to assume that the connection has been broken. Under either set of conditions, a session route will be pieced back together by the routing engine. During this time, the flow of data through TCP can become sluggish or perhaps stop altogether. When a new route is found, there is a possibility that the bandwidth and the latency will be altered. As a direct consequence of this, the TCP round-trip duration is very variable. To simulate the operation of the TCP protocol, a wireless network with a chain topology was developed. It has been reported that the throughput of TCP from end to end dramatically drops when there is a considerable reduction in the number of hops. Increasing TCP's performance over wireless networks has been the subject of discussion in a number of publications. There is a possibility that error control techniques implemented at the link layer will make wireless networks more reliable (e.g., forward error correction or automated repeat request). An example of a split TCP solution would be to utilize a landline segment and a wireless link segment to break up a long TCP connection into a number of more manageable locally based segments. Through the use of a proxy node, which acts as a bridge, it is possible to connect two neighboring segments together. The regulation of traffic flow and the alleviation of traffic congestion are handled by a variety of methods around the city. In the event that a wireless packet is lost, the explicit notification scheme class is used by routers and base stations serving as intermediate nodes to inform TCP directly.

8.5.3.2 Cross-Layer Design and Optimization

We have seen how different aspects of wireless networks may have an effect on the development and functioning of each protocol layer up to this point. Even while we still employ a tiered design, our conversations now include qualities and features drawn from more than one layer. At the physical level, packet scheduling is affected by a number of parameters, including but not limited to the dynamics of the MAC layer, the dynamics of the channel, and power control. TCP, for instance, does not function particularly well over wireless networks due to the nature of these networks. Using information at lower levels, it has been shown that TCP is able to differentiate between the many different causes of packet loss. The sharing of information across many network levels is what constitutes cross-layer design.

The goal of this design is to make the most efficient use of network resources while also granting the network a great amount of adaptability. In a cross-layer design, the formation of each layer is accomplished by using a limited set of key factors and control knobs. Other levels consult the parameters to decide how their control knobs should be adjusted in response to the present condition of the network. Parameters are utilized in this way by other levels. The formulation of cross-layer design as an optimization problem with numerous optimization variables and constraints that operate simultaneously on many levels is a standard approach to describing this kind of design. After the optimization problem has been fixed, the control knobs of a layer may have their settings adjusted so that they are at their best. Figure 8.7 depicts a different available choice. The following are some of the reasons why cross-layer design is beneficial to wireless network architecture.

The conventional method of architectural design works well inside a wired network; nevertheless, this narrows the scope of the search for the most suitable adaptation. Wireless networks must investigate a far larger optimization space comprising numerous levels to make the most of the available resources instead of just relying on a single optimization level to do so. This protocol stack was developed for networks joined by cables; it does not operate with wireless networks because of this. It may not work when applied to wireless networks because they differ in many ways. For instance, the idea of "connection" has gone through substantial development from the beginning. The distance that separates individual nodes in a network has a considerable bearing on how well they can interact with one another. Because of the one-of-a-kind characteristics of wireless networks, previously explored components must be tackled simultaneously.

Take the example of TCP running via wireless, for instance. Design components that may be found at different levels are now more closely interwoven than in wireline networks, as

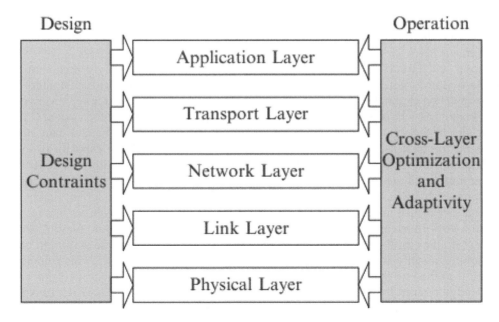

FIGURE 8.7
The cross-layer design.

was previously indicated in this chapter. Defining the "separation idea" in the context of wireless networks would be challenging. More and more people are using protocols and algorithms for wireless networking that make connecting devices on different levels easier.

This trend is expected to continue. On a few occasions, at the very least, we have seen a notable increase in overall productivity. Regarding optimization, cross-layer design presents several challenges that must be overcome. When picking parameters and weighing the potential performance benefits against the potential increase in complexity brought on by the combination of new features from many tiers, care has to be taken. Cross-layer rules have been developed.

8.5.3.3 The Need for a General Framework for Cross-layer Design in Wireless Systems

The Open System Interconnection (OSI) standard is widely used to describe the operation of several types of communication networks. The functions of an organization are layered here according to this standard. PHY layer is an example of a layer that is responsible for ensuring that bits are transferred in a consistent and efficient manner. In order to achieve this goal, data modulation and channel coding are used. Users get their share of system resources according to their MAC/Link layer membership. This objective is achieved by the use of multiplexing, scheduling, and some quality of service management. One other example of this would be the network layer, which, among other things, offers routing capabilities that improve the performance of a wireless network. Due to the fact that the MAC/Link layer and the PHY layer share the same design, we pay special attention to this particular aspect. The traditional paradigm maintains that due to the idea of layer transparency, one layer does not need to be aware of the operation of the layers below or above it in order to operate properly. The only duties that a layer is required to do are those that have been established by the layers that come before and after it. This is helpful for a multitude of reasons, so keep that in mind. If just the physical layer is modified, changing the communication medium (from copper to fiber, for example) will not have any effect on higher tiers of the system. Despite having certain appealing visual qualities, this method may not be the ideal option for wireless transmission, despite the fact that it has some of them.

Knopp and Humblet were the first to pioneer the use of cross-layer architecture in wireless communications. They did it on the uplink of a single cell with customers who often switched channels. The number of different fading channels that may be viewed by each person rises in proportion to the number of users who are using the system (MUD). If the slot in the schedule is given to the user who has the highest channel gain at each moment, the total capacity will improve (sum of simultaneous user capacities). In this strategy, the money you make is referred to as the MUD gain. An algorithm for controlling power that distributes more power to channels that are strong while delivering less power to channels that are weak is the most effective method. The power is often managed in this manner, which means that rather than raising transmit power to compensate for poor channel performance, this approach controls the power in the opposite way. A similar concept applies in the downlink, where the ideal approach for the base station to obtain maximum capacity is to always offer the strongest channel to the user. This principle is identical to the one that applies in the uplink. According to the sophisticated theoretical knowledge offered by MUD, it is essential to co-design features of the MAC/Link layer and the PHY layer in order to improve performance.

There is a large gap between information theory and networking, despite the fact that both have been used to build strategies for allocating PHY–MAC/Link resources. This book takes a look at the traditional obstacles that stand in the way of researchers in these two fields increasing their synergy with one another, as well as the tactics, models, and approaches for distributing resources across layers that are available to them. Problems such as limited battery power, restricted bandwidth, unpredictable time-varying fading, and differing protocols and standards might make it difficult to provide support for multimedia applications and services via wireless networks. This solution will also satisfy the stringent quality of service criteria that have been outlined. This chapter employs a method of cross-layer optimization to simultaneously design wireless network scheduling and power management. This approach was used since it was more efficient. We determine a system's overall performance efficiency by using techniques such as scheduling, power management, and adaptive modulation (CSI).

8.6 Conclusion

Cross-layer design is an effective technique of avoiding some of the limitations of the existing TCP/IP stack, especially in wireless network environments. This is because cross-layer design allows for the separation of different layers. Its primary objective is to do this while also enabling the coordination, interaction, and cooperative optimization of protocols across several layers. This is accomplished by preserving the functionality of the original levels. This chapter investigated the concept of cross-layering, emphasizing its virtues and disadvantages, and gave information on the ubiquity of such design paradigms in the literature as well as modern wireless standards. The most promising future research directions were also highlighted. In conclusion, cross-layer architecture is essential in wireless networks of both the present and the future. Despite the fact that there are hundreds of contributions on the issue, cross-layering research is continually exposing new ideas, notably on architectural problems and dynamic adaptation of network behaviour—but also on the trade-off between performance and interoperability. According to the research that has been done so far, cross-layer design appears to be a practical technique that can be used to make future contributions to WLANs. This technique is able to deal with the growing challenges that are caused by increasing performance, energy consumption, and mobility demands.

References

[1] Li, M., Jiang, F., & Pei, C. (2022). Improvement of triangle centroid localization algorithm based on PIT criterion (ITCL-PIT) for WSNs. *Eurasip Journal on Wireless Communications and Networking*, 2022(1). doi: 10.1186/s13638-022-02109-3.

[2] Liang, S., Cheng, J., & Zhang, J. (2023). Data aggregation algorithm of sensor network nodes for health monitoring based on fuzzy clustering. *Journal of Testing and Evaluation*, 52(1) doi: 10.1520/JTE20210459.

[3] Liu, W., & Yin, F. (2023). Anti-malicious attack of wireless sensor hospital networks based on improved particle swarm optimization algorithm. *Journal of Testing and Evaluation*, 51(1). doi: 10.1520/JTE20210455.

[4] Liu, Y., Li, C., Zhang, Y., Xu, M., Xiao, J., & Zhou, J. (2022). HPCP-QCWOA: High performance clustering protocol based on quantum clone whale optimization algorithm in integrated energy system. *Future Generation Computer Systems*, 135, 315–332. doi: 10.1016/j.future.2022.05.001.

[5] Wang, Y., Li, F., & Fang, F. (2010). Poisson versus Gaussian distribution for object tracking in wireless sensor networks. *2nd International Workshop on Intelligent Systems and Applications, Wuhan, China*, pp. 1–4. doi: 10.1109/IWISA.2010.5473254.

[6] Xu, Y., Qi, H., Xu, T. et al. (2019). Queue models for wireless sensor networks based on random early detection. *Peer-to-peer networking and application*, 12, 1539–1549. https://doi.org/10.1007/s12083-019-00759-7

[7] Polonelli, T., Magno, M., Niculescu, V., Benini, L., & Boyle, D. (2022). An open platform for efficient drone-to-sensor wireless ranging and data harvesting. *Sustainable Computing: Informatics and Systems*, 35. doi: 10.1016/j.suscom.2022.100734.

[8] Rydhmer, K., Bick, E., Still, L., et al. (2022). Automating insect monitoring using unsupervised near-infrared sensors. *Scientific Reports*, 12(1). doi: 10.1038/s41598-022-06439-6.

[9] Sedighimanesh, A., Zandhessami, H., Alborzi, M., & Khayyatian, M. (2022). Training and learning swarm intelligence algorithm (TLSIA) for selecting the optimal cluster head in wireless sensor networks. *Journal of Information Systems and Telecommunication*, 10(37), 37–48. doi: 10.52547/jist.15638.10.37.37.

[10] Singh, A.K., Pamula, R., & Srivastava, G. (2022). An adaptive energy aware DTN-based communication layer for cyber-physical systems. *Sustainable Computing: Informatics and Systems*, 35. doi: 10.1016/j.suscom.2022.100657.

[11] Tanaka, H.K.M. (2022). Cosmic time synchronizer (CTS) for wireless and precise time synchronization using extended air showers. *Scientific Reports*, 12(1). doi: 10.1038/s41598-022-11104z.

[12] Teekaraman, Y., Manoharan, H., & Manoharan, A. (2022). Diagnoses of reformed responses in curative applications using wireless sensors with dynamic control. *Sustainable Computing: Informatics and Systems*, 35. doi: 10.1016/j.suscom.2022.100677.

[13] Turcza, P. (2022). Entropy encoder for low-power low-resources high-quality CFA image compression. *Signal Processing: Image Communication*, 106. doi: 10.1016/j.image.2022.116716.

[14] Van Mao, N., & Son, V.Q. (2015). Applying queuing theory to evaluate performance of cluster wireless sensor networks. *International Conference on Advanced Technologies for Communications (ATC), Ho Chi Minh City, Vietnam*, pp. 501–506. doi: 10.1109/ATC.2015.7388380.

[15] Kumar S., & Lobiyal, D.K. (2013). Sensing coverage prediction for wireless sensor networks in shadowed and multipath environment. *Scientific World Journal*, 2013, 565419. doi: 10.1155/2013/565419.

[16] Mao, J., Jiang, X. & Zhang, X. (2019). Analysis of node deployment in wireless sensor networks in warehouse environment monitoring systems. *Eusasip Journal on Wireless Communications and Networking*, 2019, 288. doi: 10.1186/s13638-019-1615-x.

[17] Yang, R., Liu, L., Liu, Q., et al. (2022). Validation of leaf area index measurement system based on wireless sensor network. *Scientific Reports*, 12(1). doi: 10.1038/s41598-022-08373-z.

[18] Lewandowski, M., & Płaczek, B. (2021). Data transmission reduction in wireless sensor network for spatial event detection. *Sensors*, 21(21). doi: 10.3390/s21217256.

[19] Yang, Z., Wen, J., & Huang, K. (2022). A method of pedestrian flow monitoring based on received signal strength. *Eurasip Journal on Wireless Communications and Networking*, 2022(1). doi: 10.1186/s13638-021-02079-y.

[20] Singh, A.P., Luhach, A.K., Gao, X.-Z., Kumar, S., & Roy, D.S. (2020). Evolution of wireless sensor network design from technology centric to user centric: An architectural perspective. *International Journal of Distributed Sensor Networks*, 16(8). doi: 10.1177/1550147720949138.

[21] Zhiqiang, L., Mohiuddin, G., Jiangbin, Z., Asim, M., & Sifei, W. (2022). Intrusion detection in wireless sensor network using enhanced empirical based component analysis. *Future Generation Computer Systems*, 135, 181–193. doi: 10.1016/j.future.2022.04.024.

[22] Zhou, Q., Zhang, R., Zhang, F., & Jing, X. (2022). An automatic modulation classification network for IoT terminal spectrum monitoring under zero-sample situations. *Eurasip Journal on Wireless Communications and Networking*, 2022(1). doi: 10.1186/s13638-022-02099-2.
[23] Babu, M.M., Reddy, P.C., & Sam, R.P. (2022). A novel cross-layer based priority aware scheduling scheme for QoE guaranteed video transmission over wireless networks. *Multimedia Tools and Applications*, 81(20), 28129–28164. doi: 10.1007/s11042-022-12896-y.
[24] Hamrioui, S., Lloret, J., Lorenz, P., & Rodrigues, J.J.P.C. (2022). Cross-layer approach for self-organizing and self-configuring communications within IoT. *IEEE Internet of Things Journal*, 9(19), 19489–19500. doi: 10.1109/JIOT.2022.3168614.
[25] Jain, A., & Bhardwaj, A.K. (2022). Power-efficient optimized clustering method with intelligent fog computing for wireless sensor networks. *Concurrency and Computation: Practice and Experience*, 34(15). doi: 10.1002/cpe.6983.
[26] Morozs, N., Sherlock, B., Henson, B.T., Neasham, J.A., Mitchell, P.D., & Zakharov, Y. (2022). Data gathering in UWA sensor networks: Practical considerations and lessons from sea trials. *Journal of Marine Science and Engineering*, 10(9). doi: 10.3390/jmse10091268.
[27] Tran, T., Nguyen, T., Shim, K., Da Costa, D.B., & An, B. (2022). A deep reinforcement learning-based QoS routing protocol exploiting cross-layer design in cognitive radio mobile ad hoc networks. *IEEE Transactions on Vehicular Technology*, 71(12), 13165–13181. doi: 10.1109/TVT.2022.3196046.
[28] Zahid, S., Ullah, K., Waheed, A., Basar, S., Zareei, M., & Biswal, R.R. (2022). Fault tolerant DHT-based routing in MANET. *Sensors*, 22(11). doi: 10.3390/s22114280.

9

Social Impacts of Technology with the Emergence of IoT, 5G, and Artificial Intelligence

Ghazanfar Latif, Jaafar Alghazo, and Sherif E. Abdelhamid

CONTENTS

9.1 Introduction 119
9.2 Recent Studies on the Social Impact of Technology 120
9.3 Integration of IoT with 5G 123
    9.3.1 Role of Artificial Intelligence in Future Technology 123
    9.3.2 Model to integrate IoT, 5G, and Artificial Intelligence 124
9.4 Future Impacts of Technology 124
    9.4.1 Personal Privacy Impact of IoT 126
    9.4.2 Impact of Wearable IoT and 5G Devices 127
        9.4.2.1 Control of Smart Sensors 127
        9.4.2.2 Improved Patient Care 127
        9.4.2.3 More Accurate Diagnosis of Health Problems 127
    9.4.3 Economic Impacts of IoT and 5G 127
    9.4.4 Social Impact on Human Life 128
9.5 Discussion and Analysis 129
9.6 Conclusion 131
References 131

9.1 Introduction

Some scientists define technology as two main components—material and intellectual aspects—that complement and mix, so that they lead to an integrated meaning of the concept of technology, where the physical part includes all equipment and machinery, and the intellectual aspect includes the rules and knowledge foundations that lead to production. For example, 5G, a mobile technology now in use in certain areas and still rolling out globally, is sought after by many users due to its speed. However, others wonder whether this new technology will come at the expense of health. 5G comes with rates of up to 10 GB per second [1]. Another new technology called cloud computing is setting the standards for new IT systems. Cloud computing offers various services such as Infrastructure as a Service (IaaS), and Software as a Service (SAS), etc. Artificial intelligence refers to systems that work in a way that resembles the human mind.

The more technological innovation develops, the more it appears to influence human lives. In this contemporary world, the use of technology has significantly increased. Besides, there is relentless advancement of technology in all facets of life. While innovation makes life simpler for individuals, it is increasing the complexity of individuals' social life. For example, technology decreases conventional social practices. However, the social orders of today entail the necessity of technological innovation, which is a form of new information that fulfills the objectives of numerous tasks. Yet at the same time, one wonders whether the social order was brought about due the technological innovations in the first place. If all the social media platforms did not exist today, wouldn't today's generation act and interact like the generations before it? The word technology is made up of *techno* which implies application and artistry/expertise, and *logy* which means learning and science. The semantic relevance of the word innovation denotes the techniques and devices that have been created and are available for use. Technology enables the arrangement of a certain social lifestyle as it fulfills the fundamental requirements to network (socialize). The utilization of innovation in daily activities can cause dangers as well as wrongdoings. Given that humans are social beings, it is not surprising that the social media platforms took off in a way that no one had ever expected. It can be argued that online communication was a necessity for long-distance social connections, but the problem is that it is also affecting human and family relations in close proximity, sometimes even in a household [2]. For example, you might have a family in one household, each with their own digital devices and communicating on social media platforms rather than face to face. It has become a norm for people to think that the utilization of technological innovations is a prerequisite for social familiarity and networking. There are concerns regarding the speed at which current innovation spreads as well as the way it is used and its negative effects. Since instruction is a significant everyday issue, the utilization of present-day advancements makes it an important part of training, not just expansion. This exploration indicates the antagonistic impact of today's technological advances on society. It is expected that it will enhance individuals' mindfulness of the proper way of using technology.

Technology has a huge impact on what it means to be a person socially. A study of students of information technology found that 85 percent of the students who completed the survey used the Internet to access social networks [3]. Figure 9.1 shows the percentage of home Internet users globally [4]. Cars will also support the 5G network to connect them to the Internet, each other, and traffic units to form a complete network showing traffic jams, breakdowns, and even the vehicle's location if it is stolen. Home appliances will also get a share of this development. Instead of relying on HUB or Home Assistant connected to a router, then to the Internet, these devices will communicate directly with the 5G networks.

9.2 Recent Studies on the Social Impact of Technology

The industry supported by Internet of Things (IoT) technology aims to produce smart manufacturing objects that can communicate with each other to automate manufacturing logic and interconnect the various technologies. In the environments of making these interconnected technologies, effective communication is achieved from person to person, from person to machine, and from machine to machine. IoT is a new term for industrialization, such as IT infrastructure, with the goal of sharing information that has a significant

Emergence of IoT, 5G, and AI

FIGURE 9.1
World map with percentage of Internet access at home in different countries [4].

impact on the performance of the manufacturing system. The fifth-generation network is the new generation of wireless networking technology, which will provide a better experience for the individual and provide services through faster data transmission. Communication will become more reliable with 5G. For example, the fourth-generation technology allows files to be downloaded at speeds of up to 100Mbps (bits per second), while the fifth generation offers speeds of more than 10 Gbps. This means that the fifth generation will mark a fundamental change in speed compared to the previous generations. Artificial intelligence (AI) technologies support several companies and sectors. For example, in the transportation sector, AI supports transportation vehicles and helps in facilitating traffic. AI has also been introduced to the field of health care, as it develops effective new tools for diagnosis and supporting medical practitioners. Back to the social networks, massive social apps like Twitter and YouTube have changed people's perception of social interaction. Content is spreading with the press of a finger in all four parts of the planet. Even the big and important personalities are using Twitter to get closer to people, for example, Donald Trump is still tweeting that America is one of the best countries to tackle the coronavirus pandemic, and this tweet reaches millions of people who follow Trump in a fraction of a second. This type of communication has become popular with people and made them aware of the latest news and developments. One of the most important activities on social media platforms is media distribution, creativity, and education. E-commerce (electronic commerce) is considered one of the most prominent trends in society besides chatting and dating.

Vlačić et al. did a quantitative study in which they investigated the absorbing capacity of organizations when it comes to technology [5]. The researchers asked >600 Croatian firms to fill out questionnaires on technology absorption and the resultant impact. Only

103 agreed to take part in the study. Findings from the study showed that 45 of the firms had absorbed technology and were using it to make work efficient. The other findings were that 34 of the 103 firms largely exported technologies that enhanced work performance. Findings also revealed that technology resulted in enhanced innovation in the workplace. Khosravi et al. investigated the impact of technology on the isolation of senior citizens within the family context [6]. The methodology was meta-analysis, which involved looking at 6,886 relevant articles from 2000 to 2015. Findings from the study revealed that eight technologies are used to isolate the elderly, including video games, robotics, 3D virtual environment, asynchronous peer support chat groups, and social networks, among others. In as much as technology has isolated seniors, it can be used to bring them together as well.

Betts and Spenser (2017) investigated the effects that technology has on young people living in the United Kingdom [7]. A qualitative methodological approach was used in the study. The researchers collected data using a focus-group discussion with children aged 11 to 15 years. A phenomenological approach was used for the data analysis. Findings established that there are several negative effects of technology on adolescents. Findings indicated that the use of social media affects social interactions among children and their family members. Findings also established that the use of social media results in cyberbullying which affects the esteem of adolescents. From these findings, it can be deduced that technology has more negative effects among young people whose brains are not fully developed. It causes disruption and some adolescents get addicted to the technology.

Rabab studied the effect of new technologies on moral qualities of youths in Saudi Arabia, and their use as an instrument to advance virtues [8]. The principal objective gave rise to accompanying sub-objectives: checking, inspecting, and assessing the connection between youth and advanced broad communications by recognizing the power of utilizing the new media, the kinds of cooperation and inspiration, and the ramifications important to youngsters. The researchers aimed to establish the ethics of Saudi youths, who are largely Arab and Muslim.

Sutton did a qualitative study to investigate the impact of technology on education and society [9]. Data were collected through observation. Findings showed increased classroom use of technology such as the use of SMART boards and PowerPoint that enhance learning. Findings from this study also indicated that teachers are not welcoming of the use of technology in class. They are not sure how to effectively integrate technology into the syllabus. Conversely, students were receptive and excited about the use of technology in classrooms. Similarly, findings from [10], which was a meta-analysis, established that students in elementary schools enjoy the use of technology in their classes. These are young children who were born in the era of technology and have the potential to fully embrace and utilize digital learning.

Junco et al. investigated the effects that technologically mediated communication has on students' evaluation of advisors [11]. Specifically, the effects of evaluating NACADA were investigated. The researchers utilized a quantitative study with a survey design. A total of 4,500 undergraduates were included in the survey. However, the response rate was only 15 percent, with a full response from 706 students. Findings from this study indicate that there was a variance of 18 percent in people who met with their advisors. The level of student satisfaction with online advice was insignificant ($F_{(16, 490)} = 1.206$, $p = .259$). The other findings were that 61 percent of students rarely emailed their advisors. These findings indicate that communication using technology is not as effective as one on one. Similarly, Fartash et al. found that technology helps in research and the generation of new knowledge [12]. This promotes innovation in companies and enhances competencies within the market.

9.3 Integration of IoT with 5G

Over the past years, the Internet of Things has created a major change in the world of computing and sensitive applications. The great activity in product lines is expected to grow in the coming years to reach billions of devices. With the emergence of problems in devices, there is a desire to integrate sensors with physical and cyber devices. Fifth generation devices and the Internet of Things are at the forefront as the devices are expected to form a large part of the fifth generation model. Also, IoT is expected to change other technologies such as communication from one device to another with smart data analysis. It is also expected that cloud computing will extend to the emerging fog model with the use of smart devices to bring innovation into the field. There is already widespread research done on smart devices [13–14].

Communications will play an important role in the IoT paradigm in the coming decades. The IoT-5G scenario extends the sensor-based IoT capabilities in other technologies such as robots and actuators for coordination and implementation. It consists of up and downlinks for 5G-IoT connections. Programmers suggest analyzing the machine-type multicast service (MTMS). This is to derive the best design engines by analyzing various business indicators such as energy consumption. One of the most prominent problems that the Internet of Things faces is security: because of the proliferation of electronic devices in the hands of people, it requires secure algorithms. The number of IoT devices was around 7.5 billion in 2015, and it reached more than 23 billion connected devices last year [15]. The popularity of the Internet of Things has caused several complex security problems, especially the privacy concerns that make Internet users vulnerable to many risks. These risks include cyberattacks, identity theft, as well as default or persistent passwords, which can create a door for security breaches. Vulnerabilities can be exploited by cybercriminals on the dark web to gain remote access and then mess with hardware. IoT security is concerned with protecting connected devices and networks in the IoT world. It indicates steps to enhance the security of IoT devices and to reduce their vulnerability to attacks from unauthorized users.

9.3.1 Role of Artificial Intelligence in Future Technology

Technology has developed extensively in the field of medicine, as the world witnesses new procedures that earlier generations did not know through artificial intelligence. This helps doctors diagnose and detect microscopic diseases [16–18]. AI has not replaced the doctor, but rather helps the doctor to achieve the best possible medical care. Artificial intelligence provides technical services that contribute to most of the doctor's accomplishments, as it reduces the psychological pressure that the physician faces and the physical effort in trying to discover diseases. Among the positives of integrating artificial intelligence in the health field are obtaining tremendous ability and tremendous speed in collecting, storing, and easily obtaining patient information and data, in addition to complete safety and intense privacy. There is no doubt that artificial intelligence and robotics together constitute the future of the healthcare sector. Middle Eastern hospitals have demonstrated this new reality from its inception and contributed to enabling new and advanced services. For example, one of the regions of Saudi Arabia, fearing an outbreak of coronavirus, used drones to conduct a thermal survey of people and follow up on their safety. This is one of the technologies that Saudi Arabia has adopted exploiting the targeted aspects of artificial intelligence technologies [19].

9.3.2 Model to integrate IoT, 5G, and Artificial Intelligence

Technology is advancing at mind-boggling speed to the extent that technologies of today were not available a few years ago. Some examples of these technologies include Big Data, real-time machine translation, and real-time interactive chat bots that automate communication with a large population of customers. Another example is the new DeepFake technology which allows the faking of videos in ways that are undetectable [20]. The latest buzzes in the world of technology and research are AI, deep learning, and natural language processing (NLP) systems and their applications. Major technology companies depend on them to develop their services, for example Google relies on them to develop many of its services, starting with the search engine and extending to YouTube, Gmail, Maps, and even the voice assistant that recognizes speech and provides simultaneous translation. Facebook relies on them to provide recommendations appropriate for users of the application in their news feed, and Amazon relies on them to develop its voice assistant Alexa, and much more.

Deep learning, which is sometimes considered a subset of artificial intelligence, is sometimes loosely defined as automating human functions. For example, human tasks mastered through experience, such as recognizing a cat, can be done using machine vision and deep learning [21–22]. The more research is done in the field of deep learning, the more complex the tasks can be achieved. For example, a lot of research is now done on using machines equipped with computer vision and machine learning algorithms for detecting and diagnosing medical conditions. Deep learning allows for the training of machines just like you would train a human to complete a certain task. These technologies now make it possible for companies to allow their machines to make instant decisions, make a prediction, and automatically classify certain things. Speech recognition, computer vision, and self-driving cars are the most important areas in which companies use learning.

Since 2019 to the present, we have witnessed an increased dependence on artificial intelligence. Companies are keeping up with the latest research developments in the field to be the first to benefit from these technologies. In addition, many large companies have established their own research and development divisions in these fields. The ultimate goal of these companies is to produce services and products that improve the user's experience. Figure 9.2 shows a model for the integration of IoT, 5G, cloud computing, and AI.

9.4 Future Impacts of Technology

The world knows that technology is nothing but a double-edged sword. The great effect that technology has achieved by relying on machines or robots instead of manpower has resulted in decreased demand for humans. In the near future, it is expected that the use of these technologies will eliminate the human element even in the most difficult tasks such

Emergence of IoT, 5G, and AI

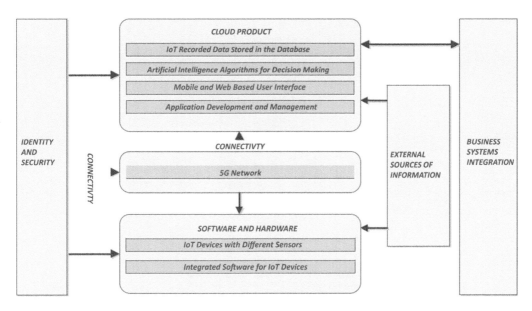

FIGURE 9.2
Model to connect IoT, 5G, cloud computing, and artificial intelligence.

as medicine and surgery. This will lead to a high rate of unemployment. It is well known that technology occupies the first position in all fields of work, and one cannot do without it. From this standpoint, scientists expect social effects caused by modern technology. First, is the high unemployment rate because companies and hospitals will dispense with the human element and replace it with robots. Second, the world fears that robots more intelligent than humans will be developed and be difficult to control. Third, is the emergence of many economic and development projects, including unprecedented qualitative developments in many areas, such as the auto industry. Fourth, future technology will move toward simulation, i.e., remote use. Fifth, technology will move toward the use of light technology, as telecommunications companies seek to dispense with copper lines and adopt optical fiber cables. Scientists are striving to develop a new computer called an optical computer. Light is made up of photons traveling at a very high speed, while electrons in electricity are somewhat slow. These changes may cause social effects for the individual, including:

- Increasing the material cost as more devices will be needed in each household.
- Increasing the consumption of electrical energy, given that most modern and advanced devices operate on electrical energy. Electricity has become the backbone of modern life.
- Technology has a major impact on the spread of violence among humans, given that most teenagers and children spend most of their time watching combat movies and playing violent games that are easily available in any home and in the hands of everyone, and this can affect the child's behavior even when he/she grows up.
- Technology is making human interaction something that can be dispensed with, since people can now interact with virtual devices such as Alexa and Siri, in addition to the fact that face-to-face interaction is no longer needed with the advances in social network applications.

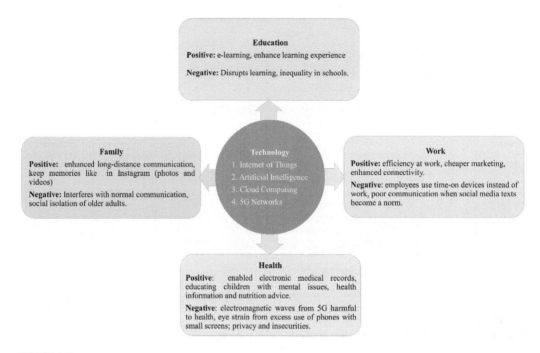

FIGURE 9.3
Overview of the positive and negative influences of technology.

Figure 9.3 shows a summary of the impact of technology both positively and negatively on the pillars of human life, including education, work, family, and health.

9.4.1 Personal Privacy Impact of IoT

IoT is the new generation of the Internet (the network), which allows understanding between devices that are connected to each other (via the Internet Protocol). What distinguishes the IoT is that it allows a person to be free from the place; that is, the person can control the tools without having to be in a specific place to deal with a specific device. It has a great impact on companies as it has the potential to stimulate the innovation sector and improve the efficiency of individual companies in pushing the market toward a clearer focus on innovation. As an example, smart doorbell technology can be likened to a strong guard dog on the one hand, and a reliable digital receptionist on the other. In addition to informing the person through a smartphone of the activity taking place at the door, the bell records all kinds of movement events in the electronic cloud, to allow viewing of these videos (along with the videos recorded directly) on digital devices at any time.

A study showed that people can be uniquely identified through e-mail, their nationalities, and their place of birth. Data that seems unidentifiable can easily be found using reverse engineering methods. Paraskevopoulos et al. described how most of the population can be identified from the behavioral patterns revealed by site data from mobile phones, by analyzing the mobile phone database of approximate sites on the basis of the closest cell tower [23].

9.4.2 Impact of Wearable IoT and 5G Devices

IoT technology is now immersed in many fields, including health. Below are some of the impacts of these technologies within the health field.

9.4.2.1 Control of Smart Sensors

One of the recent methods tried on patients who suffer from complete paralysis is the use of smartphones or modern computers to sense the thinking, ideas, and tasks they want to do through sensors that are cultivated in their brains. Stanford University announced that they had succeeded in experiments with sensors implanted in the brain. This allows such patients to think of things and the machine can directly translate or implement them. For example, they can think of words and the machine will be able to say these words, allowing these patients to communicate. They can also think of tasks such as the movement of fingers and hands and an attached robotic hand can do these tasks. There are many examples, and this field of research is making a huge impact not only on patients with complete paralysis but also on patients with other medical issues. When this technology is fully viable, it can be used by healthy individuals to implement tasks such as driving through their thoughts.

9.4.2.2 Improved Patient Care

Technological advances such as 5G speeds will allow for real-time reliable communication. This communication will allow patients to communicate with healthcare providers to have their concerns and questions answered on the spot. These advances will also allow the development of IoT devices that connect to the patient and communicate directly with the healthcare provider. These technologies will allow for home medical care which will ease the pressure on hospitals and might reduce the cost of healthcare.

9.4.2.3 More Accurate Diagnosis of Health Problems

A more accurate diagnosis of health problems is the most important goal for IoT devices measuring pressure, sugar, and heat. Early diagnosis is very important as it reduces the difficulties of surgery and enables appropriate treatment. Therefore, research is ongoing for technological devices that can accurately diagnose health problems without human intervention. This will pave the way for technological devices that will in many areas replace human medical personnel.

9.4.3 Economic Impacts of IoT and 5G

The fourth industrial revolution is high speed, which means achieving more unprecedented innovations and achievements in several vital areas, such as artificial intelligence (AI), smart cities, and the Internet of Things (IoT). This would radically transform how societies operate, advance economic development, support digital plans for countries, and contribute to creating a more connected world. One of the main factors behind this revolution is empowerment from fifth generation mobile networks. Thanks to the high speed, which is 100 times faster than the current 4G LTE network, in addition to the high reliability, mobile to mobile (M2M) communication technology, and limited energy requirements, the fifth generation networks herald a new era in the world of mobile communication. This generation's network outperforms the capabilities of previous generations by providing unparalleled speeds that allow unprecedented communication at speeds with unobservable small delays, which will enable networks to support entirely new services and applications, many of which are characterized by their ability to significantly improve the quality of life throughout the world. According to the Global System for Mobile Communications Association, it is expected that more than 1.2 billion communication channels will be operating over 5G networks worldwide by 2025 [24]. To keep pace with the ever-increasing traffic while offering higher bandwidth, 5G technology requires larger blocks of contiguous spectrum and bandwidth. Therefore, it becomes necessary to extend telecoms tower coverage in any market seeking to operate the 5G networks. With 5G network technologies expected to contribute $2.2 trillion to the global economy over the next 15 years [24], it is imperative that governments and mobile operators ensure the right infrastructure to take advantage of the huge opportunities looming. Tower companies have succeeded in overcoming the challenges facing emerging markets by keeping up with a five-dimensional model, including the participation of towers and sites, lease agreements, construction of new sites, acquisition and re-lease of existing telecoms towers, and providing coverage solutions, such as smart distribution of antennas inside buildings, in addition to connecting towers to the optical fiber network. Given the large number of towers required for the efficient operation of 5G networks, the logical step to make this possible and effective is to use the existing infrastructure. This can be achieved through tower rental agreements and the participation of other telecoms operators to increase the efficiency of the existing towers while installing additional equipment when needed to improve their network capabilities.

9.4.4 Social Impact on Human Life

It is undeniable that organizations looking to adopt IoT are driven by the need to gain a competitive edge over other players in the industry. For this to be successful, management approaches also have to be fine-tuned and innovated in such a way that communicating physical objects will go hand in hand with network management teams. According to Tarabasz (2016), network management is a formalized and constant dependency on relationships, meaning that this network will be viewed as a dynamic solution [25]. This solution would be an ad hoc creation by clients applying available products and applications of companies. Through this approach, companies will be able to introduce new and unconventional ways of thinking, thereby fostering competitive advantage through flexible management, versatility in approach to leadership, and access to new business windows and opportunities. While this line of argument does not focus on the social impacts of IoT, it dissects the management of people, which loosely borders on the assessment of people's social norms and behaviors.

Perhaps an overlooked or ignored avenue of the IoT entails the myriad ways in which it can be socially used. According to Zeeuw et al. (2019), such forms as Internet skills and capital form the basis upon which different types of uses of IoT can be attained [26]. Furthermore, the interconnectivity and open nature of IoT imply that communication is open. Hence different modes of social communication. The authors argue that the social use of IoT by different types of people informs the understanding of people's behavior in terms of creation, maintenance, and absolution of social relations in a society linked through networks. Key indicators of the abovementioned statement are entrenched in the inverse effects of income, social capital, and education on personal use, as well as information on sharing of IoT data with a third party or a partner. These varying uses and factors that determine the indicated uses highlight the existence of different social uses of IoT. The authors conclude that the different designs of the Internet and capital skills determine social use and application of the Internet of Things.

It is important to note that the social impacts of any phenomenon will eventually have an impact on the economy of a given country. The primary impact of increased consumption of IoT is an increase in Internet usage [27]. It goes without saying that in the current world of business, the Internet is a crucial component, thanks to the ever-increasing demand for convenience from consumers. The authors acknowledge the fact that an increase in Internet usage and the adoption of the IoT will pose challenges regarding data ownership and unlimited communication and intrusion. Their take is that an increase in IoT usage and overall Internet usage will result in a direct bolstering of the economy, especially with more accurate data on consumer behavior.

While an improvement in the economic status of a country or entity would likely result in a compromise in the social norms of the said country, sustained approaches of mitigation and management can be deployed to ensure that these challenges do not have long-standing effects on a social sphere. The authors hold that the five major components that make IoT tick are the storage of information, collection of information, communication, processing of information, and action performance. When these factors are seamlessly fused with the right regulatory observations, both social and economic aspects of a country are strengthened. The interconnection of homes, transport systems, business activities, communities, and the government results in the creation of a one-stop-shop ecosystem, where people can access anything they need in almost real-time. Therefore, social interactions will be boosted as economic prosperity is achieved.

Wielki (2017) claimed that the impact of IoT stretches far beyond organizational restructuring and economic improvement through an increase in sales [28]. He holds that the Internet of Things will trigger operational improvement in the sense that organizations will come up with newer business models that will be used to enhance operations and maximize profits. More specifically, the IoT will present opportunities for organizations in two major ways: the transformation of regular business processes and the enabling or creation of new business models.

The author further notes that the paramount areas of exploitation and success for the IoT will be the installation of sensors into smart products, thereby enabling organizations to track the health, usage, operation, and external environment of the product. It is imperative to note here that despite the tremendous economic improvement that this sensor installation will generate for the products of a given company, the social downside is that it will keep track of the consumer, causing unwarranted and unauthorized intrusion into the private life of the customer or client using the product. Besides, the fact that this product can be remotely controlled suggests that companies can use the product for whatever intention they may harbor.

9.5 Discussion and Analysis

Based on research and arguments brought forth by select scholars, it is evident that the IoT, 5G, AI, and cloud computing have both overwhelmingly positive and negative impacts. From a spectrum of simplicity, IoT facilitates the connection of people, objects, companies, governments, and societies through technology. The key emerging factor from the above review of research material is that economic and social impacts usually overlap. Given that it is mostly viewed from a business prism, this concept entails five major components that must work together seamlessly for efficiency: identity and security, cloud, connectivity, external sources of information, and business systems integration. A combination of these factors results in an astute IoT that can be used to bolster the economic and social facets of human life.

Based on the reviewed research, the general impacts of the IoT can be condensed into process monitoring, management of data, accurate prediction, and access from anywhere. From an economic angle, these factors facilitate proper and accurate collection of information on clients, sustained monitoring of product performance and purchase trends, near-accurate prediction of consumer behavior in forthcoming time periods, and the ability to offer customers convenience since products or services can be accessed from any given location. Based on these factors, organizations and corporations that adopt IoT stand to make massive profits because of improved operationalization, better business models, better monitoring, and increased sales because of the prediction of consumer behaviors.

Socially, the impacts of IoT, 5G, AI, and cloud computing can be broken down into positive and negative effects. On a positive scale, IoT facilitates quicker and easier communication, implying that it becomes easier for people to form social relationships even when initial interactions are business oriented. The fact that homes, schools, social amenity avenues, workstations, and the world at large can be connected, means improved and increased communication in real-time, hence better relationships. As noted above, human beings exist in an environment by forming relationships through social learning. This learning process, if done properly without unnecessary intrusions, results in strong relationships. Dating, for instance, can be massively improved through the Internet of Things, especially when sufficient data and information are collected about a person's habits, preferences, and tendencies. The Internet, through this connection, could also be an avenue for educating youth on positive social behavior and societal expectations [8]. Numerous other social concepts can benefit greatly from increased and improved connections brought about by IoT.

Conversely, the phenomenon also has far-reaching negative social impacts that can be wholly tied to data ownership, security, and privacy. For most people, individual privacy is not considered when companies collect information and data about them. Worse still, companies make products with sensors that can be controlled remotely, meaning they may easily be used for 'spying' on consumers. Given the unauthorized and unrestricted access to personal data through IoT, social relationships are highly likely to be compromised because people would rely on data rather than natural trust when interacting with one another. The foundational principles of any social relationship revolve around mutual trust, human touch, and the element of surprise. The use of IoT to study one another would mean that people perceive each other statistically or in a robotic fashion. Besides, families have been disrupted, thanks to the excessive use of technology and the Internet to the extent that normal face-to-face communication becomes scarce [29]. The inevitable result of such an occurrence would be the collapse of normal social interactions. This is in addition to the danger of some individuals' tendencies to socialize with virtual devices such as Alexa or Siri rather than socializing with real human beings. This could become a real danger in the future when such virtual machines are advanced to the extent that they can hold a real conversation. To our knowledge, this technology is still in its infant stage and the social impact remains to be seen.

A downside to technology is that for now, it cannot be successfully used to teach students with special needs such as mental illness. When it is used for regular students, it provides avenues for distraction, bullying, and other non-educational ventures. Besides, IoT can be socially used in a myriad of ways, depending on the Internet and capital ability. On this basis, it goes without saying that this connectivity could easily have negative repercussions, especially if users harbor malice. Cases of cyber-bullying could increase, online fraud, attacks, and even identity theft could be on the rise because of increased connectivity through 5G and IoT [30]. Young people engaged in such crimes would face jail terms. Therefore, it suffices to state that proper and more stringent regulatory measures have to be put in place to ensure that the feared social impacts do not materialize.

9.6 Conclusion

The modern world is on an upward trajectory in terms of consuming and relying on technology. The continuous improvements in connectivity and network enhancement through IoT, 5G, cloud computing, and artificial intelligence mean that the world is continuously becoming one tiny ecosystem. Better human interactions through real-time and accurate data mean that everyone lives in one huge bubble. As discussed throughout this chapter, this concept has numerous positive impacts, including better communication and heightened profits for businesses. However, the downside is tied to the absence of strong regulations to ensure that data security and privacy principles are adhered to. To truly realize the eye-watering benefits of IoT, governments need to ensure that personal information and data are properly protected against misuse, improper exploitation, and even identity theft. When these factors are taken into consideration, the IoT will fully revolutionize the world in a beneficial way. Till then, it is impossible to guarantee only the positive impacts of this phenomenon. On the other hand, the chapter has concentrated on the social impact of technology. The social impact of technology can either be positive or negative. Positive impact can include the use of technology in classrooms, as a means to gain knowledge, as a means to help patients, and other uses. However, there might be negative social impacts which include decreasing family and general human face-to-face interaction, changing the behavior of youth, lack of privacy, etc. The negative social impacts of technology are already seen in today's society, and it remains to be seen how this impact will increase and what its long-term effects might be. The field of IoT and new technologies is being explored in other domains for its impact on human life, such as IoT-based smart cities as well as other domains explored in recent literature. Humankind must pay attention when it comes to technology and its impact and research should be continuous in the domain of the effect of technology on all aspects of human existence.

References

[1] Kaur, K., Kumar, S., & Baliyan, A. (2020). 5G: A new era of wireless communication. *International Journal of Information Technology*, 12(2), 619–624.

[2] Anderson, M., & Jiang, J. (2018). Teens, social media & technology 2018. *Pew Research Center, 31*, 1673–1689.
[3] Simsek, A., Elciyar, K., & Kizilhan, T. (2019). A comparative study on social media addiction of high school and university students. *Contemporary Educational Technology, 10*(2), 106–119.
[4] World Bank. Households w/ Internet access TCdata360. https://tcdata360.worldbank.org/indicators/entrp.household.inet (accessed 22 May 2022).
[5] Vlačić, E., Dabić, M., Daim, T., & Vlajčić, D. (2019). Exploring the impact of the level of absorptive capacity in technology development firms. *Technological Forecasting and Social Change, 138*, 166–177.
[6] Khosravi, P., Rezvani, A., & Wiewiora, A. (2016). The impact of technology on older adults' social isolation. *Computers in Human Behavior, 63*, 594–603.
[7] Betts, L.R., & Spenser, K.A. (2017). "People think it's a harmless joke": young people's understanding of the impact of technology, digital vulnerability and cyberbullying in the United Kingdom. *Journal of Children and Media, 11*(1), 20–35.
[8] Rabab, R. (2013). *The Impact of the Use of Social Networks to Form a Pattern of Ethical Value System for Saudi Youth: An Empirical Study*. Department of Media, Faculty of Arts & Humanities, King Abdulaziz University.
[9] Sutton, B. (2015). The effects of technology in society and education. Masters dissertation, SUNY Brockport.
[10] Chauhan, S. (2017). A meta-analysis of the impact of technology on learning effectiveness of elementary students. *Computers & Education, 105*, 14–30.
[11] Junco, R., Mastrodicasa, J.M., Aguiar, A.V., Longnecker, E.M., & Rokkum, J.N. (2016). Impact of technology-mediated communication on student evaluations of advising. *NACADA Journal, 36*(2), 54–66.
[12] Fartash, K., Davoudi, S.M.M., Baklashova, T.A., Svechnikova, N.V., Nikolaeva, Y.V., Grimalskaya, S.A., & Beloborodova, A.V. (2018). The impact of technology acquisition & exploitation on organizational innovation and organizational performance in knowledge-intensive organizations. *Eurasia Journal of Mathematics, Science and Technology Education, 14*(4), 1497–1507.
[13] Latif, G., Shankar, A., Alghazo, J.M., Kalyanasundaram, V., Boopathi, C.S., & Arfan Jaffar, M. (2020). I-CARES: Advancing health diagnosis and medication through IoT. *Wireless Networks, 26*(4), 2375–2389.
[14] Latif, G., & Alghazo, J. (2021). IoT cloud based Rx healthcare expert system. In *Fog Computing for Healthcare 4.0 Environments* (pp. 251–265). Springer.
[15] Farhan, L., Kharel, R., Kaiwartya, O., Quiroz-Castellanos, M., Alissa, A., & Abdulsalam, M. (2018, July). A concise review on Internet of Things (IoT)—problems, challenges and opportunities. In *2018 11th International Symposium on Communication Systems, Networks & Digital Signal Processing (CSNDSP)* (pp. 1–6). IEEE.
[16] Bashar, A., Latif, G., Ben Brahim, G., Mohammad, N., & Alghazo, J. (2021). COVID-19 pneumonia detection using optimized deep learning techniques. *Diagnostics, 11*(11), 1972.
[17] Butt, M.M., Latif, G., Iskandar, D.A., Alghazo, J., & Khan, A.H. (2019). Multi-channel convolutions neural network based diabetic retinopathy detection from fundus images. *Procedia Computer Science, 163*, 283–291.
[18] Latif, G., Butt, M.O., Al Anezi, F.Y., & Alghazo, J. (2020, October). Ultrasound image despeckling and detection of breast cancer using deep CNN. In *2020 RIVF International Conference on Computing and Communication Technologies* (pp. 1–5). IEEE.
[19] Hassan, O. (2020). Artificial intelligence, Neom and Saudi Arabia's economic diversification from oil and gas. *Political Quarterly, 91*(1), 222–227.
[20] Güera, D., & Delp, E.J. (2018, November). Deepfake video detection using recurrent neural networks. In *2018 15th IEEE International Conference on Advanced Video and Signal Based Surveillance (AVSS)* (pp. 1–6). IEEE.

[21] Latif, G., Alghazo, J., Maheswar, R., Vijayakumar, V., & Butt, M. (2020). Deep learning based intelligence cognitive vision drone for automatic plant diseases identification and spraying. *Journal of Intelligent & Fuzzy Systems, 39*(6), 8103–8114.

[22] Latif, G., Bouchard, K., Maitre, J., Back, A., & Bédard, L.P. (2022). Deep-learning-based automatic mineral grain segmentation and recognition. *Minerals, 12*(4), 455.

[23] Paraskevopoulos, P., Dinh, T.C., Dashdorj, Z., Palpanas, T., & Serafini, L. (2013). Identification and characterization of human behavior patterns from mobile phone data. *D4D Challenge session, NetMob*.

[24] Gangfada, I.A., Abdullahi, U.S., Ugweje, O., Koyunlu, G., Suleiman, H.U., & Chijioke, C.E. (2020, June). An evaluation of the physical layer characterization of 5G networks. In *2020 4th International Conference on Trends in Electronics and Informatics (ICOEI) (48184)* (pp. 496–504). IEEE.

[25] Tarabasz, A. (2016). The impact of the Internet of Things on new approach in network management. *International Journal of Contemporary Management, 15*(2), 151–170.

[26] van der Zeeuw, A., Van Deursen, A.J., & Jansen, G. (2019). Inequalities in the social use of the Internet of Things: A capital and skills perspective. *New Media & Society, 21*(6), 1344–1361.

[27] Saidu, C.I., Usman, A.S., & Ogedebe, P. (2015). Internet of Things: impact on economy. *British Journal of Mathematics & Computer Science, 7*(4), 241.

[28] Wielki, J. (2017). The impact of the Internet of Things concept development on changes in the operations of modern enterprises. *Polish Journal of Management Studies, 15*.

[29] Battiston, D., Blanes i Vidal, J., & Kirchmaier, T. (2017). Is distance dead? Face-to-face communication and productivity in teams. CEP Discussion Papers dp1473.

[30] Raselekoane, N.R., Mudau, T.J., & Tsorai, P.P. (2019). Gender differences in cyber-bullying among first-year University of Venda students. *Gender and Behaviour, 17*(3), 13848–13857.

Part III

Advanced Wireless Sensor Networks: Power, Data Gathering Techniques, and Security

10

Power Management Strategies in Wireless Sensor Networks

Senthil Kumaran Rajendran and Ganesan Nagarajan

CONTENTS

10.1 Introduction 137
10.2 Requirements and Characteristics of WSNs..138
10.3 Related Work 138
10.4 Difficulties in WSNs..139
    10.4.1 Power Management..139
    10.4.2 Security...140
    10.4.3 Data Aggregation..140
10.5 Causes of Energy Wastage in WSNs 140
10.6 Power Gating 140
10.7 Leakage Gating and WSNs..142
10.8 Techniques of Power Matching..143
10.9 Different Types of Power Sources 144
10.10 Power Modes in DRAM 145
10.11 Power Management Strategy 145
10.12 Dynamic Duty Cycle Scheduling Scheme 146
    10.12.1 Static Power Saving..147
    10.12.2 Dynamic Power Management 147
        10.12.2.1 Sleeping Policy 148
        10.12.2.2 Awakening Policy...149
10.13 Conclusion and Future Work..149
References 149

10.1 Introduction

The sensor model comprises the sensor module, memory, network stack, sensor application interface, and the microcontroller. The power model consists of a central processing unit (CPU), a battery, and a radio model. In total, the wireless sensor network consists of four units, the power unit, sensing unit, communication unit, and processing unit. Each unit degrades a certain amount of energy, which results in higher energy consumption. Therefore, the consumption of energy should be minimized to improve the network lifetime. There is a huge consumption of energy for certain reasons, such as smaller size of nodes, variations that cannot be done manually, and some nodes that do not function properly, which leads to maximized consumption of power. Communication and processing are the two tasks performed in WSNs, and communication requires more power than processing, so the process of communication should be reduced and the same battery with the correct energy should be preferred for both operations. Another major problem is that batteries that are used cannot be replaced or recharged, so the above-mentioned processes should cooperate and manage to work together.

10.2 Requirements and Characteristics of WSNs

The major requirements of WSN technology are:

- low power consumption
- unlicensed radio band
- easy deployment and extension of network
- sufficient for lower data rate
- minimal cost and small sized devices

Wireless sensor networks have the following unique characteristics:

- dense node deployment
- battery powered device
- self-configurable
- data redundant
- change in topology

10.3 Related Work

Luo et al. designed a dynamic power management scheme for sensor networks [1]. Based on the event occurring, each node can switch on and off its components. The node awakening scheme includes three policies: event-driven, message-driven, and timer-driven. In event-driven, when an event occurs, the sensor awakens the CPU. During a message-driven state, the node goes to sleep after forwarding the message to the next node. In a timer-driven state, the deepest sleep state period was obtained. An operating system directed power management technique to enhance the energy efficiency of sensor nodes was developed. This technique shuts down the sensor nodes when not in a sensing region and wakes the sensors whenever necessary [2].

Cotuk et al. investigated the impact of transmission power control on network lifetime [3]. This was analyzed with mathematical programming models. The effect of granularity of power levels on energy dissipation characteristics through a linear programming framework was explored. Nayak and Devulapalli proposed a new methodology for clustering with the help of fuzzy descriptors [4]. The super cluster head was selected using the fuzzy inputs: remaining battery power, mobility, and distance of the node to the base station.

The base station was mobile. It is necessary to route the data sensed by the sensors to the base station via the super cluster head. The Mamdani rule was used to evaluate the output variable chance (i.e., super cluster head).

Zhang et al. developed a solution for the energy consumption problem and proposed a virtual multi-input and multi-output based cooperative routing algorithm [5]. With the help of initial energy, residual energy, and the state of the channel, the proposed algorithm solves the issues of power allocation and inappropriate routing path. This algorithm chooses cooperative relay nodes based on the VMIMO model. Giuseppe Anastasi et al. implemented a method to reduce the transceiver power consumption [6]. This proposed protocol, named Adaptive Staggered Sleep Protocol (ASLEEP), vigorously adjusts the sleep scheduling of nodes based on the traffic scenario of the network. Also, it does not require information about the topology of the network.

Buettner et al. presented a low-power MAC protocol for wireless sensor networks [7]. X-MAC transmits a preamble packet which contains the target receiver addresses. This preamble packet produces a certain delay at each node, which leads to the consumption of more energy. This is applicable to non-targeted receivers. Nodes can be turned to switch off and on (i.e., sleep and listening modes). Preamble packets received by non-targeted receivers are to be kept in sleep mode.

Ian F. Akyildiz et al. surveyed various research issues in the sensor network protocol stack [8]. The issues involved in the physical layer are modulation schemes, hardware design, and the strategies to overcome propagation effects. The major problems in the MAC layer are error control coding schemes and power saving operations. At the network layer, the protocols are developed to address network topology changes and higher scalability [9–13]. Consider the performance evaluation of two fuzzy-based cluster head selection systems for wireless sensor networks. The major problems in sensor network applications are cluster formation and cluster head selection. In the first method, the fuzzy input variables are distance from the cluster centroid, remaining battery power, and network traffic, and the output for the fuzzy system is the cluster head selection of a node. In the second method of cluster head selection, the input variables used by the fuzzy system are remaining battery power, degree of number of neighbor nodes, and distance from the cluster centroid.

10.4 Difficulties in WSNs

WSNs have faced several difficulties in real-time applications. Some of the difficulties are power management, security, and data aggregation.

10.4.1 Power Management

In general, sensors are equipped with minimal resources, especially power. In order to reduce the power usage, the sensor node can be switched to a sleep state and an active state whenever needed. Power can be consumed by three functional units: sensing, communication, and processing. At a certain instant of time, the node moves to a dead state. Data routing can be affected when a node is dead. Hence, power management must be done with utmost care.

10.4.2 Security

As data transfer takes place in wireless media, it is prone to hacking. Data transfer takes place in an unlicensed frequency band-2.4 GHz (i.e., industrial, scientific, and medicinal (ISM) bands. Due to this unlicensed frequency, hacking of data is easier. Hence, security is one of the important factors to be considered in wireless sensor networks.

10.4.3 Data Aggregation

Sensors are placed in an area to monitor the environment and surroundings. The sensed data can be routed to the base station through the intermediate sensors. The ultimate aim is to route the sensed data to the sink node efficiently. Therefore, data aggregation also plays a vital role in wireless sensor networks.

10.5 Causes of Energy Wastage in WSNs

Due to the several limitations of wireless sensor networks, lower power consumption is one of the major criteria for protocol design. Several sources of energy waste in WSNs are:

- Due to collision, retransmission of data causes energy wastage.
- Control overhead packets (request-to-send, clear-to-send, and acknowledgement) can use a smaller amount of energy.
- Overhearing refers to the nodes receiving packets that are destined for other nodes.
- Idle listening refers to listening to an idle channel in order to receive possible traffic.

While designing the WSN, many technical challenges are faced, like security problems, smaller bandwidth, design of cross-layers, in that one of the most tremendous complications is power management. The sensor nodes that are present in the WSN are dependent on batteries, which means only a certain amount of energy can be utilized. The power of the battery consists of a threshold condition, even though it is built based on the unit harvesting energy. The battery that is used in sensors should be sustainable and reliable with high performance and lifetime because the whole network density, reconfiguration, time of operation, and applications depend on the management of power. Power management deals with minimizing the power utilization with mobile hardware components such as central processing unit, random access memory, read only memory, and liquid crystal display (LCD). Another way of lowering power consumption is the protocol stack used by mobile or wireless systems. The various types of sensors with their power consumption are illustrated in Table 10.1.

10.6 Power Gating

This is a device that operates in two types of modes, known as the active and sleep modes. This power gating scheme was designed to save energy in both WSN modes. The line that supplies power and chip depends on the electronic switch for processing, which is an

TABLE 10.1

Various sensor types with their power consumption

Sensor type	Power consumption
Gas sensor	(500–800)mw
Image sensor	150mw
Pressure sensor	(10–15)mw
Acceleration sensor	3mw
Temperature sensor	(0.5–5)mw

example of this technique. The PMOS or NMOS transistor is used in the development of power gates that split the power line of the module, which results in automatic shutdown of the nodes or circuit blocks that are not needed at present. For example, while the process of network measurement is done, there is no need for a radio transmitter, so it shuts down automatically using this technique. Even though a shutdown, standby, or power saver pin is present in major WSN modules, it cannot totally power off the whole circuitry at once. In power-saving mode, it consists of a minimum quiescent current which is greater than 1μA. Most manufacturers do use this power gating technique because when the power saver pin is disabled, the WSN device can continue its normal operation. But when the module totally shuts down or powers on, this increases the delay, resulting in higher power conservation in any electronic device. This does not cause any problems in the application of WSNs, so this issue can be easily solved by using this scheme of power gating.

The power gating module is constructed using the following devices:

- processor
- sensing module
- radio transceiver

The operations which are performed in power gating and the time needed to complete the process are mentioned below:

1.	Wake up	:	–
2.	Few processing	:	1s
3.	Taking of measurement	:	5s
4.	Sends/receives data to/from sink node	:	3s
5.	Many processing	:	1s
6.	Sleep	:	–

After this process is completed, the comparison between the two schemas is made, which is given by:

- The radio modules and sensing are present in the standby and sleep pins of power gating.
- The external implementation of power gating where the devices are in direct contact with the power lines results in no loss due to the presence of voltage regulators.

Total energy consumed is calculated by the below formula:

Total energy consumed per day = active + inactive (power saving or power gating)

TABLE 10.2

Sensor operation states and their power consumption

Operations	Microcontroller unit	Sensors unit	Radio unit
Active (energy consumed)	5 mW	30 mW	350 mW
Inactive (power saving)	2 µW	5 µW	20 µW
Inactive (power gating)	2 µW	1 µW	1 µW

Table 10.2 represents the various sensor operations and the individual unit power consumption status.

10.7 Leakage Gating and WSNs

The technique of leakage gating is difficult because the design of the hardware is becoming more complicated in recent times. In this scheme, the leakage is only dependent on the power lines. For example, the connection between the two modules, like the MCU with radio transceiver, can create a point of leakage when any one of the modules shuts down. To evaluate the value of leakage current, the measurement of every leakage line should be done one after the other, which is used in determining the variation of current. The circuitry is added or connected with an analogue switch once the point of current leakage is identified. The technique of leakage gating was developed by depending on the basis of the power gating scheme. This technique of leakage gating is totally implemented using analogue switches where the PMOS or NMOS transistors are used in power gating.

The output parameters to be calculated and their range in the leakage gating scenario are given by:

• gate leakage	:	<50 nA
• delay in time	:	≥ 0.5 s
• leakage current	:	nearly 30 µA
• sleeping current	:	less than 40 times of leaked current
• duty cycle	:	minimum
• loss	:	<0.5%

The WSN is built with sensor modules and radio transceivers, both of which rely entirely on voltage regulators to function. These regulators are used to save energy, which results in high performance by enhancing the efficiency of energy by 95 percent. The microcontroller units certainly need the voltage regulators so that they can save energy, and every cell has a voltage which ranges dynamically. Even though voltage regulators play a major role in WSNs, the module can also be designed and implemented without them while we implement the technique of power gating. It reduces the unwanted power consumed by the inactive nodes but also faces a problem that degrades the reliability of the system. This drawback is overcome by using primary cells with minimized duty cycles. The output obtained by implementing this technique is represented in Table 10.3.

TABLE 10.3

Power gating output

S.No	Attributes		Values
1.	Sampling rate	:	15 minutes
2.	Overall sleeping power	:	4 µW
3.	Energy consumed by each cycle	:	1114 mWs
4.	Primary cell	:	19 Ah
5.	Average level of voltage	:	3.53 V
6.	Lifetime of system	:	2,258 days
7.	Reduction of lifetime	:	55%

10.8 Techniques of Power Matching

Power matching technique follows the rule of thumb that states that the energy capacity is indirectly proportional to the capacity of power. For example, energy capacity of a battery made of alkaline is very low but the current provided by it is very high and the process of charging the cell takes a very long time but the current needed to charge it is very low, at a range of 5–35 mA. As the battery charges very slowly, it increases the time delay, which increases the loss of energy, so the solution to avoid this issue is to make the design of the WSN module very simple.

The components used to design the WSN module using this technique are:

- SC charger
- microcontroller unit
- DC/DC converter
- load

The step-by-step process carried out in the module designed is given by the following:

1. Selection of DC/DC converter.
2. Input voltage range is chosen. (V_{SC}^{min} or V_{SC}^{max})
3. Depending on the charging time, the value of capacitance is calculated.
4. The evaluation of energy consumed by load is done.
5. The overall energy consumed is estimated. (E_{out})

$$E_{out} = \frac{E_{load}^{max}}{E_{conv}^{avg}} = \frac{2.13}{0.85} = 2.5 J$$

6. The energy stored at the process of slow charging is evaluated. (E_{in})

$$E_{in} = \frac{1}{2} C_{eq} \left(V_{SC_{max}}^2 - V_{SC_{min}}^2 \right) = \frac{1}{2}(1.1)(3^2 - 1^2) = 4.4 J$$

7. Then by depending on the condition,

$$E_{in} > E_{out} = \begin{cases} yes, proceeds\,to\,next\,step \\ no, returns\,to\,step\,3 \end{cases}$$

8. Evaluation of average energy loss is done. E_{-PM}^{total}

$$E_{-PM}^{total} = N_{cycles} * E_{-PM}^{max} = \frac{1005\,days}{15\,min}(0.55) = 53(KJ)$$

9. End or Re-evaluate.

The maximum time needed to charge the device is calculated by:

$$T_{charge}^{max} = 5\frac{VSC_{max}}{I_{charge}^{max}}\left(\frac{C}{2}\right) = 2.5C\frac{3}{0.035} \cong 214C$$

The maximum energy consumed at load is evaluated by using the formula:

$$E_{load}^{max} = P_{load} * Time_{active} = (355\,mW)(6s) = 2.13J$$

The loss of maximum energy is calculated using the equation:

$$E_{-PM}^{max} = \frac{1}{2}C_{eq}V_{SC_{min}}^2 = \frac{1}{2}(1.1)(1^2) = 0.55J$$

This technique of power matching reduces the wastage of power up to 60 percent.

10.9 Different Types of Power Sources

Power management is a computing device feature that allows the user to control the total electrical power consumed by a device with the least effect on performance. This enables the device to operate in several power modes, each with different characteristics of power related to performance. In recent years, most researchers have focused on the rapidly growing technology of wireless sensor networks. The predominant constraint of this technology is power consumption. This can be supported by battery type, energy harvesting, etc. In this hot topic of research, the ultimate goal is to save energy. A battery can be classified into:

1. Rechargeable batteries: provision for recharging the battery of the sensors.
2. Non-rechargeable batteries: does not have provision for recharging the battery.

The connection between the nodes should be independent to create a network for the needed applications so if any problem arises, such as modification in connectivity or routing issues, it can be solved by itself. The other way of reducing power consumption is by using multi-hop communication as the transmission uses low energy and it allows long-distance communication without any effect of propagating signal when compared with single-hop communication. After the process of deployment, the sensor nodes must be able to reconfigure themselves by finding the correct location or position to be placed.

10.10 Power Modes in DRAM

There are three power modes, temperature-compensated self-refresh mode, partial array self-refresh mode, and power down mode. The DRAM memory is designed such that it can operate in any one of the above-mentioned modes. The memory unit is affected by the power conservation due to the RAM timing, which is dependent on latency. The needed row and bank are selected before access is given to the memory present in a particular cell by a processor subsystem. The signal known as Row Access Strobe (RAS) is used to activate the row. This can be done till the exhaustion of data. The activation time should be kept correct because the stability of the system is totally dependent on the time set to activate the memory of a row. The time allotted can be in any range because the access is done in a sequential and insignificant manner. Only one row can be activated at a time. If we have to activate another node that is not active, the present active node should be deactivated first, which is known as the process of refreshing, where two important operations are done, such as reading and writing. When the clock logic is used to access RAM, the nearest clock cycles are rounded up. Column Access Strobe (CAS) is another method of activating memory cells that results in latency due to the delay of the CAS signal and the presence of valid data. High performance and energy consumption are due to lower CAS latency. Therefore, to save even more power, dynamic power management (DPM) is preferred because it totally closes the sensor node when not in use.

10.11 Power Management Strategy

In general, the sensor node's power consumption is a function of time and the event. The term "event" is used to identify whether any event is happening or not. As shown in Figure 10.1, the sensor node has three states: excited, active, and dormant.

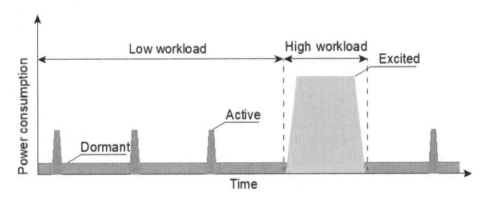

FIGURE 10.1
Various power states: dormant, active, and excited.

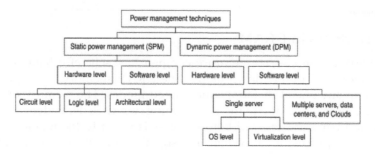

FIGURE 10.2
Power management techniques.

Power transmission is mainly affected by two factors: distance and link factor. The power management scheme (see Figure 10.2) uses energy harvesting schemes from different nodes with sleep/awake methodology. Most of the energy-efficient algorithms improve the power efficiency and the battery lifetime.

Normally, not all blocks of a system participate in performing different functions and it is useful to shut down inactive blocks to reduce power consumption.

10.12 Dynamic Duty Cycle Scheduling Scheme

High latency leads to larger end-to-end delays where multi-hop transmission of a packet is used. If a sensor wakes up, numerous sender nodes that had delayed the transmission to the receiver may contend with each other to access the channel. This contention upturns when the sleep latency leads to collisions. In order to reduce the sleep latency, the duty cycle (I_{dc}) should be kept as small as possible. The duty cycle can be referred to as the wake-up interval of sensor nodes. Due to smaller (I_{dc}) values, residual energy distribution is uneven for all sensor nodes. Due to the various characteristics (position, energy source) of the sensor nodes, they do not have the same residual energy. Hence, (I_{dc}) can be adjusted based on the value of residual energy to achieve the following goals:

- to reduce the sleep latency
- to balance energy consumption

In energy harvesting WSNs, two novel dynamic duty cycle scheduling schemes are used:

- In the first scheme, based on the value of recent residual energy, I_{dc} can be computed. Due to harvesting, the nodes' residual energy can be increased with time. But this scheme considers the current residual energy cannot capture such a prospective increase.
- In the second scheme, the sensor node calculates the prospective increase in residual energy with respect to time. This reduces the duty cycle (I_{dc}). Thus, network lifetime can be improved with an increase in residual energy.

10.12.1 Static Power Saving

The two main factors that depend on the conservation of energy are throughput and lifetime of the network, so the methodology of static power saving uses Energy Aware Routing (EAR) Protocol, which creates more paths to reduce the consumption of power because one faultless path uses more power. This process of finding paths is done by flooding, which finds a route from source to destination. It nearly works as the directed diffusion protocol where the probability factor is used to maintain the routes. The lesser amount of energy consumed by each path decides the probability function. This protocol is built depending on the main element, which is known to be the survivability of the network. The type and position of a node are determined by using the class-based addressing scheme. Then, the creation of the routing table is done to disable the higher power-consuming paths. Localized flooding is performed by the destination node to maintain the paths that are not disabled. The setting of the path becomes difficult due to the data gathering of position and when the steering mechanism is activated. In this, the subsystem of nodes is given by power, processor, communication, active memory, BUS, and RAM timing. Battery and DC/DC converters are involved in the process of power subsystems. The best example that can be given for the processor subsystem is a device that is efficient to operate at various modes of power, like microcontrollers (example: STM32S0–5 power modes–stop, stand-by, sleep, low power sleep, and low power run modes), clock frequencies, and at different voltages. Power conservation, wake up time, and wake up source are the three elements that are used to evaluate the performance of the above-mentioned five modes [14]. When the mode changes from one to another, it consumes some amount of energy and latency on its own.

The state of operational mode that is very active and cannot be identified easily because of the reduction in the rate of power transmitted does not minimize the power consumed by the transmitter. The power used for transmitting data and the heat that is released due to dissipation of power are the result of a tradeoff between the above two processes taking place at the power amplifier. The dissipation and transmission power are indirectly proportional to each other. This is due to the fact that only at some specified levels of power does the transmitter work efficiently. The use of low-performing transceivers results in wastage of high-input DC power in the form of heat even when the mode of transmitting power is high, that is, above 60 percent. Now, the interaction between one and another subsystem takes place in the process of BUS timing, so when the internal high-speed buses involve maximum power, which is dependent on frequency and bandwidth, the timing of the bus protocol is used to evaluate the frequencies of the bus, and the interaction type is used in configuring both bandwidth and frequency. The performance is determined by the bus control drivers when there is a change in the frequency of buses. Power is received by every subsystem from the power subsystem. The capacitor-transmitter pairs are used to design the active memory (DRAM), which consists of rows and columns. Every row acts as an individual bank of memory which can store data, and it is rechargeable. One of the most important factors is the refreshing interval, where the number of rows and columns present should be refreshed, where the refresh intervals and clock frequencies are indirectly proportional to each other. For example, 2K uses maximum power but operates very fast at a lesser interval by refreshing many cells. 4K uses minimum power but operates at a frequency that is slower by refreshing fewer cells.

10.12.2 Dynamic Power Management

An asynchronous awakening scheme is proposed based on the observed event with a tradeoff between communication cost and power consumption. In this scheme, each node is free to switch on or off its components. Residual power helps to determine the sleep state period. A dynamic power management scheme allows the sensor to be placed in low power sleep modes when the system is inactive. This technique minimizes the power of every single node by providing only a certain amount of energy to each system to perform the operation, and when there is no operation to be processed, it shuts down the system. The network wide sleeping state reduces the power consumed by the overall network. There are two types of sleeping schedules, synchronous and asynchronous. The first step, which is done by these two processes, is defining their own individual schedule for sleeping mode by each node. In the synchronous process, the sensing by coordination and communication by inter-node is done efficiently by the splitting of schedules to the neighbour nodes from the main node. The drawback of this operation is that it consumes a lesser amount of power and the synchronization time of the neighbor nodes should be maintained. In the process of asynchronous scheduling, the receiving partner will send an acknowledgement to the node that is responsible for communication. Until it receives the acknowledgement, that node does the transmission of preamble. During the transmission of data, it will give rise to the side effect of latency, which is the main issue faced in asynchronous scheduling. Then, upon identification of the node that is ready to communicate and the nodes that are waiting in queue, the individual node performs periodic wait up in both the processes. The important process while designing the wireless sensor network is to calculate how the lifetime of the network is affected by the rate of overall power dissipated by every single node. The unwanted processes should be eliminated, and lastly, the work to be done should be estimated, i.e. budgeting the amount of power needed for each process that takes place. The classification of nodes into subsystems is demonstrated in DPM by dynamic scaling, task scheduling, and modes of dynamic operation which involve the cost of transition. The conceptual architecture is used in DPM, which exhibits two approaches, central and distributed, to enhance power management.

10.12.2.1 Sleeping Policy

In the deepest sleep state, the sensor node can't detect the event or receive messages from other nodes. Hence, the probability of deep sleeping mode is a major concern in WSNs. The chance of the deep sleeping mode was determined by the "n-duplicate-covered" method. The term "n-duplicate-covered" node refers to the fact that more than (1+n) sensors are in the area monitored. Due to this, the sleep state has been classified into four states (s1, s2, s3, and s4). The probability of entering shallower (s1, s2, and s3) and deep (s4) sleep modes is 1/(n+1) and n/(n+1), respectively. After a sleep period, a deep sleep state is stimulated with the help of the clock counter. The deep sleep period 'T' can be denoted in Eq (1) as,

$$T = \mu \times e^{\left(\frac{V_{st}}{V_{ct}}\right)} \tag{1}$$

where,
V_{st} = battery operating voltage
V_{ct} = present battery voltage

For a lithium battery, V_{st} = 3.6V and V_{ct} = 2.8 to 4.2V.

$$P_i(T_{th}, 0) = e^{-\lambda_i T_{th}} \tag{2}$$

where,
λ_i = event generation mean rate value
T_{th} = threshold value

If $P_i(T_{th}(k), 0) > P$ (fixed value), then the node moves into sleep state k.

10.12.2.2 Awakening Policy

To awaken the sensor node, three mechanisms, namely event-driven, message-driven and timer-driven, are followed:

Event-driven: this policy operates in a shallower sleep state. If there is a sudden variation in temperature or a signal generated by a moving object, an interrupt is produced and it awakens the CPU. Using data fusion techniques, the CPU processes the signals and moves to sleep mode.

Message-driven: this policy operates with k = 1 or 2 due to the receiver being ON in these sleep states. In a sleeper mode, if a sensor node 'i' needs to transmit packets to node 'j', it will initialize a message to wake the node 'j'. After the awake message is received, node 'j' will verify the sleeping time 't' with the threshold value $T_{th}(k)$. If the condition is satisfied, then the node instantly wakes up; otherwise it will awaken the node till the condition is satisfied. When the node receives an acknowledgement, it will route the packets to node 'j', otherwise it will initialize the awakening message. This scheme avoids the higher energy consumption of failed packet transmission.

Timer-driven: this scheme uses the counter interrupt to awaken the node from the deeper sleep state. Based on the battery status and the parameter (m), a deep sleeping state period can be obtained.

Dynamic power management is based on the event generation probability, the transmission of a packet, the coverage area, and the value of the battery.

10.13 Conclusion and Future Work

A review of the study of various power management techniques is presented. Due to the power management problem in WSNs, a comprehensive approach to energy saving is essential. This chapter briefly explains static and dynamic power management in wireless sensor networks. Also, dynamically managing the sensor node operations to reduce the power consumption after deployment is discussed. In the future, the combined techniques from all the existing sources of power from batteries and the surrounding environment can be considered to improve the network lifetime of WSNs. This review will be helpful to researchers who are involved in academic and industrial research in power-aware wireless sensor networks.

References

[1] R. C. Luo, L. C. Tu, and O. Chen (2005). "An Efficient Dynamic Power Management Policy on Sensor Network," *Proceedings of the 19th International Conference on Advanced Information Networking and Applications (AINA '05)*, Taiwan, pp. 1–4.

[2] A. Sinha and A. Chandrakasan (2001). "Dynamic Power Management in Wireless Sensor Networks," *IEEE Design and Test of Computers*, vol. 18, no. 2, 62–74.

[3] H. Cotuk, K. Bicakci, B. Tavli, and E. Uzun (2014). "The Impact of Transmission Power Control Strategies on Lifetime of Wireless Sensor Networks," *IEEE Transactions on Computers*, vol. 63, no. 11, 2866–2879.

[4] P. Nayak and A. Devulapalli (2016). "A Fuzzy Logic Based Clustering Algorithm for WSN to Extend the Network Lifetime," *IEEE Sensors Journal*, vol. 16, no. 1, 137–144.

[5] J. Zhang, D. Zhang, K. Xie, H. Qiao, and S. He (2017). "A VMIMO-Based Cooperative Routing Algorithm for Maximizing Network Lifetime," *China Communications*, vol. 14, no. 4, 20–34.

[6] G. Anastasi, M. Conti, and M. di Francesco (2009). "Extending the Lifetime of Wireless Sensor Networks through Adaptive Sleep," *IEEE Transactions on Industrial Informatics*, vol. 5, no. 3, 351–365.

[7] M. Buettner, G. V. Yee, E. Anderson, and R. Han (2006). "X-MAC: A Short Preamble Mac Protocol for Duty-Cycled Wireless Sensor Networks", *4th International Conference on Embedded Networked Sensor Systems (SenSys'06)*, pp. 307–320.

[8] I. F. Akyildiz, W. Su, Y. Sankarasubramaniam, and E. Cayirci (2002). "A Survey on Sensor Networks," *IEEE Communication Magazine*, vol. 40, 102–114.

[9] R. S. Kumaran and G. Nagarajan (2017). "Energy Efficient Clustering Approach for Distributing Heavy Data Traffic in Wireless Sensor Networks," *AMSE Journal*, vol. 22, no.1, 98–112.

[10] H. Azarhava and J. M. Niya (2020). "Energy Efficient Resource Allocation in Wireless Energy Harvesting Sensor Networks," *IEEE Wireless Communications Letters*, vol. 9, no. 7, 1000–1003.

[11] S.K. Rajendran and G. Nagarajan (2022). "Network Lifetime Enhancement of Wireless Sensor Networks Using EFRP Protocol," *Wireless Personal Communications*, vol.123, no. 2, 1769–1787.

[12] J. Singh, S.S. Yadav, V. Kanungo, Yogita, and V. Pal (2021). "A Node Overhaul Scheme for Energy Efficient Clustering in Wireless Sensor Networks," *IEEE Sensors Letters*, vol. 5, no. 4, 1–4.

[13] R.S. Kumaran and G. Nagarajan (2022). "Mobile Sink and Fuzzy Based Relay Node Routing Protocol for Network Lifetime Enhancement in Wireless Sensor Networks," *Wireless Networks*, vol. 28, no. 5, 1963–1975.

[14] N. Qi, K. Dai, F. Yi, X. Wang, Z. You, and J. Zhao (2019). "An Adaptive Energy Management Strategy to Extend Battery Lifetime of Solar Powered Wireless Sensor Nodes," *IEEE Access*, vol. 7, pp. 88289–88300.

11
Power Management in Wireless Sensor Networks

Prasanta Pratim Bairagi, Kanojia Sindhuben Babulal, and Mala Dutta

CONTENTS

11.1 Aim of this Study 152
11.2 Introduction 152
    11.2.1 About Wireless Sensor Networks 152
    11.2.2 Types of Wireless Sensor Network 153
    11.2.3 About the Sensor Node and Its Architecture 153
11.3 Power Consumption in WSNs 155
    11.3.1 About Power Consumption 155
    11.3.2 Measurements of Power Consumption 155
        11.3.2.1 Power Consumption at Node Level 155
        11.3.2.2 Power Consumption at Network Level 156
        11.3.2.3 Power Consumption at Software Level 157
11.4 Sources of Power Waste 158
11.5 Power Management in WSNs 158
    11.5.1 About Power Management 158
    11.5.2 Power Management Approaches 158
        11.5.2.1 Management at Node Level 158
        11.5.2.2 Management at Network Level 159
        11.5.2.3 Management at Software Level 159
11.6 Power Conservation Techniques in WSNs 160
    11.6.1 About Power Conservation 160
    11.6.2 Challenges in Power Conservation 160
        11.6.2.1 Energy Dissipation 160
        11.6.2.2 Quality of Service 160
        11.6.2.3 Transmission Mode 160
    11.6.3 Different Power Conservation Techniques 161
        11.6.3.1 Efficient Node Deployment 161
        11.6.3.2 Data Reduction 161
        11.6.3.3 Duty Cycling 161
        11.6.3.4 Mobility Based Power Conservation 162
        11.6.3.5 Power Efficient Routing Protocol 162
    11.6.4 Analysis of Power Efficient Routing Protocol 163
        11.6.4.1 Network Parameters 164
        11.6.4.2 Simulations Results 164
11.7 Conclusion and Future Work 167
References 169

DOI: 10.1201/9781003326205-14

11.1 Aim of this Study

The study includes the analysis of power consumption at various levels of a network and different power efficient approaches to minimize consumption to extend the network's lifetime and manage power efficiently. The aims of this study are:

a. To provide an overview of the entire process of power consumption at different levels of WSNs.
b. To provide a brief list of possible ways to manage power consumption at different levels of WSNs.
c. To provide a detailed analysis of different power conservation techniques available for WSNs.

11.2 Introduction

11.2.1 About Wireless Sensor Networks

Wireless sensor networks (WSNs) are comprised of a huge quantity of autonomous sensor nodes that monitor environmental conditions like temperature, pressure, and sound, etc. [1]. WSNs are a distributed type of network which is equipped with battery powered sensor nodes. Sensor networks are normally used in the fields of engineering, health, and military operations etc. to monitor various environmental parameters. Apart from monitoring data from its surroundings, a sensor network is also responsible for processing the data as well as communicating the data. The primary component of a WSN is sensor node with the help of which the sensor network functions. A sensor node is comprised of different individual components: sensors for sensing data, a processing unit to process the data, and a power supply unit which is equipped with batteries [2]. A sensor node has limited power and is basically responsible for sensing data from its surroundings, processing it, and sending it to its destination.

The working principle of WSN is:

a. sensing data
b. processing data
c. communicating data

A wireless sensor network has at least one sink node or base station which serves as a destination node for a particular region within the sensor network. Normally the base station node has higher power as compared to the other nodes of the network. The major portion of power consumption happens within this node. The lifespan of a WSN is highly dependent on the power of the sensor node. The higher the power, the longer the lifetime of the network. As a result, power in a WSN is a critical resource that must be handled carefully.

Apart from sensor nodes, routing also plays a significant role in WSNs. During the routing of data in a sensor network, power consumption also occurs. Routing protocols are used for performing the routing operations. Routing protocols are in charge of transferring data from one sensor node to another [3]. Along with the sensing capability of a

Power Management 153

sensor node, performance of the routing protocols will also determine the quality of the WSN. The selection of best-in-class routing protocols is necessary for performing these functions efficiently.

In WSNs, there are three forms of communication that might occur [4].

a. **Communication inside the cluster:** When all communication within a cluster, such as sensor nodes, senses information of interest and reports it to the CH, this is referred to as intra cluster transmission.
b. **Communication between clusters:** When two CHs communicate instead of sending data straight to a sink or base station, this is known as inter cluster transmission.
c. **Communication between the cluster and the base station:** In this case, the CH communicates directly with the sink or base station.

The major advantages of a wireless sensor network are:

a. It supports the autonomous configuration system of nodes which enables easy entry and exit of nodes within the network.
b. It ensures the establishment of a communication as well as sharing of information even where the establishment of a traditional network is not possible.

In WSNs, the major issue is limited power resources. The quantity of energy used by a WSN is directly proportional to its lifespan. As a result, the energy used in a WSN is a valuable resource that must be carefully managed.

11.2.2 Types of Wireless Sensor Network

WSNs are classified into numerous varieties based on the environment and the method of node deployment, allowing them to be used in a variety of settings. Table 11.1 shows some of the most important characteristics of various types of WSNs [5].

The several variations of WSNs are:

a. terrestrial WSNs
b. underground WSNs
c. underwater WSNs
d. multimedia WSNs
e. mobile WSNs

11.2.3 About the Sensor Node and Its Architecture

WSNs are made up of a variety of homogeneous or heterogeneous sensor nodes that gather and analyze sensed data for a specific application. The major goal of sensor nodes is to convert environmental values like light, temperature, and pressure into a signal that humans and communication devices can understand. Each sensor node in a WSN must follow the same predetermined rules, which start with detecting data from its neighborhood and conclude with transferring data to the sink.

A sensor node is comprised of different individual units: a sensor unit, a central processing unit, and a power unit [6]. The internal components of a sensor node are shown in Figure 11.1. The sensor unit is comprised of some kind of sensors which are mainly responsible for sensing data from its surrounding, and an analog to digital converter (ADC) to convert the

TABLE 11.1
Characteristics of various types of WSNs

Sl No.	Type of WSN	Characteristics
1	Terrestrial	a. Nodes deployed in pre-planned or ad-hoc manner.
		b. Solar cells are used as a secondary power source.
2	Underground	a. Nodes are deployed underground to determine the underground condition.
		b. Additional sink node is required above ground.
3	Underwater	a. Acoustic waves are used for communication.
4	Multimedia	a. Nodes equipped with cameras and microphones are deployed in pre-planned manner.
		b. High bandwidth demand, high energy consumption, QoS are the challenges.
5	Mobile	a. Nodes that can move on their own and have the ability to reposition and organize themselves in network.
		b. High coverage, better battery efficiency.

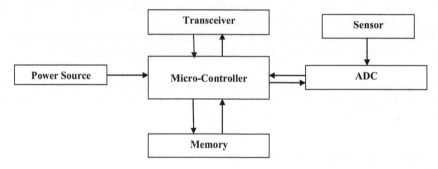

ADC- *Analog Digital Converter*

FIGURE 11.1
Sensor node components.

physical information to signals. The central processing unit can be further classified into two subunits: a processing unit consisting of microcontrollers which perform the necessary processing of sense data and make it ready for communication; and a communicating or transmitting unit which is involved in the transmitting of data from one sensor to another. The processing unit also has a small amount of memory where it can store the information collected by the sensor. The third unit, the power unit, is responsible for supplying the power to each and every part of the sensor node so that they can perform their operations.

In WSNs, sensors can be categorized based on a variety of factors, including technological features, detecting mechanisms, output signals, sensor materials, and application field. Although a separate categorization is required when considering its use, it may be divided into the following groups [7]:

a. **Active sensors:** In order to make measurements, active sensors stimulate the surroundings.
b. **Directional sensors:** They can monitor the environment without producing any interference.
c. **Narrow beam sensors:** In order to assess the surroundings, this sort of passive sensor needs a defined orientation.

11.3 Power Consumption in WSNs

11.3.1 About Power Consumption

In WSNs, the main component is the sensor nodes and they are powered by small batteries. Sensors are typically used in inaccessible locations where battery replacement is not possible. The use of a battery, on the other hand, limits the sensor's lifetime and has an impact on WSN design and operation. Power is a valuable resource for any sensor network, and it must be intelligently used to extend the network's lifespan.

Apart from the restricted battery, there are a number of other elements that influence a node's power usage and network longevity. These factors are:

a. duration of activeness of nodes
b. coverage of nodes
c. connectivity among the nodes
d. deployment environment of nodes
e. type of application running on the nodes

In a wireless sensor network, the power consumption happens at three different levels [8]. Figure 11.2 depicts the classifications of power consumption into various levels.

11.3.2 Measurements of Power Consumption

In WSNs, power consumption is a significant factor in determining the success of sensor node deployment as well as the performance of the network. To maximize the lifespan of the network, it is essential that the utilization of power should be done in an efficient way. To ensure the efficient utilization of power, it is mandatory to measure the power consumption at each level of the network.

The power consumption that occurs at each level of the network is detailed below.

11.3.2.1 Power Consumption at Node Level

As depicted in Figure 11.1, WSN nodes are made up of many modules: sensor module, processing module, communication module, and power supply module [9]. To make the sensor perform, all the components must work together, and during execution each component will also consume power. To analyze a WSN node's energy consumption, it is necessary to look at the energy consumption of its components.

The power consumption that occurs at each module of the node level is detailed below.

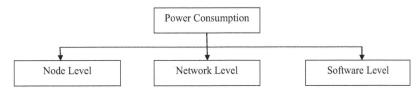

FIGURE 11.2
Classification of power consumptions at different levels.

TABLE 11.2

Power consumption of different sensors [11]

Sl No.	Sensor type	Energy consumption (in mW)
1	Gas	500–800
2	Image	150
3	Pressure	10–15
4	Acceleration	3
5	Temperature	0.5–5

a. **Sensor module:** In the sensor module, the major portion of power consumption happens because of its sensing operation. Apart from this, the sensor module is also responsible for executing other operations like sampling of signals, conversions of signals, and modulation of signals.
b. **Processing module:** In this module, the major portion of power consumption happens because of different operations like controlling of a sensor and information processing.
c. **Communication module:** The communication module mainly consumes energy during its operations, such communication between the nodes, communication between the protocols, etc.
d. **Power supply module:** The power supply unit is responsible for supplying the power to each and every part of the sensor node. To perform this operation this module consumes some power.

It has been revealed that transmitting data from a node to its appropriate neighbors consumes more energy than processing data at the node itself [10]. Numerous sensors are accessible in a sensor network, each of which may be identified by their usefulness and power consumption during transmission.

Some of the popular sensors' power consumption per unit may be calculated as in Table 11.2.

11.3.2.2 Power Consumption at Network Level

The network level is made up of multiple layers, each of which performs a particular operation, resulting in power consumption at each layer. Individual network characteristics such as latency, transmission delay, delivery ratio, and throughput may be evaluated at the network level to determine the performance of the WSN [12]. Figure 11.3 highlights the different layers available at network level and Table 11.3 highlights the major operations which cause power consumption in each layer.

The power consumption that occurs at each layer of the network level is detailed below.

a. **Application layer:** The application layer is primarily responsible for collecting data from its neighboring nodes [13]. It eliminates redundant data, compresses it, and sends it to the designated recipient. Due to this activity, power consumption happens in this layer.
b. **Transport layer:** The transport layer ensures that data is delivered reliably between nodes [14]. Congestion occurring during packet transmission is the primary source

Power Management

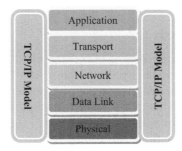

FIGURE 11.3
Layer of network level.

TABLE 11.3

The major operations which cause power consumption in each layer

Sl No.	Layers	Operations
1	Application layer	Data redundancy
2	Transport layer	Congestion
3	Network layer	Routing
4	Data link layer	Collision
5	Physical layer	Data transmission

of power consumption at this layer. Flow control, multiplexing etc. are also responsible for consumption of power within this layer.

c. **Network layer:** The major operation performed in the network layer is data routing [15]. The data routing happens in the form of single-hop routing or multi-hop routing. In this layer the major portion of power consumption occurs due to the routing operations.

d. **Data link layer:** In this layer, collisions between radio transmissions of nodes and error-checking are a key source of power consumption [16]. Apart from collisions, there are a number of additional reasons for power consumption in this layer, such as nodes remaining in an unneeded active state rather than idle when no transmission is happening.

e. **Physical layer:** The physical layer is in charge of radio channel monitoring, sensing activities, and signal processing [17]. In a network this layer provides the necessary communication link in the form of a radio channel, using which a node can transfer its data. The transmission or reception of data through the radio channel consumes power at this layer.

11.3.2.3 Power Consumption at Software Level

A sensor node often has communication drivers, data processing software, and an operating system [18]. An operating system consumes a huge amount of power at the software level. The power consumption of an operating system can be calculated based on its size, processing power, and scheduling capabilities.

11.4 Sources of Power Waste

For sensor systems, power is a limited resource that must be properly managed in order to extend the life of the sensor nodes for the duration of a mission. A sensor node's power consumption might come from either "useful" or "wasteful" sources. Useful power consumption of a node happens when a node performs operations like sensing, processing, sending, and receiving data in order to achieve the application's objectives. When it comes to wasteful consumption, there are several sources or procedures that cause power waste: overhearing, idle listening, interference etc. [19]. Given the foregoing information, a well-constructed procedure must be implemented in order to avoid these power wastes.

a. **Idle listening:** Because a node cannot predict when it will get a message, it must continually listen to the media and hence remain idle, causing the node to consume power.
b. **Overhearing:** Because of the shared nature of the wireless media, when a sender sends a packet to the next hop, it is received by all neighbors of the source, even if it is only meant for one of them. When the node is a one-hop neighbor of the sender but not the destination, the energy expended is called overhearing.
c. **Interference:** Each node between the emitter and the interference range receives the packet, but it cannot be decoded.
d. **Collision:** When a node gets several packets at the same time, the packets collide. When packets collide, they are retransmitted, resulting in energy waste. This can be saved by preventing collisions in the first place, saving network energy.

11.5 Power Management in WSNs

11.5.1 About Power Management

Power management is a very important and crucial factor in WSNs. To enhance the lifetime of a network it is very important to utilize all the resources in the most appropriate way so that maximum output can be generated from it. Therefore it is very much essential to manage the power resources efficiently. In order to regulate the power use in WSNs, different approaches are available.

11.5.2 Power Management Approaches

Like power consumption measurement, power management can also be done at different levels of the network.

11.5.2.1 Management at Node Level

A sensor unit serves as a connection between the physical environment and the logical processing unit setup. The communication module, which is in charge of transferring data across sensor nodes, will use more energy than executing the computation within each node [10]. In WSNs, base stations are generally connected to the main power source, while nodes in the network are powered by batteries. As a result, it is necessary to select power-efficient node hardware and various efficient operation methods in order to make the network more energy efficient.

11.5.2.2 Management at Network Level

The network layer is mainly in charge of routing data across the sensor network, determining the most efficient path for packets to take on their journey to a certain destination. The network level plays a vital role in managing the total power consumption of the network. Below are the techniques to manage energy use at various levels of this level.

   a. **Management at application layer:** A sensor node collects information from its neighbors. Following that, the data is validated, compressed, and transmitted to its intended destination via the application layer. Several energy-efficient data collecting strategies are often used to gather data from surrounding nodes [20]. This layer is capable of managing power usage during transmission by removing unnecessary data.
   b. **Management at transport layer:** This layer is responsible for successful delivery of packets between the nodes. It also takes care of congestion during the delivery of packets. In this layer, by maintaining low input flow of data as compared to output flow at the communication window, power consumption can be managed [21].
   c. **Management at network layer:** At this layer, different aspects such as topology selection and routing have a key role in managing power consumption across the network [22]. The basic goal of network topology is to create a strong network. On the other hand, routing is responsible for selecting the best possible route between source and destination. By selecting a topology which is much more compatible with network and a power efficient protocol, we can manage the power consumption at this level.
   d. **Management at data link layer:** Packet exchange through a channel is another major source of power consumption in WSNs [23]. In this layer, by establishing a secure communication channel we can manage the power consumption.
   e. **Management at physical layer:** The radio signals utilized for communication at this layer have three modes: idle, sleep, and active [24]. We may regulate the power use in this layer by managing the radio channel.

11.5.2.3 Management at Software Level

At this level, the operating system consumes a significant amount of power. The operating systems which are specially designed for sensor networks can be classified mainly in two categories: event based operating systems and multi-threaded operating systems [25]. Some of the most popular operating systems are SOS, Mantis, Contiki, and TinyOS. In general, the power consumption of an operating system depends on its size, processing power, and scheduling capabilities. Table 11.4 discuss the major power management approaches at different levels.

In WSNs, by using the following mechanism we can manage power consumption at software level.

   a. by using a lightweight operating system
   b. by managing scheduling operations efficiently

TABLE 11.4

Major power management approaches at different levels [26, 27]

Sl No.	Levels	Power management approach	Examples
1	Node level	By using low power consumption microcontrollers	MSP430
2	Network level	By using power efficient routing protocols	LEACH, GEAR
3	Software level	By using lightweight operating system	Mantis

11.6 Power Conservation Techniques in WSNs

11.6.1 About Power Conservation

The goal of power conservation techniques is to maximize the utilization of limited power sources available within the network to extend the network's lifetime. Because the network's life-period is limited by the amount of power dissipated by nodes, conservation of power is extremely important. There are numerous approaches available for reducing power consumption. Some researchers aim to reduce a node's power consumption at the computation or data transmission level [28], others aim to reduce power consumption in the entry/exit functions [29], while others try to construct sensor networks based on network architecture and efficient routing principles [30].

11.6.2 Challenges in Power Conservation

The WSN faces a number of challenges in terms of power utilization. These are energy dissipation, quality of service, transmission mode, etc.

11.6.2.1 Energy Dissipation

Energy is transferred and some of it is dissipated whenever there are activities happening in a system. The term "dissipation" is frequently used to describe how energy is wasted. Wasted energy is the portion of the total energy that is not utilized to perform necessary operations. During communication between nodes in WSNs, energy dissipation happens and it needs to be managed. Energy should be conserved so that batteries are not quickly exhausted or discharged, as these are not easily recharged in surveillance applications [31].

11.6.2.2 Quality of Service

Effective communication within the provided time is ensured by service quality. For every sort of traffic distribution, protocols should verify for network stability, and redundant data should be transferred through the network. It must also adhere to specific resource constraints, such as bandwidth, memory buffer capacity, and processor power [32].

11.6.2.3 Transmission Mode

In wireless sensor networks, the transmission method is very crucial. Nodes can communicate or send data to other nodes in the network using single-hop or multi-hop network topologies, depending on the kind of network architecture chosen [33].

11.6.3 Different Power Conservation Techniques

Power is a prominent resource of WSNs because it is directly related to the lifespan of the network. Due to the architecture and power breakdown of wireless sensor networks, numerous strategies to conserve power consumption must normally be implemented.

11.6.3.1 Efficient Node Deployment

The deployment of nodes is the foundation of any WSN application. The effective deployment of nodes, as well as the determination of deployment cost, coverage, connection, and other factors, all contribute to the network's longevity [34]. The two main node deployment techniques available in WSNs [35] are:

  a. **Deterministic deployment:** Manually placing nodes in the sensing region is used in this approach. There is no need to initiate the localization process because the nodes' positions are known in advance. Deterministic deployment can transform the problem into a mathematical model in order to identify a node deployment solution. The nodes in a deterministic deployment are often situated in a way that allows for quick event reporting and a short data routing path.
  b. **Random deployment:** In this approach nodes are randomly distributed around the monitored region. When coverage requirements are not restrictive, this deployment method can be used. Clusters are constructed by randomly deploying nodes to provide maximum network connectivity with the cluster head being the node with the most energy. The role of a cluster head is to collect data from its cluster's nodes and transfer it to the base station.

11.6.3.2 Data Reduction

The goal of data reduction methods is to decrease the quantity of data that has to be sensed, processed, and given to the sink. They compress a vast quantity of data into a smaller functional entity that can be decoded with little data loss and lower power usage. Data compression and prediction are the two primary approaches for reducing data [36].

  a. **Data compression:** In data compression, the sense data is compressed by the node and only the data that is required is sent. Lossless compression, dual Kalman filters, deterministic compression approaches, and adaptive model selection are some of the ways used to perform data compression.
  b. **Data prediction:** The goal of data prediction is to create an abstraction of a perceived fact. In order to respond to future requests, data prediction algorithms describe the sensing process. In the network, two model instances are created, one in the sink and the other in the source node. The sink node responds to any current requests without requiring contact with other nodes, lowering communication effort and, as a result, lowering power consumption.

11.6.3.3 Duty Cycling

Node activity scheduling approaches are often known as duty cycling strategies [37]. They are also referred to as the percentage of time that nodes are active throughout the course of their lifespan. Sleep and activity patterns for nodes should be synchronized and designed to meet the needs of certain applications. In a node, operations like sending, receiving, and idle listening drain the limited battery power, and only the sleep state assures power savings. Two alternative and complementary means can be used to accomplish duty cycling. They are:

a. **Topology control protocols:** The primary concept underlying topology control protocol is to take care of network redundancy in order to extend the life of the network [38]. It is also defined as the process of determining the best selection of nodes for ensuring connection. Topology control protocol allows nodes that aren't in use in the transmission to go to sleep mode and conserve power. This technique takes into account the active/passive status of the nodes in order to create a power management system.

b. **Sleep/wake-up protocols:** The sensor node's radio subsystem defines sleep/wake-up protocols that reduce the length of time a node remains idle. The three most popular forms of sleep/wake-up protocols are on-demand protocols, scheduled rendezvous protocols, and asynchronous protocols [39].

In this category, the most fundamental strategy to reduce power usage is to use on-demand protocols. The essential principle of on-demand protocol is that a node should only wake up when another node requests communication. A scheduled rendezvous strategy is an alternative solution to duty cycling. The entire idea of scheduled rendezvous systems is that each node should awaken at the same time as its neighbors. Nodes usually wake up on a timetable and stay active for a short period of time to interact with their neighbors. They then go to sleep till the next encounter happens. Use of asynchronous protocol is the third type of duty cycle technique. Asynchronous protocols allow a node to wake up whenever it wishes while maintaining communication with its neighbors. The primary goal of this approach is to eliminate the need for nodes to be synchronized together. This technique allows each node to determine its own sleep/wake schedule independently, which helps to reduce conflict and load.

c. **MAC protocols:** In WSNs, power efficient MAC protocols are essential for lowering power consumption. TDMA and CDMA are two of the most widely used MAC protocols for reducing transmission power consumption [40]. To save energy, the TDMA protocol employs two distinct radio channels for appropriate synchronization of data and control messages. The CDMA protocol assists in collision avoidance and supports bounded delay.

11.6.3.4 Mobility Based Power Conservation

Packets transferred from a node to the sink in a sensor network utilize a multi-hop communication channel. Due to this communication mechanism, some routes may be overloaded. Normally in WSNs, nodes closer to the sink must relay more packets, which causes them to consume more power [41].

Node mobility is a relatively recent way to gather data in wireless sensor networks while conserving energy. In reality, there are two ways to deploy mobile nodes. The first one uses a mobilizer on the node, whereas the second places the node on a moveable element [42].

11.6.3.5 Power Efficient Routing Protocol

Routing protocols are mainly in charge of transferring data from one sensor node to another. Additionally, routing protocols are also responsible for maintaining best possible

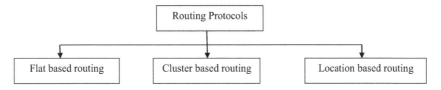

FIGURE 11.4
Classifications of routing protocol.

route between source and destination so that data can be transferred most efficiently. The amount of energy consumed by a routing protocol is particularly essential in improving the longevity, connectivity, and stability of WSNs.

Different routing techniques have been developed to address the communications of WSNs. In WSNs, based on the network architecture, the routing protocols are mainly classified into three categories [43] (Figure 11.4). They are:

a. **Flat based routing protocol:** Flat routing is a kind of multi-hop routing in which all nodes are active at the same moment [33]. Despite the fact that the network is often huge, all nodes are assigned the same sensing assignment. As a result, because all nodes send data, redundancy is probable, resulting in excessive energy usage. WSN supports a number of different types of flat based routing protocols, e.g., EAR, Directed Diffusion, SPIN, etc.

b. **Cluster based routing protocol:** In this routing, different clusters and cluster heads are established and assigned. High-energy nodes are picked randomly for data processing and transmission, while low-energy nodes are employed to sense and transmit data to cluster heads [44]. Hierarchical-based routing is the best fit when network scalability and efficient communication are required. WSN supports a number of different types of cluster based routing protocols, e.g., LEACH, PEGASIS, etc.

c. **Location based routing protocol:** In location based routing, the location of nodes is used to determine routing rather than the network address. A source node knows the destination's geographical position and transmits a message to the destination [27]. The position of nodes is determined using a low-power GPS system in this technique. WSN supports a number of different types of cluster based routing protocols, e.g., GAF, GEAR, DREAM, etc.

Selection of efficient routing protocols is critical in terms of performance of the network. Routing protocols are primarily responsible for gathering various routing information such as source and destination node addresses, node deployment type, packet size and type, and network bandwidth, etc. Once the selection is done correctly, the performance as well as the lifetime of the network will increase automatically.

11.6.4 Analysis of Power Efficient Routing Protocol

The management of routing information within a network is also very much essential in term of enhancing the lifetime of the network. Routing protocols are mainly responsible for collecting different routing information, such as address of source and destination node, type of deployment of the nodes, packet size and type, and bandwidth of the network, etc. Based on this information, the routing protocols try to establish the best possible route from the sender to receiver through the established network. Once the management of all this routing information is done efficiently, the lifetime of the network will increase automatically.

11.6.4.1 Network Parameters

In order to obtain the best quality of service, network parameters help determine which routing protocol is most suitable or the best performer for the network. To evaluate the performance of a protocol, different network parameters are used. These parameters are [45, 46]:

a. **Network delay:** This is the entire amount of time it takes for a data packet to travel from its source to its destination. In the instance of network delay, we might state that a lower number indicates higher performance.
Network delay = receiving time - sending time.
b. **Data delivery ratio:** The ratio of total packets delivered to total packets sent from a source node to a destination node in a network is known as PDR. In the instance of PDR, we may argue that a higher PDR number indicates greater performance.
Data delivery ratio = total number of data received / total number of data sent.
c. **Throughput:** Throughput refers to the average rate of successfully sent messages through a network connection. We may say that the greater the value, the better the performance.
Throughput = total packets received / total time.
d. **Packet loss:** This is the ratio of total packet losses to total packets delivered. In the instance of PL, we may argue that a lower value indicates higher performance.
Packet loss = total packets lost / total packets sent.

11.6.4.2 Simulation Results

Simulation is the method of creating a digital model of a real-world system for use in behavioral assessments. Simulation is normally used to test the performance of a complex network before it is built. The simulation was done with varying numbers of nodes to investigate the performance of these protocols thoroughly. The placement of nodes in the network is depicted in Figure 11.5.

In WSNs, several power efficient protocols like GEAR, LEACH, and DREAM, etc. are available which are responsible for performing the routing operations [47, 48].

For this analysis, we have taken five different sizes of network consisting of 20, 40, 60, 80, and 100 nodes, keeping all the network parameters constraints, and the results are reported in terms of network delay, data delivery ratio, packet loss, and throughput.

a. **Network delay:** Table 11.5 and Figure 11.6 highlight the network delay values against each network size.

Observation: From Figure 11.6 it may be observed that the overall performance of GEAR is better in terms of network delay.

b. **Data delivery ratio:** Table 11.6 and Figure 11.7 highlight the data delivery ratio against each network size.

Power Management 165

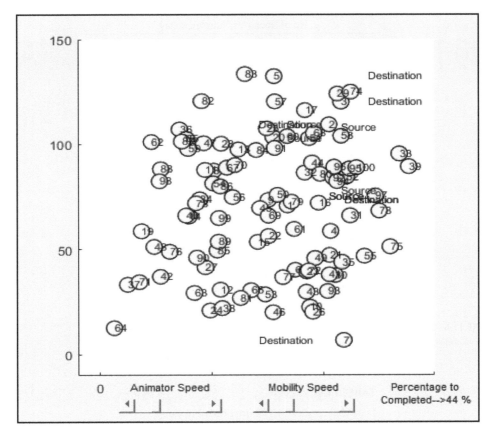

FIGURE 11.5
Node deployments in WSNs.

TABLE 11.5

Network delay values against no. of nodes

		\multicolumn{5}{c}{No. of Nodes}
Sl No.	Protocol	20	40	60	80	100
1	LEACH	174	108	83	71	61
2	GEAR	14	31	2	12	33
3	DREAM	42	82	122	162	202

Observation: From Figure 11.7 it may be observed that the average performance of the GEAR protocol is better in terms of data delivery ratio as compared to the other two.

c. **Throughput:** Table 11.7 and Figure 11.8 highlight the throughput values against each network size.

Observation: From Figure 11.8 it may be observed that the overall performance of the LEACH protocol is better in terms of average throughput.

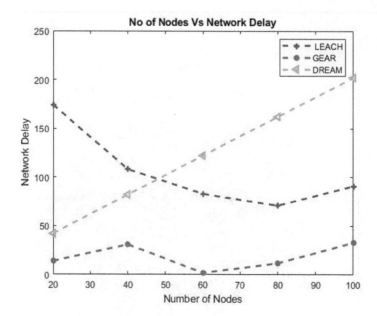

FIGURE 11.6
Network delay values against no. of nodes.

TABLE 11.6

Data delivery values against no. of nodes

Sl No.	Protocol	\multicolumn{5}{c}{No. of Nodes}
		20	40	60	80	100
1	LEACH	67	67	70	71	71
2	GEAR	109	105	106	96	92
3	DREAM	100	90	92	99	99

d. **Data loss:** Table 11.8 and Figure 11.9 highlight the data loss values against each network size.

Observation: From Figure 11.9 it may be observed that the overall performance of LEACH is better in terms of data loss.

Selecting the most efficient routing protocols is crucial in terms of network performance. Routing protocols are in charge of collecting information such as source and destination node addresses, node deployment type, packet size and type, and network bandwidth, among other things. The network's performance and lifespan will immediately improve when the selection is accurate. From the above simulation results, we can say that the performances of the selected protocols are different from one parameter to another. We can also say that the performances vary depending on the network parameters. In this simulation, the performance of the LEACH protocol is better in terms of packet loss and

Power Management

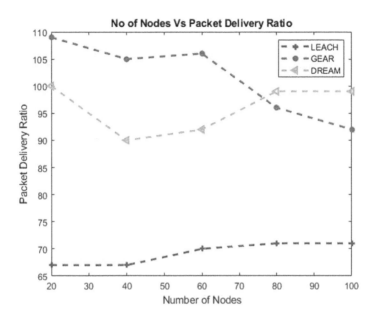

FIGURE 11.7
Data delivery values against no. of nodes.

TABLE 11.7

Throughput values against no. of nodes

Sl No.	Protocol	\multicolumn{5}{c}{No. of Nodes}

Sl No.	Protocol	20	40	60	80	100
1	LEACH	533	1023	1590	2095	2739
2	GEAR	116	316	261	132	121
3	DREAM	172	56	178	191	106

network throughput. On the other hand, GEAR gives better performance in parameters like network delay and data delivery ratio.

11.7 Conclusion and Future Work

This study highlights different areas where power consumption happens within the network. It also explains different approaches which can be taken to minimize power consumption within the network so that the lifetime of the network can be maximized. Despite the fact that a sensor network may employ a variety of power management measures to maximize its lifetime, a number of challenges still persist. At the network level, particularly in routing, a WSN has a large degree of power dissipation. As a result, there is much more scope for improvement in this field.

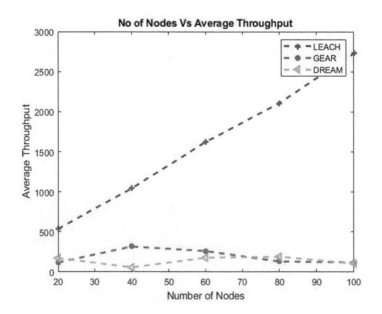

FIGURE 11.8
Throughput values against no. of nodes.

TABLE 11.8

Data loss values against no. of nodes

Sl No.	Protocol	\multicolumn{5}{c}{No. of Nodes}
		20	40	60	80	100
1	LEACH	0	0	0	0	0
2	GEAR	2	6	0	9	0
3	DREAM	5	6	13	28	2

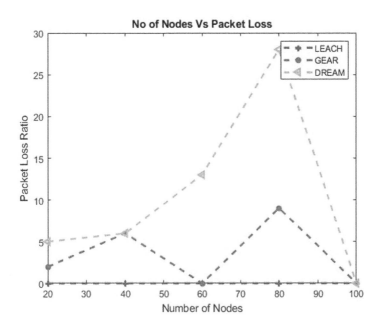

FIGURE 11.9
Data loss values against no. of nodes.

References

[1] V Perumal and K. Meenakshi (2017). "The Comparison of Energy Efficient in Wireless Sensor Network Using Various Cluster Methods and Different Protocols," *IJRDR*, Vol. 5, Issue 2, 1974–1978.

[2] D. Prasad and V. Padmavati (2017). "An Approaching of Energy Management Routing Protocols in Wireless Sensor Network," *IRJET*, Vol. 4, Issue 5, 1060–1064.

[3] Anjali, Shikha, and M. Sharma (2014). "Wireless Sensor Networks: Routing Protocols and Security Issues," 5th ICCCNT July 11–13, 2014.

[4] G. Acs and L. Buttyabv (2007). "A Taxonomy of Routing Protocols for Wireless Sensor Networks," BUTE Telecommunication Department.

[5] K. Mehta and R. Pal (2017). "Energy Efficient Routing Protocols for Wireless Sensor Networks: A Survey," *International Journal of Computer Applications*, Vol. 165, Issue 3, 41–46.

[6] K. Sindhuben Babulal and R. Ranjan Tewari (2010). "E2XLRADR (Energy Efficient Cross Layer Routing Algorithm with Dynamic Retransmission for Wireless Sensor Networks)," *International Journal of Wireless & Mobile Networks*, Vol. 2, Issue 3, 167–177.

[7] T. He et.al (2006). "Achieving Real-Time Target Tracking Using Wireless Sensor Networks", *Proceedings of the 12th IEEE*, Vol. 4, Issue 7, 37–48.

[8] P. P. Bairagi and M. Dutta (2021). "Various Energy-Saving Approaches in Wireless Sensor Networks: An Overview," in *2021 10th IEEE International Conference on Communication Systems and Network Technologies (CSNT)*, pp. 499–504. IEEE.

[9] K. Sindhuben Babulal and R. Ranjan Tewari (2011). "Cross Layer Energy Efficient Cost Link Based Routing for Wireless Sensor Network," in *2011 World Congress on Information and Communication Technologies*, pp. 804–809. IEEE.

[10] V. Bharghavan, A. Demers, S. Shenker, and L. Zhang (1994). "MACAW: A Media Access Protocol for Wireless LAN's," in *Proceedings of the ACM SIGCOMM Conference on Communications Architectures, Protocols and Applications*, pp. 212–225. ACM.

[11] M. AboZahhad, M. Farrag, and A. Al (2015). "A Comparative Study of Energy Consumption Sources for Wireless Sensor Networks," *International Journal of Grid Distribution Computing*, Vol. 8, Issue 3, 65–67.

[12] W. Heinzelman (2000). "Application Specific Protocol Architectures for Wireless Networks." Ph.D. thesis, Massachusetts Institute of Technology.

[13] J. N. Al-Karaki and A.E. Kamal (2004). "Routing Techniques in Wireless Sensor Networks: A Survey". *IEEE Transactions on Wireless Communications*, Vol. 11, 6–28.

[14] A. Belen, G. Hernando, J. Fernan, M. Ortega, J. Manuel, L. Navarro, A. Prayati, and L. Redondo-Lopez (2008). *Problem Solving for Wireless Sensor Networks*. Springer.

[15] Y. L. Tan and Z. Zhang (2011). "Performance Requirement on Energy Efficiency in WSNs," in *3rd International IEEE Conference on Computer Research and Development (ICCRD)*. IEEE.

[16] S. Xu and T. Saadawi (2001). "Does the IEEE 802.11 MAC Protocol Work Well in Multihop Wireless Ad Hoc Networks?" *Comm. Mag.*, Vol. 39, 130–137.

[17] V. Raghunathan, C. Schurgers, S. Park, and M.B. Srivastava (2002). "Energy Aware Wireless Microsensor Networks," *IEEE Signal Processing Magazine*, Vol. 19, Issue 2, 40–50.

[18] J. Hill, R. Szewczyk, A. Woo, S. Hollar, D. Culler, and K. Pister (2000). "System Architecture Directions for Networked Sensors," *ACM SIGOPS Operating Systems Review*, Vol. 34, 93–104.

[19] P. Minet (2009). "Energy Efficient Routing," in H. Liu, X. Chu, and Y.-W. Leung (eds.), *Ad Hoc and Sensor Wireless Networks: Architectures: Algorithms and Protocols*, pp. 49–68. Bentham Science.

[20] P. Mohanty, S. Panigrahi, N. Sarma, and S.S. Satapathy (2010). "Security Issues in Wireless Sensor Network Data Gathering Protocols: A Survey," *Journal of Theoretical and Applied Information Technology*, March, 14–27.

[21] B. Scheuermann, C. Lochert, and M. Mauve (2008). "Implicit Hop-by-Hop Congestion Control in Wireless Multihop Networks," *Ad Hoc Networks*, Vol. 6, 260–286.

[22] Y.Xu, S. Bien, Y. Mori, J. Heidemann, and D. Estrin (2003). *Topology Control Protocols to Conserve Energy in Wireless Ad Hoc Networks*. Center for Embedded Networked Computing.

[23] T. van Dam and K. Langendoen (2003). "An Adaptive Energy-Efficient MAC Protocol for Wireless Sensors Networks," in *SenSys '03: Proceedings of the 1st International Conference on Embedded Networked Sensor Systems*, pp. 171–180. ACM.

[24] V. Raghunathan, C. Schurgers, S. Park, and M.B. Srivastava (2002). "Energy Aware Wireless Microsensor Networks," *IEEE Signal Processing Magazine*, Vol. 19, Issue 2, 40–50.

[25] J. Hill, R. Szewczyk, A. Woo, S. Hollar, D. Culler, and K. Pister (2000). "System Architecture Directions for Networked Sensors," in *ACM SIGOPS Operating Systems Review*, Vol. 34, 93–104.

[26] R. Lajara, J. Pelegrí-Sebastiá, and J.J.P. Solano (2010). "Power Consumption Analysis of Operating Systems for Wireless Sensor Networks," Sensors, Vol. 10, no. 6, 5809–5826.

[27] W.R. Heinzelman, A. Chandrakasan, and H. Balakrishnan (2000). "Energy-Efficient Communication Protocol for Wireless Microsensor Networks," in *33rd Annual Hawaii International Conference on System Sciences*. IEEE.

[28] R. Min, M. Bhardwaj, S.-H. Cho, E. Shih, A. Sinha, A. Wang, and A. Chandrakasan (2001). "Low-Power Wireless Sensor Networks," in *Fourteenth International Conference on VLSI Design*.

[29] K.M. Alzoubi, P. Wan, and O. Frieder (2002). "Distributed Heuristics for Connected Dominating Sets in Wireless Ad Hoc Networks," *Journal of Communications and Networks*, Vol. 4, 22–29.

[30] R.C. Shah and J.M. Rabaey (2002). "Energy Aware Routing for Low Energy Ad Hoc Sensor Networks," in *IEEE Conference on Wireless Communications and Networking*.

[31] M. Busse, T. Haenselmann, and W. Effelberg (2006). "TECA: A Topology and Energy Control Algorithm for Wireless Sensor Networks," in *Proceedings of ACM/IEEE International Symposium on Modeling, Analysis and Simulation of Wireless Mobile Systems*. ACM.

[32] J. Kulik, W.R. Heinzelman, and H. Balakrishnan (2002). "Negotiation-Based Protocols For Disseminating Information In Wireless Sensor Networks," *Wireless Network*, Vol. 8, 169–185.

[33] C. Intanagonwiwat, R. Govindan, and D. Estrin (2000). "Directed Diffusion: A Scalable and Robust Communication Paradigm for Sensor Networks," in *Proceedings of ACM MobiCom*, pp. 56–67. ACM.
[34] A. Seetharam, G. Balasubramanian, A. Hossain, and S. Chakrabarti (2007). "Energy Efficient Deployment and Scheduling of Nodes," in Proceedings of the 10th International Symposium on Wireless Personal Multimedia Communications.
[35] J. Mao, X. Jiang, and X. Zhang (2019). "Analysis of Node Deployment in Wireless Sensor Networks in Warehouse Environment Monitoring Systems," *EURASIP Journal on Wireless Communications and Networking*, Article No. 288. https://doi.org/10.1186/s13638-019-1615-x
[36] E. Fasolo, M. Rossi, J. Widmer, and M. Zorzi (2007). "In-Network Aggregation Techniques for Wireless Sensor Networks: A Survey," *IEEE Wireless Communications*, Vol. 14, Issue 2, 70–87.
[37] D. Ganesan, A. Cerpa, W. Ye, Y. Yu, J. Zhao, and D. Estrin (2004). "Networking Issues in Wireless Sensor Networks," *Journal of Parallel and Distributed Computing*, Vol. 64, 799–814.
[38] A. Mainwaring, J. Polastre, R. Szewczyk, D. Culler, and J. Anderson (2002). "Wireless Sensor Networks For Habitat Monitoring," in *Proceedings of the ACM Workshop on Wireless Sensor Networks and Applications*, pp. 88–97. ACM.
[39] G. Anastasi, M. Conti, M. Francesco, and A. Passarella (2009). "Energy Conservation in Wireless Sensor Networks: A Survey," Ad Hoc Networks, vol. 7, 537–568.
[40] I. Kurtis Kredo and P. Mohapatra (2007). "Medium Access Control in Wireless Sensor Networks," *Computer Networks*, Vol. 51, 961–994.
[41] I. F. Akyildiz and I. H. Kasimoglu (2004). "Wireless Sensor and Actor Networks: Research Challenges," *Ad Hoc Networks Journal*, Vol. 2, Issue 4, 351–367.
[42] J. Li and P. Mohapatra (2007). "Analytical Modeling and Mitigation Techniques for the Energy Hole Problem in Sensor Networks," *Pervasive Mobile Computing*, Vol. 3, Issue 3, 233–254.
[43] P. Bairagi and L. Saikia (2018). "A Comparative Study on Location Based Routing Protocols in Wireless Sensor Network," *International Journal of Computer Sciences and Engineering*, Vol. 6, 1060–1064.
[44] H. Cheng, G. Yang, and S. Hu (2008). "NHRPA: A Novel Hierarchical Routing Protocol Algorithm for Wireless Sensor Networks," *Journal of China Universities of Posts and Telecommunications*, Vol. 15, Issue 3, 75–81.
[45] S. Vanthana, V.Sinthu, and J. Prakash (2013). "Comparative Study of Proactive and Reactive AdHoc Routing Protocols Using Ns2," in *2014 World Congress on Computing and Communication Technologies* Institute of Electrical and Electronics Engineers. IEEE.
[46] M. Dua, V. Ranga, K. Mehra, P. Kardam, and S.M. Bahsakhetre "Performance Evaluation Of AODV, DSR, DSDV Mobile Adhoc Protocols on Different Scenarios: An Analytical Review," Department of Computer Engineering, National Institute of Technology, Kurukshetra (India).
[47] S. Ito and K. Yoshigoe (2007). "Consumed-Energy-Type-Aware Routing for Wireless Sensor Networks," Master's thesis, Donaghey College of Information Science and System Engineering.
[48] K. Akkaya and M. Younis (2005). "A Survey on Routing Protocols for Wireless Sensor Networks," *Ad Hoc Networks*, Vol. 3, Issue 3, 325–349.

12
Wireless Sensor Networks: Power Management

Ciro Rodríguez and Isabel Moscol

CONTENTS
12.1 Introduction 173
12.2 Wireless Sensor Network 174
      12.2.1 Requirements for Implementing a WSN 174
12.3 Energy Harvesting Methods for Sensor Nodes 175
      12.3.1 Electric Batteries 175
      12.3.2 Rechargeable Batteries 175
      12.3.3 Energy Harvesting 175
      12.3.4 Hybrid Energy Harvesting (HEH) 175
12.4 Analysis of Power Consumption in Wireless Sensor Networks 176
12.5 Energy Management 176
12.6 Node Architecture 177
12.7 Conclusions 178
References 178

12.1 Introduction

The application of wireless sensor networks has been of great efficiency for many companies for the performance of their own products. These sensors were inspired by military applications, especially in the surveillance of conflict zones [1]. On the other hand, they have also been of great use in automotive production, environmental monitoring, health care, and industrial automation [2].

Wireless sensor networks have the functionality to monitor in any environment, no matter how inaccessible and remote the environment may seem, thanks to the capacity of the sensors and the functions that each sensor node fulfills [3]. That is why wireless sensor networks are becoming more popular in applications that need continuous monitoring, as some of the main causes of their popularity are their low cost of access, installation, and maintenance, in addition to their flexibility and scalability [4].

The use of sensors is expanding in different areas of our lives. Wireless communication has allowed these sensors to be applied in more and more innovative and useful fields. However, a problem that comes with wireless communication between nodes is the energy they require to continue working properly. This chapter will give a broad overview of the sources of energy for wireless sensor networks as well as studies on how to optimally manage that energy.

DOI: 10.1201/9781003326205-15

12.2 Wireless Sensor Network

Sensor networks can be considered a specific field within ubiquitous computing. The sensor network is formed by intelligent nodes, which self-organize and manage the network (see Figure 12.1), i.e., it is a type of ad-hoc network. This allows them to cooperate with each other, with the ability to measure a physical parameter of the environment and as a result to process the information and circulate it through the network [5].

The sensor nodes have a structure consisting mainly of a wireless transceiver mechanism, a processing unit, usually a microcontroller, sensors or actuators, and a power system for power supply, usually comprised by a battery. Sensor nodes have the ability to monitor physical conditions of the environment, such as light intensity, temperature, humidity, pressure, chemical information, among other factors [6].

12.2.1 Requirements for Implementing a WSN

There are several areas of application of wireless sensor networks, the most important of which are environmental areas, for example, to monitor the seismic activity of an active volcano, detection of anomalies associated with earthquakes, detection of forest fires, etc. In agriculture, they are involved in pest prevention and crop irrigation. Other applications include monitoring the oceans. In home automation they are involved in temperature control and security systems against intruders. In health, they can monitor different physical aspects of patients with diseases that require permanent supervision.

System requirements vary by application, e.g., in discrete manufacturing and asset monitoring, low power consumption is required as may be battery energy storage and wireless power transfer [7]. However, both cases differ in that discrete manufacturing requires low latency time, which is typically a few tens of milliseconds; while in asset monitoring the latency time is less critical, where although the latency time depends on the asset being monitored, it is common for update times to be on the order of minutes or even hours [7].

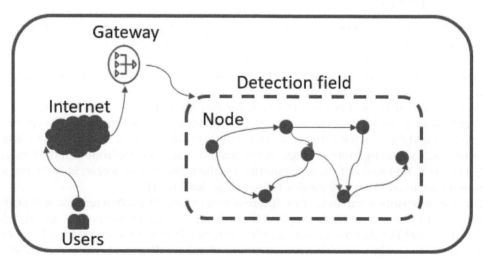

FIGURE 12.1
Wireless sensor network model.

Regardless of the application, it will always be important to work with a modular design (hardware and software) that allows the efficient maintenance of the asset, since often the system will be in permanent operation fulfilling monitoring functions—even in sensitive cases—where downtime can raise costs or lead to damages of high magnitudes, such as in an industrial plant or in the detection of earthquakes, respectively.

12.3 Energy Harvesting Methods for Sensor Nodes

12.3.1 Electric Batteries

Electric batteries generate electrical energy after a chemical reaction. Primary batteries do not recharge once their energy is depleted, but secondary batteries can be recharged. Wireless sensor nodes that use primary batteries have problems due to the short life of the power source, so when placing the sensors in remote locations, battery replacement is difficult not only because of the time involved, but also because of labor and maintenance costs.

12.3.2 Rechargeable Batteries

Rechargeable batteries are typically used to store energy from solar or a renewable energy source. There are currently five types of rechargeable batteries, which have several advantages and disadvantages. For example, nickel-cadmium batteries have a very fast discharge rate, while lithium-ion batteries have a low self-discharge rate but are expensive due to their complex circuitry. On the other hand, nickel-metal hydride batteries have a lower charge and discharge cycle. Other types of rechargeable batteries are sealed lead acid and polymeric lithium-ion.

12.3.3 Energy Harvesting

Energy harvesting consists of obtaining energy from environmental sources. This technique originated as a solution for sensors to find their own energy source to be self-sufficient and not have the complications of working with non-rechargeable batteries. Some of the most researched and currently implemented are:

Vibrational energy: can be caused by humans or machines. The movements produced by the vibrations in the capacitor induce the variation of the capacitance, thus generating electrical energy.

Thermal energy: produced by the potential difference when two different materials with different temperatures are interconnected by a metal.

Photovoltaic energy: the voltaic cell is responsible for harvesting solar energy, which is one of the most abundant sources on our planet. Electrical energy is generated by the release of electron flow after the reception of a large amount of photons of light.

12.3.4 Hybrid Energy Harvesting (HEH)

This increases reliability by combining two power sources, so that one fills in for the active source when needed. By having more than one data source, more energy is collected, which increases the lifetime of the sensor. Some of the most used scenarios are mentioned below:

- Two different energy harvesting mechanisms on the same platform.
- Electronic switch/multiplexer for switching between energy harvesting sources.
- Individual power converter for each source.
- Direct connection of power sources in parallel/series configuration.

12.4 Analysis of Power Consumption in Wireless Sensor Networks

There are small electronic boards that contain all the included components. These are referred to as motes, which consist of stand-alone electronic boards that include parts of a sensor node. The reason for using motes instead of ad-hoc designs is the speed of development and reuse of third-party code [8]. Among the main functions of the software running the motes is power control. The motes are organized in time slots to detect which paths are available for communication.

Additionally, speed in the development and reuse of code is achieved by the implementation of operating systems as base software; some of these are TinyOS v1, TinyOS v2, Contiki, and Nano-RK. The operating systems aim to control the power consumption of each part of the sensor node and increase its efficiency. To reduce the power of each part of the node, it turns it off or puts it into a low-power mode, or *sleep*, when not in use and when required *wakes* it up or turns it on. Reducing the power consumption of each part of the node reduces the overall consumption and, therefore, extends the battery life of the motes.

12.5 Energy Management

Wireless sensor networks consist of sensors/actuators, a power supply, and a wireless communication system or ready-made system, such as motes. One of the major problems of wireless sensor networks is the availability of energy, so it is of particular importance to maximize the amount and sources of supply and properly monitor its flow throughout the system through energy management from hardware, firmware, and protocols (see Figure 12.2). Additionally, the consumption of a node within a sensor network is defined by the lifetime of its battery. Some recommendations to decrease the energy consumption of sensor nodes within a wireless sensor network are:

- Minimize listening times of wireless communication transceivers.
- Hibernate components or sensors when they are not needed.
- Maximize data packet routing performance.
- Use an energy regeneration system, such as solar, wind, etc.
- Minimize data collisions in the transceiver channel.
- Execute the routing protocol considering the geographical locations of the nodes, as well as the capabilities of the wireless transceivers used.

Hardware control: this is manual monitoring and control, by turning off devices that are not in use, managing low power modes and, in devices that do not have this mode, using

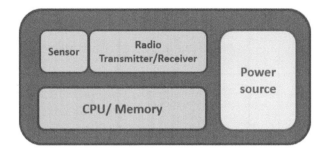

FIGURE 12.2
Autonomous device of a wireless sensor network.

switches for power supplies. These techniques are aimed at microcontrollers since they are in charge of managing the sensor/actuator nodes.

Firmware control: it is recommended to have a high impedance in the lines connected to the circuit that are not currently active, and to have variable clock frequency management since the nodes of the sensor networks usually have periodic work cycles. In this way, wear and tear is reduced, which extends the useful life of the sensors.

Protocols: protocols are implemented according to the type of layer.

- The SMAC protocol works with the awake/sleep scheme. When it is in an *awake* state it is in activity; when its activity is not requested it will remain in *sleep* or low energy consumption mode.
- The BMAC protocol uses a central node that varies its state according to a certain period.
- The WiseMAC protocol is derived from BMAC. It dynamically adjusts the duration so that awake times are reduced.

The proper handling of energy management can be achieved with algorithms to manage the timing of the states in each of the elements that make up the node, based on the information collection time.

12.6 Node Architecture

The nodes must have characteristics that allow them to operate in a wireless sensor network. In addition, the nodes that are within the sensor network must have the functionality to perform the tasks that have been assigned to them as these come configured within the computational framework, memory, volatile or non-volatile, and in the energy storage [9]. Nodes must have the following characteristics (see Figure 12.3):

- Have a processing element (MCU).
- Have a wireless communication element (radio frequency transceiver).
- Have peripheral communication elements (I/O ports). These ports will be used to obtain information from sensors and/or transducers.

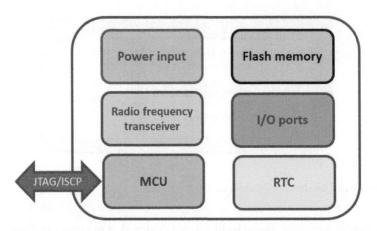

FIGURE 12.3
Elements of a sensor node.

- A system for power supply to the sensor node. In this case it will be an autonomous photovoltaic system.
- Real-time clock (RTC). This element is very important for the implementation of the energy efficiency algorithm.
- Programming or debugging interface (JTAG/ISCP).

12.7 Conclusions

Good energy management or administration of the wireless sensor network is important as this will allow a quality lifetime. Also, sensor nodes have an important feature and that is that they can remain in operation without battery recharge for years, which can be enhanced with hardware control, firmware, protocols, and finally with logical decision algorithms.

Thanks to the characteristics of wireless sensors, the office and consumer products sector is one of the main drivers of wireless technologies. On the other hand, special attention should be paid to the integration of wireless components in industrial devices since the lifetime of industrial devices is much longer than that of consumer products.

References

[1] J.A. Rapallini and H.H. Mazzeo (2014). "Wireless Sensor Network Systems," Universidad Tecnológica Nacional, Centro de Codiseño Aplicado (CODAPLI) Facultad Regional La Plata, Buenos Aires, Argentina. https://adut.frlp.utn.edu.ar/wp-content/uploads/2021/02/Sistemas-de-Redes-de-Sensores-Inalambricos.docx.pdf

[2] N. Chio-Cho, D. Tibaduiza-Burgos, L. Aparicio-Zafra, and M. Caro-Ortiz (2009). "Wireless Sensor Networks," presented at the International Congress of Mechatronics Engineering-UNAB.

[3] D. Gascón (2010). "Wireless Sensor Networks, the Invisible Technology," *Technology and Society*, 180–181, pp. 53–55.
[4] L. R. Amondaray, F. J. A. Fuentes, and C. A. Calderón (2020). "Software Defined Wireless Sensor Networks: State-of-the-Art Review," *RIELAC*, 41, pp. 39–50.
[5] E. E. Lopez, A. M. Sala, J. V. Alonso, and J. D. Jimenez, "An Introduction to Wireless Sensor Networks," p. 3. https://repositorio.upct.es/bitstream/handle/10317/337/2004_AI_6.pdf.pdf?sequence=1&isAllowed=y
[6] J. Del-Río (2018). "Electronic Design with Solar Panel to Extend the Lifetime of a Node in a Wireless Sensor Network (WSN)," graduate thesis, Fundación Universitaria Los Libertadores, Colombia.
[7] N. Aakvaag and J. -E. Frey (2006). "Wireless Sensor Networks: New Interconnection Solutions for Industrial Automation," *ABB*, 2, p. 4.
[8] J. Taboada (2020). "Comparative Study of SC DC-DC Topologies and Design of a Series-Parallel Topology for Energy Harvesting Purposes," graduate thesis, Pontificia Universidad Católica del Perú.
[9] D. M. Delgado (2019). "Energy Management Algorithm Applied within WSNs," Revista Nthe, 29, pp. 19–30.

13
Security Enabling for IoT and Wireless Sensor Network Based Data Communication

Ghazanfar Latif, Jaafar Alghazo, and Zafar Kazmi

CONTENTS
13.1 Introduction 181
13.2 IoT and Wireless Sensor Network Based Embedded System 182
13.3 Data Security for IoT and Wireless Sensor Networks 183
       13.3.1 Importance of Data Security 184
       13.3.2 Data Security for Internet of Things (IoT) 184
       13.3.3 Different Methods of Data Security 185
       13.3.4 Role of Cryptography for Data Security 186
       13.3.5 Types of Data Attacks for IoT Based Devices 186
       13.3.6 Literature Solutions for Secure IoT Data Communication 187
       13.3.7 Data Security Goals for IoT Devices 187
13.4 Security Issues in the Architecture of IoT and WSN 188
13.5 Proposed Model for Security of IoT Devices 189
13.6 Benefits of the Proposed Solution 191
13.7 Conclusion and Future Work 192
References 193

13.1 Introduction

The emergence of Internet of Things (IoT) technologies is an integral part of our lives. It has led to many security issues which affect the network and the operating system. IoT is a complex and open system that communicates with any endpoint, thus making security a grave concern. Since IoT devices are connected to the Internet, they can accept connections from any anonymous source. This multiplies the risk of a security breach. With IoT communicating over the internet with any incoming connection, security parameters should be kept in place to ensure the integrity of the system. With so many IoT devices connected to the internet, a bigger picture has emerged: Big Data. Millions and billions of devices are connected to the internet and huge numbers of terabytes of data are being transferred over the network, and it requires analysis and science to make sense of this huge volume of data. This is done through Big Data analytics. This study focuses on the network communication of the IoT and the use of Big Data. We try to understand the patterns and the factors that affect the network communication of IoT-based devices.

IoT devices are used in every known industry for automation processes and to provide real-time results. IoT devices can be used in home automation, smart cities, smart grids, and for automation processes in industries. What makes IoT devices fascinating is that real-time sensors can communicate data to the cloud and make it easily viewable through an app. This allows devices to be remotely managed over the internet: for example, controlling the accessories in your smart home, like switching on the air conditioner, turning off the light, or closing the door of the parking garage without the involvement of any human physical presence. These devices can transfer data over the internet without the need for any third-party interaction [1]. With so many data points at play, we have seen the emergence and advancement of fields of technology like machine learning, data science, embedded systems, and wireless sensor networks. Some of the commercial benefits of IoT devices in different areas include healthcare systems, medical imaging, transportation, smart home design, agriculture, and industrial automation. IoT has provided real-time automation, reducing both labor and cost.

Any device such as an IoT device that is connected to the internet can be vulnerable to an attack. These security risks could range from compromising the IoT device to the leak of Big Data. Information such as customer data can be a valuable commodity that could be sold for money to third parties. The manufacturing companies require diagnostic scans of their product for research or re-usability. They build sensors into their products that could relay information to their cloud server. This data can be misused by manufacturing companies [2]. One of the many concerns which are associated with IoT devices is privacy issues. Since IoT devices are embedded, when compromised they could be used as weapons in cyber warfare. It is also stated in [3] that IoT devices store massive volumes of data which could be used by countries for spying. Such huge data leaks could be used by malicious parties to sabotage trust in a government and damage its reputation. Securing such a large volume of data is a challenge. A lot of research is being done on the use of IoT devices in facets of life such as healthcare [4–5], oil and minerals exploration [6–7], smart cities [8–9], and retail business [10–11], among others. The integration of machine learning with IoT and Big Data is also an open field of research.

This chapter outlines the security risk involved in IoT and wireless network architecture when it comes to data sharing. This chapter will propose a comprehensive method of security enhancement for data sharing and storage with the help of a proposed diagram that will enhance the security of IoT-based devices and WSNs.

13.2 IoT and Wireless Sensor Network Based Embedded System

Embedded devices like IoT devices use microcontrollers and microprocessors and sensors that collect real-time data. This real-time data is then fed into the system, which uses artificial intelligence algorithms to process the data. After the processing is done, based on the output, the required action is taken. Examples of these type of embedded devices are ATMs and washing machines. The embedded system has a certain degree of restriction on both hardware and software, which only perform the intended function as required. The hardware of embedded systems runs using a microprocessor and microcontroller, which is the central processing unit (CPU) that is combined with an external device and memory along with the microprocessor, and it uses a specific type of chips for peripheral interfaces and storage.

Embedded devices consist of both hardware and software which work together. Embedded devices are now a part of our daily lives. As technology evolves, there are many challenges and constraints that the embedded devices will have to face. In this chapter, we will outline some of the issues, such as real-time response, recovery from failure, working with multi-vendor and distributed architecture, flexibility, timing, security, and power optimization. This chapter will discuss some of the solutions and how these solutions will be beneficial to counter some of the emerging security challenges for embedded devices. It will also discuss different data-sharing techniques from endpoint to endpoint and endpoint to IoT cloud servers.

Considering the net worth generated through innovation, with new technology markets opening, it is estimated that IoT produced $14.4 trillion in net worth between 2013 and 2022 [12]. Enterprises and companies ranging from small and medium to large have taken advantage of IoT devices to enhance automation and increase their profit. IoT devices have provided a platform to accumulate large volumes of data and analyze this data to discover patterns and trends. This poses a great challenge for the manufacturing industries and suppliers. Big Data will overcome this challenge, helping us to examine the data and find both relevant and irrelevant data. Huge data alludes to accumulations of data collections with sizes past the capacity of generally utilized programming devices. Enormous data is portrayed by "4 Vs," that is, volume, variety, velocity, and veracity. Personal data collection sizes are increasing and range from terabytes to petabytes with the sole aim of data collection [13].To accommodate the need to take care of such a huge amount of data, a foundation for "Big Data" devices has been developed. Big Data carries huge benefits for the organization that is ready to utilize it.

13.3 Data Security for IoT and Wireless Sensor Networks

The internet is an integral part of our life that comes with undoubted benefits. But with these benefits, there are associated security risks, such as hacking of online accounts, data breaches, and privacy breaches. There is a need to ensure that the privacy of the data is maintained, whether on a physical server or while passing through an unsecured network. To provide confidence to internet users, the issue of security needs to be addressed, assuring that no unauthorized user will have access to anyone else's data. This means that each party can communicate securely without losing data or fear of hacking. IoT may be the new frontier in the phase of the evolution of technology from analog to digital, but the lack of security measures in place can't be ignored. The main challenge for IoT is security, considering the fact that IoT has facilitated data mining and improved decision making using artificial intelligence algorithms. To address the issue of IoT security, we need to understand it in two broad senses: namely, handling the issue of data mining, and secondly, understanding the IoT device's communication capabilities and the risk associated with them.

As IoT technology is advancing, the number of hacking incidents is also increasing. These incidences are usually associated with loss of privacy, data leaks, and compromising of a system. Some of the common attacks include hijacking IoT devices, home intrusion, privacy leaks, and remote vehicle hijacking. A distributed denial of Service (DDoS) was waged by a botnet named Mirai in 2016 which disrupted many networks and websites [14]. Airbnb was also subjected to investigation after a hidden camera was discovered in one of the rooms in 2019 [15]. The healthcare system has been a point of attack for cyberattacks targeting critical health systems which have embedded devices integrated into them. Research is underway for the protection of health records and health data through the use of various techniques [16]. It is assumed that about a million IoT devices, which include doorbells, IP security cameras, and baby monitors, have been hijacked and used by the hacker for spying [17]. There is no critical patch or firmware update provided to close these security flaws, according to researchers. The vector used to target these IoT devices is generally on a network using a peer-to-peer (P2P) approach, which allows the intruder to get access without a manual setup. Therefore, it is important to highlight the challenges and solutions to arrive at a valid conclusion.

13.3.1 Importance of Data Security

Data security is one of the aspects of security whose importance cannot be denied, as it protects against data breaches and leakage of data. It is crucial to safeguard confidentiality of government data and sectors like banking and the military against any data breach. Data breaches like these can affect people's lives and put them in harm's way. They can affect businesses and the private sector. These data breaches can be used by hackers for monetary gain or for blackmailing. Data security is the key concern for every user of IoT devices and if the device fails to protect user data, people will stop using these devices [18]. Clients only trust companies that protect their data [19].

Confidentiality and privacy are fundamental rights of every user. Some IoT devices come with a built-in camera and audio sensors which record. If this recording falls into the wrong hands, they can use it to exploit the user through social engineering [18]. In the world of smart cities, IoT devices are built into many systems and record a massive amount of confidential data that can be misused by the hacker. It is believed that a lot of data losses occurred when the Mirai botnet attack took place. Hence it is important to safeguard the data on IoT devices [20].

13.3.2 Data Security for Internet of Things (IoT)

IoT devices consist of many sensors which collect data in real-time and these devices need to be kept secured. Some of the common issues with IoT arise because when the data is collected through the sensor it is stored on a third-party server. For example, IoT devices for fog detection are used to collect data and store it on a third-party server, which can be risky as a hacker can steal and exploit the data [21]. If the data is stored on a third-party server, the third-party company should be reliable and trustworthy. Identity theft is an example where a hacker can break into an embedded device and gather user data such as age, name, and location. These embedded devices are a point for collecting data [22].

The intensity of the privacy and safety threats associated with IoT devices is still unknown to most of the public. Whenever a computer device shares data over a Wi-Fi, it is breaching protection of one sort or another. When an intruder compromises a computer system, the intruder has access to the files system and custom data [23]. Obtaining a secure IoT device is challenging as there are many factors in play such as price, existing goods, and not having adequate knowledge of security. Research estimates global IoT markets have doubled from 235 billion dollars in 2017 to about $520 billion in 2021. Such reports indicate that the Internet of Things is growing and is becoming a prominent part of the world of technology. According to another survey carried out by 451 researchers, 55 percent of IT professionals have listed IoT protection as their top priority [24]. There are many endpoints through which a hacker can gain access to an IoT device, like exploiting the operating system, through the cloud server, or social engineering the device to access the data. This means we have to list all possible countermeasures and protect the IoT device from such attacks [2].

Since IoT devices produce data in real-time and this data needs to be shared in real-time with the server, faster networks like fiber channels are required to avoid any delay on the network. Next, a large storage server and a great bandwidth are needed to cope with the high internet traffic. Currently, there is no open platform that can allow these apps to talk to one another. Microsoft, Google, and Android use their private interoperability network for their own devices. This raises one more big challenge which is the integration of multiple security solutions [25]. With the emergence of IoT devices, the demand for IP addresses grows and cannot be handled by the current IPv4 and needs to be replaced by IPv6 [23]. Understanding the significance of the IoT devices, data needs to be secured by using different protocols [26].

Security for billions of devices connected to the Internet of Things will be a great challenge. More innovative technologies will likely emerge providing long-lasting solutions [27]. Controlling authentication on a large scale is a huge challenge that cannot be effectively met by the media communications industry. Numerous security strategies have been proposed in the course of the most recent 15 years, ranging from cryptographic procedures to absent information structures that conceal information to information anonymization systems that change the information to make it increasingly hard to exploit. Be that as it may, numerous such procedures do not scale to enormous datasets and do not explicitly address the issue of accommodating security with protection, which is a significant concern while sending Big Data [28]. Cryptographic methodologies, for example, secure set crossing point conventions, and may ease such concerns. In any case, these systems do not scale for enormous datasets. Ongoing methodologies dependent on information change and mapping into vector spaces, and a blend of secure multiparty calculation (SMC) and information cleansing methodologies, for example, differential protection and k-obscurity, have tended to be versatile.

13.3.3 Different Methods of Data Security

Some of the methods which are used to protect data are cryptography, access controls, organizational standards, next generation firewall, and using a complex password. One of the common security mechanisms deployed to protect the system is encryption. Data should be encrypted using multiple algorithms, for example RSA encryption [29].It is a recommended practice that each user change his password after a certain interval of time and use a complex password. IoT devices need to be connected to the internet to work properly. These devices store data on the cloud server, which should be done through encryption. While encrypting the main data connection it is recommended and important to secure the secondary communication, which is generally used for maintenance and update. Some of the IoT devices use a web interface for the convenience of the client, and this communication should be decoded by default. If this is not the case, then it becomes very easy for any other person to hack the usernames and passwords or use assembly data to pretend that the logged-in accounts are controlled by these devices which are using the same network [39].

13.3.4 Role of Cryptography for Data Security

Cryptography is an essential part of securing the data. The stronger the cryptographic algorithm, the more difficult it becomes to crack the password. The most widely used algorithm for commercial purposes is the asymmetric encryption algorithm called RSA based on public-key cryptography. Some of the powerful hashing algorithms which are used to hash the password are SHA-256, MD5, and blowfish. Cryptography secures the communication, authentication of credentials, and firmware in IoT devices. It is preferable to utilize public-key encryption with message authentication code (MAC). As per a recent study, cryptography is an essential element to protect IoT devices, which it is mainly used to secure communication channels. IoT-centric communication allows developers to use Transport Layer Security (TLS) such as MQTT and AMQP which have the authority to make sure that all the data that has traversed through the network should be incomprehensible to unknown parties. TLS is considered a deserving successor to the better-known standard known as Secure Sockets Layer (SSL), which has remained the long-time standard for web encryption (such as HTTPS) but is now considered unreliable [31].

Public key encryption is asymmetric encryption, which uses a pair of keys, namely the public key and private key. If the data is encrypted using a private key, then it must be decrypted using a public key. Generally, the private key is kept private, which allows it to communicate with the outside world and validate the foreign machine. This specific function of cryptography is best for some aspects of IoT infrastructure [29].

13.3.5 Types of Data Attacks for IoT Based Devices

IoT can be subjected to different types of attacks as the hacker can exploit more than one loophole. These types of attacks involve botnet, the man-in-the-middle attack (MITM), and social engineering attacks. An analysis of data security and potential threats from IoT devices for middle card players for both individuals and businesses was done in [32]. Depending on the hackers, they can use any conventional method to exploit the loopholes. IoT devices are connected to mobile phones, laptops, and PCs and can be exploited as a tool by hackers to create a botnet. Mirai botnet proves that botnets are an extreme threat to IoT devices. As per recent studies, the Mirai botnet stole data from approximately 2.5 million devices, such as routers, printers, etc. [33]. For initiating service outbreaks on IoT devices, botnets are especially used by attackers. These botnets help the attackers in launching very serious cyber-attacks against vulnerable IoT devices. According to [34], man-in-the-middle attacks can be used in real-time to attack a wide range of IoT apps. With man-in-the-middle, attackers may interact with multiple IoT devices, which results in a critical malfunction. For example, an intruder may manipulate intelligent home accessories, such as bulbs, using the MITM to change or turn on and off its light. These attacks can have catastrophic impacts on industrial equipment and medical applications involving the Internet of Things [35]. The concern of IoT companies is that hackers can easily hack their devices, such as smartwatches, smart meters, smart home devices, etc. so that they can steal information about specific users and organizations. After getting the user's personal information, hackers can even perform identity thefts. With social engineering, hackers trick people into giving their personal information, such as bank account details, passwords, etc. Social engineering also helps cyber criminals to approach a system by downloading malicious software privately on the IoT devices. Mostly, social engineering attacks are conducted through malware emails, in which the hacker establishes a bond with a person to convince them. Despite this, social engineering attacks can be much easier to execute in the case of IoT devices [36].

13.3.6 Literature Solutions for Secure IoT Data Communication

In [37], authors investigated the ways to secure IoT technologies through both qualitative and quantitative research. The findings showed that IoT devices can be secured with real-time monitoring of response and recovery. They revealed that using weaker credentials can lead to security issues. Therefore, the authors presented various ways to enhance security. Blockchain ciphers offer blockchain-based solutions for securing IoT devices. This research discussed blockchain-based IoT design for enhanced security using secondary sources. In [34], authors investigated the IoT attack vectors and then proposed a solution based on security information and event management (SIEM). The technique used for this research is experimental and it showed that SIEM can secure the IoT ecosystem.

Different types of cyber-attacks are classified in [38].The authors highlighted the challenges faced by IoT with secondary sources. The research paper further presented the blockchain-related solutions for two of the more common IoT and industrial IoT (IIoT) applications. The authors subsequently created a taxonomy and the correct solutions for the safety research areas of IoT and IIoT. The IoT security scheme for symmetric IoT data encryption was proposed in [39]. The authors offer simple concepts of safety to tackle this issue, present a new construction, and give safety evidence of the degree of construction protection. Quality statistics are also given for proof-of-concept implementation. The results showed that the new program provides a good compromise between the protection of identity and complexity.

Authors in [40] highlighted solutions to protect IoT information retrieval. The proposed scheme is based on private information retrieval (PIR). It saves the data to various servers and recovers the requested piece of data without revealing its identity. The information is encrypted in this method until it is sent to the cloud servers. The experimental research on several different configurations supported the feasibility and the efficiency of the proposed scheme with exceptional results. A new approach is presented to secure IoT devices with the help of blockchain in [22]. The authors conducted both secondary and experimental research. The results showed that the IoT device become secure to a great extent. Similarly, in [14] authors proposed various solutions to reduce the security attacks in IoT technologies. One such technique is NIST RMF. This improved the results and enhanced the outcomes to stop IoT attacks. The research is limited to only secondary data, and it only covers abstract details. It has not backed up arguments properly with authentic sources to show that the research is authentic, credentialed, and reliable. Authors in [14] take into consideration the physical factors and does not provide strong evidence. Apart from this, the research in [40] is not extremely secure, having loopholes in it, while authors in [22] did not provide any new ideas and required a lot of resources. They explained things in detail and proposed a solution, which is good for small IoT devices where data is in less quantity. Authors in [18] did not focus enough on the how factor, and their SIEM-based approach is vague and not very descriptive. Similarly, in [23], the authors provided security tips but did not propose any practical internal approaches.

13.3.7 Data Security Goals for IoT Devices

The key goal of data security is to protect the security principles for IoT devices. These principles include confidentiality, privacy, authorization, authentication, etc. IoT devices collect a lot of private data, which should be protected [15]. The users should set complex passwords so that the attacker will not be able to break the credentials within a reasonable time. Furthermore, eavesdropping should not be allowed through any medium: only an authorized person should have access to the data. The privacy of the individual needs to be preserved by preventing unauthorized access.

Businesses use connected devices to advance their business gain, keeping in mind that security cannot be ignored. Every organization should protect user data and prevent identity theft. For any business model, it is an essential need to protect user data and safeguard privacy. Mobile devices that are provided to employees need to be tested and locked. Strong password and biometrics authentication must be used in case a tablet gets stolen. Using a protection system on IoT devices limits the number of apps that can be run on company-based IoT devices, differentiating between personal and cooperate devices and removing company data if stolen [39]. People should change new devices' default passwords. Also, individuals should not use the same computer password again. Because it is powerful, it is suggested not to connect an intelligent computer to the internet. It is better to first test which features are available without connecting to the internet on the computer. One may discover that one's smart device offers good features without internet access. The computer should be used offline in this case. This is a smart way to safeguard security without any expenses. A truly open ecosystem must be designed with standard application programming interfaces, which allow interoperability with a stable, automated system of patching. Devices must be designed to protect against standard security abuse by the best practices on the market. Devices on wired networks must be well secured [15].

13.4 Security Issues in the Architecture of IoT and WSN

The Internet of Things paradigm is becoming a more and more important and promising area of technology, and it will change our ways of communication. It encompasses many technologies such as radio frequency identification (RFID), cloud services, wireless sensor networks (WSNs), etc. The mentioned technologies are becoming a factory or base of application domains including connected industry, environmental monitoring, smart cities, and healthcare.

WSN is a subset of IoT, and the difference is that IoT exists at a higher level than WSN. In simple words, WSN is a technology of IoT, and it uses a protocol that configures the network. The protocol of WSN collects data and information packets from several sensors in a specific environment. Altogether, security issues must be considered and fixed to avoid any future problems. The IoT architecture is complex since it deals with millions if not more of sensors that interact with each other or external entities. The current architecture has many drawbacks and challenges. One of the challenges that the current architecture has is confidentiality challenges. This challenge is the most complex one because information and communication need to be confidential. This problem can be solved using standard encryption functions such as a common encryption algorithm. Another challenge for the current architecture is source authentication. It is crucial to ensure data authentication since WSN uses a shared wireless communication medium, which makes it a difficult task. Thirdly, there is the data integrity challenge, and fourthly, availability challenges. The WSN can be attacked through its sensor nodes, affecting the network.

Security Enabling

The challenges are not limited to the aforementioned four. There are more and more challenges such as data privacy and data security which have already been discussed. Another common challenge nowadays is fault performance. This can be caused by physical failures such as power, physical damage, etc. We also have the challenge of scalability. Scalability means that number of routing schemes must be scalable enough to the number of sensors. Since there are large numbers of sensor nodes in a single network, the cost of these nodes is very important. Further, one challenge is the operating environment. Also, quality of service is a big challenge of current WSNs architecture because of the quality of service that the application requires.

13.5 Proposed Model for Security of IoT Devices

By taking into account the above-mentioned factors, a secure model has been proposed as shown in Figure 13.1. It is a hybrid approach in which multiple security measures have been gathered to provide a secure solution as shown in Figure 13.2. Initially, the data will be acquired through sensors and actuators, then that data will be aggregated as multiple sensors will be used. Afterward, the analytics will be applied, and the data will be encrypted and various security mechanisms, such as blockchain and AI, will be used, then the data will be stored on the cloud. Whenever the data is retrieved, the user will be authenticated, then the data will be decrypted and shown.

The layered architecture of the IoT frameworks contains three layers, i.e., the application layer, network layer, and physical layer. The hackers can target any layer; hence, we propose that communication at all these layers should be encrypted as shown in Figure 13.3. At the application layer, secure communication protocols should be used, and end-to-end device encryption should be utilized. In this way, the attacker will not be able to target the application layer of the IoT devices. At the network layer, the IoT devices comprise all the networking information. Protocols, such as IPSec, should be used to encrypt data, as there are various attacks that can target network information for exploiting the network

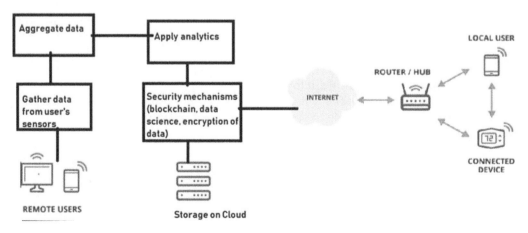

FIGURE 13.1
Proposed secure IoT model for data communication.

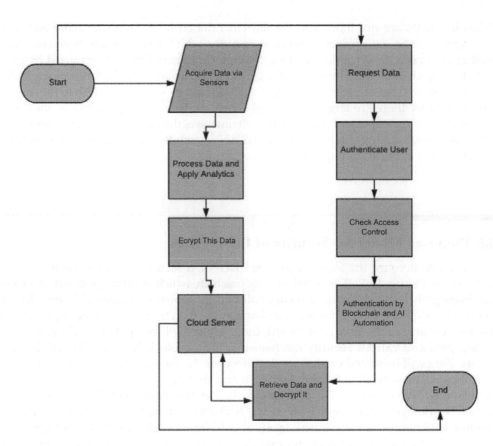

FIGURE 13.2
Descriptive flow of the proposed approach.

layer and entire devices. Furthermore, at the network layer, a firewall should be enabled so risky traffic can be blocked. Also, it is preferable to utilize intrusion detection software as the data within the IoT devices is highly confidential. For safe and secure communication, IPSec, a firewall, and intrusion detection should be utilized. At the physical layer, there are sensors, RFID sensors, and cameras that record a massive amount of data, and most of it is personal. Thus, at this layer, software-based cryptographic algorithms should be used. As per the studies, it is difficult to break RSA within a reasonable time; thus, adding hashing with salt along with RSA will make it difficult for the attacker to access this data. For further security, it is suitable to store this encrypted data on a private cloud or on a trustworthy database or public cloud. Most of the third parties are risky; hence, the right one should be chosen.

To further explain the internal working of the model, Figure 13.4 shows the proposed IoT framework. This ensures end-to-end security at all layers of IoT. The IoT devices will interact with the controllers, and they will further interact with the gateway. It is better to enable a firewall and have access control. Only the authorized user should be able to access the network traffic. The data attained from the IoT devices should be encrypted end to end before it is stored on the cloud server.

Security Enabling

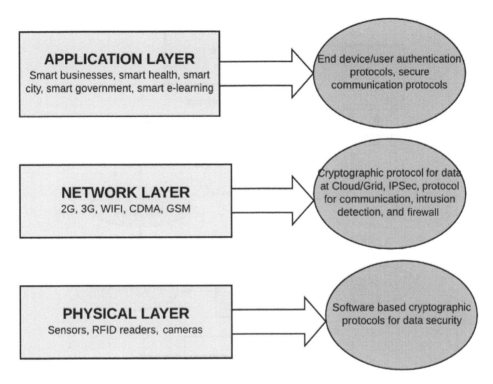

FIGURE 13.3
Securing layer architecture of Internet of Things (IoT).

As per the proposed model, the data acquired through the sensors should be encrypted using RSA, and then it should be hashed with MDF along with salt. This will make it difficult for the attacker to steal the data. At the network level, TLS encryption should be enabled. The data should be stored on a private cloud as third-party cloud services are less trustworthy. Furthermore, for providing access to the customer, biometric authentication should be conducted with behavioral aspects, such as mouse characteristics, keystroke dynamics, voice recognition, etc. The complete flow is shown in Figure 13.5.

13.6 Benefits of the Proposed Solution

Data security is a key aspect of data management. The more the organization is in control of its data security measures, the more it stands a chance of advertising guaranteed data security. There have been developments that have enhanced the way data management and security are handled. Soon, there will be a need to ensure that before any data is stored in any system, there is adequate evidence to prove that the system cannot be easily penetrated. It is clear that a huge number of organizations have been able to introduce a raft of measures that will ensure that data security is always maintained. It has also been established that training of employees to be in line with data security management

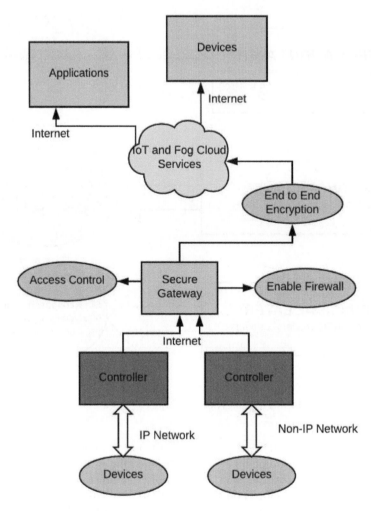

FIGURE 13.4
IoT security framework architecture.

requirements will be a key aspect when it comes to ensuring there is no external intrusion into the organization's systems.

13.7 Conclusion and Future Work

The significance of a secure IoT device cannot be denied. With the advancement in information technology, the Internet of Things (IoT) has been used globally for automating entire operations and functionalities. IoT technologies offer multifarious benefits, such as automation of routine tasks, smart homes, smart grids, smart cities, connected health, etc. IoT devices store a massive amount of data and most of this data is confidential. If this data is compromised, it will put the individual at risk. This data leakage can cause loss

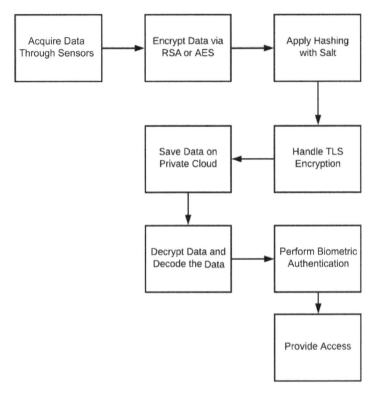

FIGURE 13.5
Flow diagram of the proposed data security Internet of Things model.

of information, loss of money, loss of reputation, or other losses. Hence, it is important to secure this data at any cost. If the data is compromised, people will stop buying such IoT applications. The current IoT technologies are facing a lot of security challenges. Recently, a Mirai botnet exploited IoT devices to steal massive amounts of data. The goal of data security in IoT devices should be to provide privacy, confidentiality, and full proof of security. All these goals are not met yet. Hence, this chapter proposed a solution to protect data within the IoT devices at all layers. It is proposed to enable firewalls, intrusion detection systems, blockchain, AI, and secure cloud servers. For cryptography, RSA should be utilized with hashing with salt. All these measures will ensure security.

References

[1] El Kafhali, S., Chahir, C., Hanini, M., & Salah, K. (2019, October). Architecture to manage internet of things data using blockchain and fog computing. In *Proceedings of the 4th International Conference on Big Data and Internet of Things* (pp. 1–8). ACM.
[2] Folk, C., Hurley, D.C., Kaplow, W.K., & Payne, J.F. (2015). The security implications of the Internet of Things. *Fairfax: AFCEA International Cyber Committee*.
[3] Lu, Y., & Da Xu, L. (2018). Internet of Things (IoT) cybersecurity research: A review of current research topics. *IEEE Internet of Things Journal*, 6(2), 2103–2115.

[4] Latif, G., Ben Brahim, G., Iskandar, D.N.F., Bashar, A., & Alghazo, J. (2022). Glioma tumors' classification using deep-neural-network-based features with SVM classifier. *Diagnostics*, 12(4), 1018.

[5] Bashar, A., Latif, G., Ben Brahim, G., Mohammad, N., & Alghazo, J. (2021). COVID-19 pneumonia detection using optimized deep learning techniques. *Diagnostics*, 11(11), 1972.

[6] Latif, G., Alghazo, J.M., Maheswar, R., Sampathkumar, A., & Sountharrajan, S. (2020). IoT in the field of the future digital oil fields and smart wells. In *Internet of Things in Smart Technologies for Sustainable Urban Development* (pp. 1–17). Springer.

[7] Latif, G., Bouchard, K., Maitre, J., Back, A., & Bédard, L.P. (2022). Deep-learning-based automatic mineral grain segmentation and recognition. *Minerals*, 12(4), 455.

[8] Mahmoud, A.A., Alawadh, I.N.A., Latif, G., & Alghazo, J. (2020, April). Smart nursery for smart cities: Infant sound classification based on novel features and support vector classifier. In *2020 7th International Conference on Electrical and Electronics Engineering (ICEEE)* (pp. 47–52). IEEE.

[9] Latif, G., Khan, A.H., Butt, M.M., & Butt, O. (2017). IoT based real-time voice analysis and smart monitoring system for disabled people. *Asia Pacific Journal of Contemporary Education and Communication Technology*, 3(2), 227–234.

[10] Hossain, M.S., Chisty, N.M.A., Hargrove, D.L., & Amin, R. (2021). Role of Internet of Things (IoT) in retail business and enabling smart retailing experiences. *Asian Business Review*, 11(2), 75–80.

[11] Latif, G., Alghazo, J.M., Maheswar, R., Jayarajan, P., & Sampathkumar, A. (2020). Internet of Things: Reformation of garment stores and retail shop business process. In *Integration of WSN and IoT for Smart Cities* (pp. 115–128). Springer.

[12] Espinoza, H., Kling, G., McGroarty, F., O'Mahony, M., & Ziouvelou, X. (2020). Estimating the impact of the Internet of Things on productivity in Europe. *Heliyon*, 6(5), e03935.

[13] Yadav, P.K., Sharma, S., & Singh, A. (2019, September). Big Data and cloud computing: An emerging perspective and future trends. In *2019 International Conference on Issues and Challenges in Intelligent Computing Techniques (ICICT)* (Vol. 1, pp. 1–4). IEEE.

[14] Shah, S.H., & Yaqoob, I. (2016, August). A survey: Internet of Things (IOT) technologies, applications and challenges. In *2016 IEEE Smart Energy Grid Engineering (SEGE)* (pp. 381–385). IEEE.

[15] Ali, B., & Awad, A.I. (2018). Cyber and physical security vulnerability assessment for IoT-based smart homes. *Sensors*, 18(3), 817.

[16] Chakrabortty, S., Aich, S., & Kim, H. (2019). A secure healthcare system design framework using blockchain technology. *21st International Conference on Advanced Communication Technology (ICACT), 2019*, 02 May 2019, pp. 260–264.

[17] Alani, M.M. (2018, December). IoT lotto: Utilizing IoT devices in brute-force attacks. In *Proceedings of the 6th International Conference on Information Technology: IoT and Smart City* (pp. 140–144). ACM.

[18] Díaz López, D., Blanco Uribe, M., Santiago Cely, C., et al. (2018). Shielding IoT against cyber-attacks: An event-based approach using SIEM. *Wireless Communications and Mobile Computing*, 2018.

[19] Zolanvari, M., & Jain, R. (2015). IoT security: A survey. *Computer Scientists & Computer Engineers at WashU*.

[20] Oracevic, A., Dilek, S., & Ozdemir, S. (2017, May). Security in Internet of Things: A survey. In *2017 International Symposium on Networks, Computers and Communications (ISNCC)* (pp. 1–6). IEEE.

[21] Minoli, D., & Occhiogrosso, B. (2018). Blockchain mechanisms for IoT security. *Internet of Things*, 1, 1–13.

[22] Shafagh, H., Burkhalter, L., Hithnawi, A., & Duquennoy, S. (2017, November). Towards blockchain-based auditable storage and sharing of IoT data. In *Proceedings of the 2017 Cloud Computing Security Workshop* (pp. 45–50). ACM.

[23] Bertino, E., & Islam, N. (2017). Botnets and Internet of Things security. *Computer*, 50(2), 76–79.

[24] Kuusijärvi, J., Savola, R., Savolainen, P., & Evesti, A. (2016, December). Mitigating IoT security threats with a trusted Network element. In *2016 11th International Conference for Internet Technology and Secured Transactions (ICITST)* (pp. 260–265). IEEE.

[25] Routray, S.K., Jha, M.K., Sharma, L., Nyamangoudar, R., Javali, A., & Sarkar, S. (2017, May). Quantum cryptography for IoT: A perspective. In *2017 International Conference on IoT and Application (ICIOT)* (pp. 1–4). IEEE.

[26] Kshetri, N. (2017). Can blockchain strengthen the Internet of Things? *IT professional*, 19(4), 68–72.

[27] Adat, V., & Gupta, B.B. (2018). Security in Internet of Things: Issues, challenges, taxonomy, and architecture. *Telecommunication Systems*, 67(3), 423–441.

[28] Vijaithaa, R., & Padmavathi, G. (2017). A study on device oriented security challenges in Internet of Things (IoT). *International Journal of Advanced Networking and Applications*, 8(5), 3224–3231.

[29] Razouk, W., Sgandurra, D., & Sakurai, K. (2017, October). A new security middleware architecture based on fog computing and cloud to support IoT constrained devices. In *Proceedings of the 1st International Conference on Internet of Things and Machine Learning* (pp. 1–8). ACM.

[30] Maheshwari, N., & Dagale, H. (2018, January). Secure communication and firewall architecture for IoT applications. In *2018 10th International Conference on Communication Systems & Networks (COMSNETS)* (pp. 328–335). IEEE.

[31] Singanamalla, S., Jang, E.H.B., Anderson, R., Kohno, T., & Heimerl, K. (2020, October). Accept the risk and continue: Measuring the long tail of government https adoption. In Proceedings of the ACM Internet Measurement Conference (pp. 577–597). ACM.

[32] Kandasamy, K., Srinivas, S., Achuthan, K., & Rangan, V.P. (2020). IoT cyber risk: A holistic analysis of cyber risk assessment frameworks, risk vectors, and risk ranking process. *EURASIP Journal on Information Security*, 2020(1), 1–18.

[33] Sagu, A., & Gill, N.S. (2020). Machine learning techniques for securing IoT environment. *International Journal of Innovative Technology and Exploring Engineering (IJITEE)*, 2278–3075.

[34] Díaz López, D., Blanco Uribe, M., Santiago Cely, C., et al. (2018). Shielding IoT against cyber-attacks: An event-based approach using SIEM. *Wireless Communications and Mobile Computing*, 2018.

[35] Angelova, N., Kiryakova, G., & Yordanova, L. (2017). The great impact of Internet of Things on business. *Trakia Journal of Sciences*, 15(1), 406–412.

[36] Mahmoud, R., Yousuf, T., Aloul, F., & Zualkernan, I. (2015, December). Internet of Things (IoT) security: Current status, challenges and prospective measures. In *2015 10th International Conference for Internet Technology and Secured Transactions (ICITST)* (pp. 336–341). IEEE.

[37] Lu, Y., & Da Xu, L. (2018). Internet of Things (IoT) cybersecurity research: A review of current research topics. *IEEE Internet of Things Journal*, 6(2), 2103–2115.

[38] Sengupta, J., Ruj, S., & Bit, S.D. (2020). A comprehensive survey on attacks, security issues and blockchain solutions for IoT and IIoT. *Journal of Network and Computer Applications*, 149, 102481.

[39] Gehrmann, C., & Gunnarsson, M. (2019, December). An identity privacy preserving IoT data protection scheme for cloud based analytics. In *2019 IEEE International Conference on Big Data* (pp. 5744–5753). IEEE.

[40] Riad, K., & Ke, L. (2018). Secure storage and retrieval of IoT data based on private information retrieval. *Wireless Communications and Mobile Computing*, 2018.

14
Wireless Sensor Network Security

Ciro Rodríguez and Isabel Moscol

CONTENTS
14.1 Introduction	198
14.2 Security in Wireless Sensor Networks	199
14.2.1 Primary Security Objectives	199
14.2.1.1 Confidentiality	199
14.2.1.2 Authentication	199
14.2.1.3 Integrity	199
14.2.1.4 Availability	199
14.2.2 Secondary Security Objectives	200
14.2.2.1 Data Freshness	200
14.2.2.2 Traceability	200
14.2.2.3 Secure Location	200
14.3 Security Attacks on WSNs	200
14.3.1 Based on the Attacker's Capability	200
14.3.1.1 Internal and External Attacks	200
14.3.1.2 Active and Passive Attacks	201
14.3.2 Based on the Protocol Stack	201
14.3.2.1 Physical Layer	201
14.3.2.2 Link Layer	202
14.3.2.3 Network Layer	202
14.3.2.4 Transport Layer	204
14.3.2.5 Application Layer	204
14.4 WSN Monitoring Systems	205
14.4.1 Indoor Environmental Monitoring	206
14.4.2 Outdoor Environmental Monitoring	206
14.4.3 Monitoring in Agriculture	206
14.4.4 Health Monitoring	206
14.5 Security Mechanisms for Sensor Networks	207
14.6 Wireless Security Protocol	207
14.6.1 WEP	207
14.6.2 WPA	207
14.6.3 WPA2	208
14.6.4 WPA3	208
14.7 Attacks against TCP/IP Networks	208
14.7.1 Network Listeners	208
14.7.2 IP Fragmentation	208
14.7.3 Denial of Service Attacks	208

DOI: 10.1201/9781003326205-17

14.7.4 Footprinting 209
 14.7.5 Fingerprinting 209
14.8 TCP/IP Network Security 209
 14.8.1 Generic Vulnerabilities 209
 14.8.2 Network Security Scanners 210
 14.8.3 VPN 210
 14.8.4 Firewall 210
14.9 Existing Solutions 211
 14.9.1 Data Encryption Standard (DES) 211
 14.9.2 Node-level Security Solution 211
14.10 Conclusions 211
References 212

14.1 Introduction

The current wireless communication systems have been experiencing notoriously revolutionary progress in recent years. The use of wireless networks is nowadays an alternative for organizations of all kinds to be competitive. Multiple areas, from market solutions companies to academic research centers, have benefited from the changes resulting from 5G, which promotes virtualization, software architectures composed of networks, multiple input multiple output (MIMO) systems, new frequency bands, among others. Streamlining and automating processes save computational cost, labor, materials, and pre-operational and operational time. Wireless networks are an essential pillar in the transformation to Industry 4.0, providing profitability and sustainability to businesses.

In wireless communication sender and receiver are not linked by a physical propagation medium; it instead implements the regulation of electromagnetic waves in space. Therefore, physical equipment is only required in the transmitters and receivers of signals, among which are antennas, laptops, PDAs, cell phones, etc.

Wireless sensor networks have become a crucial tool ranging from military to civilian applications and everyday data usage techniques. Due to the importance of these networks in large sectors such as industry, it is equally important to employ security mechanisms that provide data integrity and authenticity. In recent times, multiple security techniques have been applied that are not only approved and verified but are also used to date in different branches of applications that make use of these wireless sensor networks.

An important resource in sensor security is the use of directional antennas, which arises from the need to save energy consumption, increase the quality of transmissions, or reduce the number of jumps due to the large transmission coverage.

Although security is of high importance, it is arduous to carry out methods that provide the proper security. Various methods that are still in an initial stage provide results that are encouraging with regard to the integrity of the information in the sensor nodes. After this encouraging look at the development of new techniques, we must look to a new horizon of quality and data integrity and expand strategies to reduce the risk of sensor nodes that, as well as generating the need to own and use them, must provide security methods.

14.2 Security in Wireless Sensor Networks

WSNs are a special type of network whose security services must protect the information that is transmitted through the network, since in the environments in which it is used there are data breaches, audio leaks, and information theft due to insecure networks. Security services in a WSN must protect the information through the network, the resources, and the behavior of the nodes. It should be noted that within the security objectives there are the primary ones, known as standard security objectives, such as confidentiality, integrity, authentication, and availability (CIAA). Additionally, the secondary objectives of data freshness, self-organization, time synchronization and secure location should be considered.

14.2.1 Primary Security Objectives

14.2.1.1 Confidentiality

It should be guaranteed that sensitive data is accessed and viewed only by those who are authorized to see it. Security mechanisms are responsible for ensuring that a message on the network is not understood by someone other than the receiver of the message. However, the confidentiality of the data is breached when a node replication attack is launched, as the initially compromised nodes are cloned by completely replicating their characteristics. The cloned nodes may have all the data containing trade secrets, secret government information, private medical information, or any kind of confidential information.

14.2.1.2 Authentication

This ensures the identity of the nodes, i.e., validates who is communicating the information. An attacker can inject a flow of false packets, so it is very important that the receiver has a verification mechanism to ensure that the information is received from an authentic source.

During the transmission process, in case of a node replication attack—the creation of apparently legitimate cloned nodes—the secret credentials of the captured node increase the complexity of authenticity verification. However, this aspect is covered under the verification of the issuing source of the information, with blocks avoiding possible damage to the network.

14.2.1.3 Integrity

Complementing the authentication, only authorized persons should have the permissions to update, delete, create, or extract information from the nodes. In addition to the abovementioned [1], integrity guarantees that the data is kept intact during communication. It ensures that the sent message is received unadulterated, not accidentally or maliciously altered, or destroyed.

14.2.1.4 Availability

Availability guarantees access and usability of information on demand. High availability of information must be maintained through mechanisms that ensure the service even under attack.

In case of a node replication attack, an attacker can compromise the availability of the WSN by releasing a denial-of-service (DoS) attack, which can severely obstruct a network's processing. By interfering with genuine signals, the availability to authorized parties is also affected.

14.2.2 Secondary Security Objectives

14.2.2.1 Data Freshness

Proper latency and updating are necessary for the proper functioning of a sensor system. This requirement is especially significant when nodes use shared keys since a potential attacker can perform a replay attack from old messages.

Sensor nodes operate autonomously, for instance their local clocks are not synchronized with each other; this can lead to uncertainty or even ambiguity in the detected data. For time-critical applications, time-accurate operation is mandatory. However, in some cases energy efficiency takes on greater importance than data freshness, so dependence on this the synchronization algorithm should be implemented.

14.2.2.2 Traceability

Audit information must be stored and protected so that actions that affect the security of the system can be traced back to the responsible party. Also, information must be irrefutable and verifiable by a third party so that a sender or receiver cannot deny having transmitted or received a message.

14.2.2.3 Secure Location

Usefulness of a WSN depends on its ability to locate each sensor automatically and accurately in the network. A network designed to locate errors will need accurate location information. However, it is difficult to locate and manage them when nodes are deployed in a random fashion and in the absence of a supporting infrastructure [2]. Solving the localization problem consists of knowledge about the physical location of the deployed nodes, through discovery algorithms, that must be distributed, accurate, robust, and scalable.

14.3 Security Attacks on WSNs

14.3.1 Based on the Attacker's Capability

14.3.1.1 Internal and External Attacks

Internal attacks occur when genuine WSN nodes perform in an unauthorized or unintentional manner, while external attacks come from non-WSN nodes. In the first case, internal nodes can be affected by the execution of malicious code because attackers have obtained keys, code, and data from legitimate nodes. The attacker can use nodes whose characteristics are like those belonging to the network and can amplify the damage by using devices such as laptops. If using individual nodes, the attack range would be limited to their interaction with adjacent or nearby nodes, while using more powerful devices the entire network can be damaged. Laptop attacks have greater computational capacity and a wider transmission range with better processing capacity and more energy, causing greater impact on the WSN.

14.3.1.2 Active and Passive Attacks

An active attack alters the data stream or creates a fake data stream. It includes:

- Sensor network routing attacks, which act on the network layer.
- DoS attacks, which cause failure of certain network nodes.
- Node subversion attacks, which capture and reveal node information, such as cryptographic keys.
- Fake node attacks with the intention of injecting false and/or malicious information into the network.

Passive attacks include eavesdropping and monitoring by unauthorized agents on packets within the WSN. They mainly involve eavesdropping and monitoring of packets within the network, data traffic analysis, and node infiltration.

In a WSN all traffic is directed to a base station, so communications follow a many-to-one or many-to-few pattern. It could provide information about the network topology and the location of the base station and other strategic nodes by means of traffic patterns.

14.3.2 Based on the Protocol Stack

14.3.2.1 Physical Layer

This layer is responsible for signal detection, data modularization, and encoding. Some invasive attacks allow the adversary to get complete control over a sensor node by physically accessing it, which enables the extraction of cryptographic primitives and unrestricted access to data stored in memory, destruction of the node, modification of a program code, or even replacing it with a malicious node to cause substantial damage to the entire system. Non-invasive attacks, on the other hand, rely on data gathered from the physical cryptosystem, as opposed to weaknesses in algorithms.

Jamming

This is a type of denial-of-service attack, which occurs due to an inherent weakness in the physical layer that directly affects the transmission medium and can be either intentional or accidental. This attack consists of the transmission of radio signals that interfere with the radio frequency that the sensor network is using. In industrial environments, the possibility of interference due to machinery and other devices is very high. Therefore, it is important to implement a mechanism that guarantees a noise-free environment. On the other hand, the intentional emission of interference can cause the batteries of the nodes to run down due to the attempted retransmission of the frames.

Tampering

This attack consists of unwanted receivers intercepting and accessing messages that correspond to another destination. Wireless communications are susceptible to this type of attack due to the broadcasting of messages. All the attacker needs to do is tune in to the appropriate frequency.

14.3.2.2 Link Layer

This is responsible for multiplexing data flows, data frame detection, medium access and error control. Insecurities can occur through radio signals or manipulation of efficiency measures within protocols, which can lead to a power drain leaving the entire network inactive [3].

Collision

It occurs when two nodes try to transmit simultaneously on the same frequency, with the consequence that when packets collide, they must be discarded. This happens because the packet data changes, causing errors in the checksum.

Depletion or Continuous Access to the Channel

Some link layer protocols try to retransmit repeatedly in the face of a collision. A malicious node can continuously interrupt the communication between two nodes by forcing the sender to retransmit continuously. This can cause indefinite postponement in channel usage and also consumes the energy of nodes that repeatedly try to retransmit.

Interrogation

This type of attack manipulates the RTS/CTS (Request To Send / Clear To Send) handshake used by several MAC (media access control) protocols to mitigate the hidden station problem. A node's resources can be fatigued by repeatedly sending RTS messages to produce CTS responses from a neighboring target node. To prevent this attack a node can restrict the number of connections it accepts from the same identity or use anti-repetition protection with strong authentication at the link layer.

Denial-of-Sleep Attack

Due to the energy constraints that exist in the sensor nodes, it is usual to periodically switch to sleep mode, to extend the network lifetime. This way of conserving energy is mainly based on MAC protocols, since one of the main reasons for energy decrease is overhearing, which means that a node receives packets directed to other nodes, having its radio transmitter on for a longer time. Most WSN MAC protocols attempt to reduce overhearing by ensuring the node is only awake when there is associated traffic.

14.3.2.3 Network Layer

Routing and data forwarding functions are vital activities of the sensor nodes, where each one acts as a WSN router. Routing protocols must be efficient with respect to energy and memory consumption without neglecting security and prevention of attacks and node failures. This implies that authorized receivers receive all the messages sent and can check their integrity and validate the identity of the sender. Some attacks related to the network layer are mentioned below.

Blackhole Attack

A malicious node discards messages from its child nodes. If this is done with all packets it can leave a destination unreachable or the network unusable [4].

Spoofed, Replayed, or Altered Attack

This type of attack seeks to store and forward old messages so that, when an outdated (or non-existent) topology is described, it causes an update with false data in the routing tables of the nodes.

Selective Forwarding

A fake node is introduced to the WSN to discard certain packets to prevent their arrival at predefined destinations. This attack becomes more damaging when malicious nodes are within the path of the WSN data flow [4].

Sinkhole Attack

This attack attempts to attract as much traffic as possible to compromised nodes. It attempts to deceive the neighboring nodes of a compromised node by making it look attractive with respect to the routing algorithm, becoming an information sink.

Wormhole Attack

This is a complex type of attack to solve, since it does not require the impersonation or infection of the network nodes. It works with external hardware such as laptops or any external network device. The attack consists of storing traffic from one region to forward it to a different region. It usually involves two distant malicious nodes sending packets to each other using another channel called wormhole.

Acknowledgement Spoofing

Its objective is to impersonate a downed link or a dead node using the ACKs (Acknowledgements) of the link layer. When the attacker hears a packet that is for a node that cannot respond, he replies with an ACK as if he were that node, thus making the victim believe that the node is alive.

Fake Node Replication

The vulnerability of a node's information can result in its replicability, which would facilitate a fake node's insertion in the network by supplanting a legitimate node by having the same identifier, as well as information on the keys that allow it to authenticate itself as a member of the network.

Sybil Attack

A malicious node illegitimately takes multiple identities. In many cases, nodes in a WSN need to work together to perform a task, therefore, subtasks are distributed among the nodes [4]. In this situation a node can simulate multiple nodes using other identities and thus harm the network, for example by advertising a false identity or by taking the identity of other legitimate nodes, invalidating the information of the legitimate nodes and modifying the routing information.

Homing

Using traffic pattern analysis, it is possible to identify nodes with special responsibilities, for example the head of a cluster or the nodes in charge of managing cryptographic keys. Then other attacks can be performed such as jamming or physical access to destroy them. To prevent this, encryption is generally used at the head.

Jellyfish Attack

The jellyfish attack delays packets for a period of time before forwarding them. It can also change the order in which these packets are received and sent randomly, instead of in a FIFO (first in–first out) order of packets received. This affects the flow control mechanisms' ability to perform a reliable transmission. This attack causes a significant increase in end-to-end latency, degrading the quality of service and affecting real-time applications.

Rushing Attack

In this attack, duplicate packets are multicast at points where there is more than one path to follow. This system conserves network bandwidth by sending a single data stream to multiple receivers. An attacker can exploit the duplicate suppression mechanism by quickly forwarding route discovery packets to gain access within a route.

Blackmailing

Attacking nodes accuse an innocent node of being a malicious node. This attack can be effective against distributed protocols that establish a list of good and bad nodes based on an evaluation of the nodes participating in the network. That is, a vote is taken where the majority decides whether a node is malicious and therefore should be excluded from the network.

Neighbor Attack

Upon receiving a packet, the receiving node writes its identifier before forwarding it to a neighboring node. If an attacker forwards a packet without writing its identifier, it would make two nodes outside the same communication range believe that they are neighbor nodes, resulting in a false route.

14.3.2.4 Transport Layer

Security problems in the transport layer are mostly due to flaws in the protocols, therefore, an efficient design of transport layer protocols can prevent threats. Some of the attack systems are mentioned below.

Flooding

The malware repeatedly generates new connections until the resources required by each connection are exhausted or until the system's maximum connection limit is reached.

Desynchronization Attack

In this attack, messages are repeatedly sent to one or both nodes within a communication, forcing them to request retransmissions of the lost frames. As a result, these messages are retransmitted. This causes the nodes to be unable to exchange information, using up their energy reserves.

14.3.2.5 Application Layer

Overwhelm Attack

An attempt is made to overwhelm the WSN nodes with sensor stimuli by generating a high volume of base station traffic sent over the network. The purpose of the attack is to consume network bandwidth and energy resources of the nodes.

Reprogramming Attack

Network scheduling systems allow nodes to be reprogrammed remotely. If this reprogramming process is not secure an attacker can use it to take control of a large part of the network.

Attacks on Clock Synchronization Protocols

Clock synchronization protocols provide mechanisms to synchronize the local clocks of each node within the sensor network. This synchronization is very important within a set of applications and services within a sensor network. They are also used by certain mechanisms to detect and prevent some attacks, such as wormhole. These attacks can be performed by physically capturing some nodes and injecting fake clock synchronization update messages.

Software Attacks

The attack attempts to manipulate the code or to take advantage of its vulnerabilities. A recurring example of software attacks is buffer overflow, which occurs when a process tries to store data that exceeds the fixed length of the buffer, resulting in the additional data overwriting adjacent memory locations.

14.4 WSN Monitoring Systems

MEMS (micro electronic mechanical systems) technology is progressing dramatically today along with wireless communication. In localization, the data collected from the sensor node is useless if the location is not accurate. To fix this localization problem, one could theoretically equip each sensor node with GPS, which is practically impossible due to increased cost and power consumption.

The sensor node of the self-localization method uses the localization protocol to estimate the position, where the Beacon nodes are responsible for estimating the location. Most of the distance and angle approximation techniques used to link two sensor nodes are time difference (TDOA), time (TOA), and angle (AOA) of arrival, as well as received signal strength indicator (RSSI) and hop count between the two nodes. The position of the unknown node is estimated based on the distance and angle and the positions of the reference nodes. Commonly used localization techniques are triangulation, bounding box, probabilistic approach, and fingerprinting.

WSNs can collect data from nodes in three different modes: event capture, periodic collection, and on-demand reporting. In the event capture reporting mode, sensor nodes only detect and report data if an event occurs in their environment. For example, if a WSN is deployed in a forest to monitor a forest fire, the sensor nodes will only report to the base station if a fire occurs in the forest. Periodic reports are used when data is not urgent or needs to be received at regular intervals, e.g., an electrocardiogram monitoring a patient. The sensed data report is submitted to the base station every predefined interval. In on-demand capture, requests for information are sent to sensor nodes to send data from their sensors. For example, an application that monitors chemical contaminants in water can send a request to the sensor nodes to detect the level of contaminants and report the values to the base station when required.

14.4.1 Indoor Environmental Monitoring

The ability to use sensors to measure temperature, lighting, door and window status, air flow, and air pollution can be exploited for optimal control of the enclosed environment. Sensors can be deployed to help use radiators, fans, and other devices wisely to avoid unnecessary costs, for example heating or cooling a room unnecessarily. Other applications where sensors can be used are for fire and earthquake mitigation. Today, smoke detectors are common in buildings and can trigger alarms. On the other hand, WSN can be integrated with other components such as the Geographic Information System (GIS) that allows the obtaining of more information about the real situation.

14.4.2 Outdoor Environmental Monitoring

Another application area of WSNs is outdoor environmental monitoring, for example, for animal tracking [5]: an application with a 150-node WSN is deployed to study the nesting of a particular species of birds. In other cases, WSNs were used to study wolves. A node is placed on each animal to collect information about its living conditions and behavior. Similarly, in reserves in Kenya, WSNs have been deployed to study zebras. Their goal is to place sensors on different terrestrial animals to understand their interactions and influences, and to understand wildlife migration patterns, and how they are affected by climate variability and human influences. Other applications related to environmental monitoring that have been developed are applications for observing the environment, studying natural phenomena, and forecasting weather. For example, there is the ALERT (Automated Local Evaluation in Real Time) system developed by the National Weather Service, a precursor to modern sensor-based monitoring networks for predicting rainfall and flooding in California and Arizona.

14.4.3 Monitoring in Agriculture

WSNs plays a crucial role in crop monitoring. Its benefits are seen in factors such as labor efficiency, crop growth rates, reduced costs, reduced environmental impact, and increased product quality. WSNs are involved in monitoring fields, vineyards, and orchards, helping farmers prevent damage and increase yields. Automation in agriculture has made a fundamental contribution, giving rise to what is now known as precision agriculture. This is the technique of using the elements (water, fertilizers, pesticides, etc.) in the right place at the right time to increase production and improve the quality of the final product, while protecting the environment. A practical case would be the use of WSN to monitor relevant parameters (soil moisture and air temperature) and transmit them wirelessly to the location of the farmer to implement the appropriate measures, as in an irrigation system where the correct amount of water is supplied according to the data obtained.

14.4.4 Health Monitoring

Healthcare and scientific research systems can also benefit from WSNs, as they provide more accurate, detailed, and real-time information for decision-making. One of the specific applications in the medical field is called BAN (Body Area Network). The BAN consists of a set of sensors that collect information about the patient and a device that acts as a base station, such as a cell phone, which sends the information collected to information analysis servers to look for possible anomalies and alert situations to which you should react. The information collected by these devices enables physicians to make more accurate diagnoses and provide better treatment. It includes the detection of physiological variables that are analog signals corresponding to the physiological activities of the human body or the body's actions, such as electrocardiogram, electroencephalogram, electromyography, blood glucose, blood pressure, accelerometer, etc. All this information can be transmitted, processed, analyzed, and stored in the patient's medical record. In case of emergency, alarm messages can be sent to the appropriate person.

14.5 Security Mechanisms for Sensor Networks

Techniques to protect WSNs prevent malicious attacks for usurpation and improper data manipulation. They also provide data privacy, legitimate identity verification, and network irregularity detection mechanisms. Some of the main defense mechanisms are described below.

WSNs are normally in continuous use, so it is necessary to apply techniques to ensure their availability. One way to achieve such a pattern is to back up the energy sources of the sensor nodes and their subunits. A battery-aware multipath construction method can be used to avoid overlapping paths to prolong the network lifetime. Availability should be high compared for critical metering systems; regardless of the accuracy of data collection, there should not be a power drain problem. Different techniques to support network availability are: providing redundancy for backups and failures, power routing protocols, efficient routing protocols, emergency power backup, and network protection.

14.6 Wireless Security Protocol

With the idea of getting rid of a series of interconnected computers with cables, wireless technology emerged, which allows us to communicate without the need to be connected to a physical medium as a data exchange channel. This gives us the advantage of dispensing with a specific fixed location to perform our communication process with other terminals, but like any technology that promises to be a great solution, it brings with it a number of risks in terms of information security, thus a series of protocols and security measures are necessary to keep these potential risks at bay.

14.6.1 WEP

Developed for wireless networks, its primary objective was to offer the same degree of security offered by wired networks, but at present WEP is considered easy to break and difficult to configure, and after many attempts to improve it, this protocol was dropped by the Wi-Fi alliance in 2004.

14.6.2 WPA

Wi-Fi Protected Access emerged to improve the security weaknesses offered by WEP, but when tested it also proved to be quite vulnerable to intrusion. The configuration was not as complex as that of WEP, but maintained a high degree of vulnerability.

14.6.3 WPA2

This protocol is an improvement on the previous WPA version. The most important improvement is the Advanced Encryption Standard (AES) that serves to encrypt sensitive information for the government or other classified information. It was considered safe enough to be used in home networks. The main weak point of this protocol is the WPA2 vulnerability which indicates that an intruder who has accessed the network can access some keys of the device that is vulnerable. This option is dangerous in companies but it was considered safe to be used in home networks since it does not represent as much risk as in a company.

14.6.4 WPA3

This is the new promise as far as security protocols are concerned. This protocol protects the networks from intruders who often breach the network by using dictionaries to guess the password. The Wi-Fi Alliance decided to implement these protocols by the end of 2019 while upgrading the WPA2 security of routers gradually. This technology will make home networks and networks in general more difficult to breach.

14.7 Attacks against TCP/IP Networks

As a result of the cold war, DARPA, the agency of advanced research projects of the United States, thinking that an attack could occur to its communications network, developed with the help of various universities a highly decentralized computer network. From these studies ARPANET was born. This network tolerated failures very well, and a little later when it was rethought with regard to different networks, the set of protocols denominated TCP/IP was born [6].

14.7.1 Network Listeners

This type of attack, called network eavesdropping, occurs on the first two layers of the TCP/IP model. It is highly efficient because it allows the attacker to access much sensitive information. This type of attack operates as follows: a series of applications called Sniffers intercept information contained in the TCP/IP packets [6].

14.7.2 IP Fragmentation

This form of attack involves breaking into the traffic through the network. The IP protocol is responsible for choosing the path of the packets when passing to the lower layer, the frames are packaged depending on the physical network that has been used, and the size of the packets that travel varies from network to network. In this way we have a variable number of maximum packets per type of network. An intruder could intercept any of these networks and insert a script in the frames that could violate the network.

14.7.3 Denial of Service Attacks

This type of attack involves not being able to access the resources of a network, so that someone who violated the network appropriates some resource or service within the network. In a more technical sense, this attack can be interpreted as the denial of services to a terminal within the network, which cannot connect to it. The problems associated with this type of attack can be the reduction of bandwidth as the usurpers take advantage of it to search for movie or music files (to give a less harmful example). Someone more malicious may have premeditated this type of violation to spoil the system. Another type of this category of attack is access to the system by falsifying the address of origin to avoid being discovered. This is a form of impersonation.

14.7.4 Footprinting

This type of attack involves a general survey of the network and a report is made of its current state. This initial information could lead subsequently to a better planned attack, such as the direct application of a script to some vulnerability found. The search may be for information that is publicly available and in the first instance does not involve any crime. The process involves the following steps: application of information scanning, analysis of vulnerabilities found, and reporting to determine the state of the system or network.

14.7.5 Fingerprinting

There are two methods of fingerprinting: active and passive. Active fingerprinting involves analyzing the server's response when TCP and UDP packets are sent to it. This method allows for more extensive and detailed information to be acquired, but it is easy to detect. Passive fingerprinting, on the other hand, captures data packets from a remote system, which requires both systems (sender and receiver) to be connected to the network. This method can access information such as the IP address of the computer [7].

When this information is used in a cyber-attack, it involves collecting personal data from a user, such as their name, account information, and passwords, in order to gain access to private accounts while the owner is away or on vacation. Even seemingly innocuous information, such as the date and time of the last login, could provide useful information to organize an identity theft.

14.8 TCP/IP Network Security

In this section we intend to cover the aspects related to security in the TCP/IP protocol group, such as the incorruptibility of the data transferred between the different layers of the TCP/IP model (Figure 14.1). At the time of their creation, networks had few security protocols, since the primary intention, rather than repelling attacks, was to interconnect the equipment within the network [3].

14.8.1 Generic Vulnerabilities

To ensure the security of networks, it is important to take into account the threats that may arise, so we shall catalog the main threats that concern TCP/IP protocol family networks.

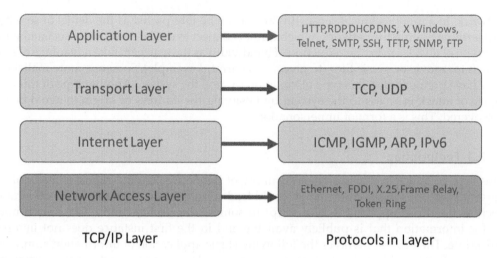

FIGURE 14.1
TCP/IP model.

14.8.2 Network Security Scanners

A network security scanner is a software tool that probes an entire network and looks for possible vulnerabilities that may arise in the network. This solution evaluates the current state of the network and each node and displays a report of the degree of security of the network. The administrator in charge of the network makes use of these scanning tools. The main elements subject to the scanning process are firewalls, routers, client computers, servers. The main vulnerabilities found are open ports, attempted vulnerabilities in passwords, OS control. In this way the tool provides a final report containing the degree of vulnerability of the network, associated vulnerabilities, and IT assets.

14.8.3 VPN

A VPN is a LAN network that can be accessed from the Internet. Network administrators are responsible for monitoring and implementing rules to control access to properly authenticated users. This network protocol is widely used by companies seeking to make available the local resources of a company through the Internet, but with the proper security that this protocol implies.

14.8.4 Firewall

This protection technique is a protocol that controls the network traffic. The usual way to explain the behavior of a firewall is stopping or allowing the network traffic. In this way the network information is protected from information that comes from external networks or the internet. As summarized points for this protection we have the following: the firewall must only allow the data defined in the protocol, all traffic must pass through the firewall, and the firewall must be the most protected part of the network.

14.9 Existing Solutions

14.9.1 Data Encryption Standard (DES)

DES is a low-level encryption technique, which transforms unformatted information into 64-bit data blocks with a 56-bit key. The encryption process operates in the following three stages:

- Permutation in 64 bits to produce the permutation input.
- Iteration of functions with random keys in 16 rounds.
- Finally, it is permuted in reverse to generate the desired cipher block.

On the other hand, decryption consists of reversing the subkeys. It is worth mentioning that this technique would not provide optimal results for large-scale applications. Nevertheless, it is a feasible and cost-effective solution for medium-scale applications [3]. Thus, the Advanced Encryption Standard (AES) emerged, which works with higher bit encoding, from 128 to 256 bits.

14.9.2 Node-level Security Solution

This implements data validation through a sampling mechanism at the base station. However, a disadvantage is that it does not identify the root cause. Therefore, a second approach emerges that deals with application-specific situations, where each node is in charge of monitoring its neighboring nodes. Upon any unusual or abnormal activity, an alert is sent to the respective neighboring nodes or to the sinkhole.

14.10 Conclusions

The demand for promising applications in sensor networks is increasing. It will be a trend in the coming years and their use will become more and more widespread, and one of the reasons is the low cost of installation compared to wired sensors. In addition to this we can add the advances in sensor technology. Better security techniques, privacy, power, computing capacity, and scalability are needed.

Energy harvesting is of great importance because it reduces the cost of new batteries, and reduces downtime and alerts about the condition of the environment, thus promoting the saving of economic resources, increasing energy efficiency, and improving system reliability. In the healthcare area, the quality of life of patients is being improved by using tiny wireless sensors that can monitor vital signs, environmental sensors, and location sensors that can be integrated into a WWBAN (Wearable Wireless Body Area Network). Biomedicine is providing new solutions with the use of WSN technology, reducing medical errors, increasing the quality of medical care, improving the efficiency of caregivers and the comfort of patients. In the area of building maintenance, wireless sensors also play an important role by helping in the monitoring and proper functioning of automated systems, and once possible current and future problems are detected, they can provide automated maintenance. Wireless sensors improve agriculture by making it more accurate, thus saving time and cost in production.

A versatile WSN architecture must consider quality of service, attacks, and encryption algorithms. The integration of software-defined networks into WSNs is reshaping the entire architecture of legacy WSNs. Attention also needs to be paid to the benefits and challenges of cloud services and virtualization technology. There is also a need to develop secure routing algorithms that provide not only coverage but also fault tolerance and preemptive security, both in data aggregation algorithms and data flow privacy.

Another area of WSN security seeks to increase the capacity of nodes to withstand physical attacks by controlling security infrastructure processes through resource optimization (computational consumption and time), agile detection and response to DoS attacks, and increasing the social privacy of sensor networks. There are underdeveloped areas where security indicators could be improved, such as the protection of sensor networks with mobile nodes or multiple base stations, or the measurement of trust between nodes.

References

[1] T. Bala, V. Bhatia, S. Kumawat, and V. Jaglan (2018). "A survey: Issues and challenges in wireless sensor network," *IJET*, vol. 7, no. 2, 53–55.

[2] Y.A. Bangash, Q. Abid, A. Ali, and Y. Alsalhi (2017). "Security issues and challenges in wireless sensor networks: A survey," *International Journal of Computer Science*, vol. 44, no. 2, 15.

[3] H. Radhappa, L. Pan, J.X. Zheng, and S. Wen (2018). "Practical overview of security issues in wireless sensor network applications," *International Journal of Computers and Applications*, vol. 40, no. 4, 202–213.

[4] G. Oreku and T. Pazynyuk, *Security in Wireless Sensor Networks*, vol. 1. London: Springer, 2016.

[5] H.C. Chaudhari and L.U. Kadam (2011). "Wireless sensor networks: Security, attacks and challenges," *Wireless Sensor Networks*, vol. 1, no. 1, 4–16.

[6] J. Joancomartí, J. García, and X. Perramón (2011) *Advanced Aspects of Network Security*. Barcelona: Editorial UOC.

[7] C. Avenía (2017). *Fundamentos de seguridad informática*, 1st ed. Bogotá: Areandino.

Part IV

Advanced Wireless Sensor Networks: Applications, Opportunities, Challenges, and Simulation Results

15

Advanced Wireless Sensor Networks: Applications and Challenges

Vandana Roy, Shyam S. Gupta, and Binod Kumar Soni

CONTENTS

- 15.1 Introduction 216
- 15.2 WSNs' Possibilities for Use 216
  - 15.2.1 Applications for the Military 216
  - 15.2.2 Monitoring the Health of the Structure 216
  - 15.2.3 Monitoring of the Environment 217
  - 15.2.4 Monitoring of Medical Care 217
  - 15.2.5 At-Home Apps 217
  - 15.2.6 Commercial Uses 217
- 15.3 Various Kinds of Mobile Networks 218
  - 15.3.1 WPAN 218
  - 15.3.2 WLAN 218
  - 15.3.3 WMAN 218
  - 15.3.4 WWAN 219
  - 15.3.5 WGAN 219
- 15.4 Classification of Wireless Sensor Networks 219
- 15.5 The IEEE 802.15.4 Expertise 219
- 15.6 Design Challenges in WSNs 220
  - 15.6.1 Scalability 221
  - 15.6.2 Culpability Tolerance 221
  - 15.6.3 Cost of Production 221
  - 15.6.4 Hardware Limitations 221
  - 15.6.5 The Transmission Media 222
  - 15.6.6 Energy 222
  - 15.6.7 Harsh Environment Conditions 222
  - 15.6.8 Self-Management 222
  - 15.6.9 Heterogeneity 222
  - 15.6.10 Redundant Data 222
  - 15.6.11 Event-Driven Challenge 223
  - 15.6.12 Quality of Service (QoS) 223
  - 15.6.13 Deployment 223
  - 15.6.14 Localization 224
  - 15.6.15 The Consumption of Power 224
- 15.7 Conclusion 224
- 15.8 Conclusion and Future Scope 224

DOI: 10.1201/9781003326205-19

Acknowledgments 225
References 225

15.1 Introduction

There has been a tremendous growth in wireless networking technology in recent years. The integration of sensors with the environment, machinery, and humans, as well as the efficient transmission of sensed information, has the potential to produce enormous societal advantages. Natural resource conservation, enhanced disaster response, increased industry productivity, and enhanced homeland security are just a few of the potential advantages. A sensor network's relevance lies in its low power feasting, squat cost, adequate intellect for signal treating, self-establishing capabilities, and data collection and enquiring aptitude [1, 2].

Many layers of WSN research have been active, including component, system, and application levels. An individual sensor's sensing, computing, and communication capabilities are the focus of component-level research. The routing method for interacting and organizing sensors in ascendable and energy-effectual ways is the subject of investigation at the system level. Research at the presentation level is concerned with the treating of sensor-generated data, based on the application's goal. We hope to lay the groundwork in this chapter for the creation of routing protocols for any application that requires the deployment of large sensors with limited power. As a result, the sensor network must be created in a generic, interoperable, and modular manner. As a result of this technique, any novel solicitation on top of the existing network can be reinforced [3].

15.2 WSNs' Possibilities for Use

Currently, wireless sensor networks are applied in a wide range of presentations, from medicinal to army to domestic and commercial. Researchers increasingly relying on dependable sensor networks will find this a useful resource. They can be used in a wide range of applications, some of which will be mentioned here to demonstrate how wide this application area is [4].

15.2.1 Applications for the Military

Military command, control, communication, and intelligence systems are increasingly relying on wireless sensor networks (WSNs). More and more military systems are relying on sensor networks because they are easier to set up and operate because of their self-configuration, fault tolerance, and unattended operation [5].

15.2.2 Monitoring the Health of the Structure

Maintenance costs can be reduced and failures can be avoided by sensors integrated in machines, which allow for condition-based maintenance and inspection at consistent intervals when the sensors designate that there may be a concern [6].

15.2.3 Monitoring of the Environment

Using sensor networks in this way was one of the first uses of the technology. It is applied to observe the health of animals or to do precise farming in the wild. It can also be used to detect natural or man-made calamities in a designated area. Sensors can be placed across a forest to detect fires or floods, for example [7].

15.2.4 Monitoring of Medical Care

Patients and the elderly can be monitored and tracked using WSNs in the healthcare industry. They provide interfaces for the diagnosis, management, and observing of humanoid biological data such as BP, heart speed, and oxygen extent.

15.2.5 At-Home Apps

WSNs can be utilized to create an intellectually alive surrounding for humans that is more convenient and efficient. Domestic appliances (washing machine, refrigerator, VCR) can be equipped with sensors that communicate with one other and with an internet-connected home network[8].

15.2.6 Commercial Uses

It may be used to monitor the wear and tear of materials in smart office spaces, to make virtual keyboards, and to monitor automobile thefts. It can also be used to manage robots in an automated manufacturing setting.

The bandwidth and distance between connecting nodes offered by wireless machines vary dramatically in numerous dimensions. The EMFs they emit and the amount of electricity they use are also crucial to mobile nodes [9-11]. Authors have listed the four major wireless technologies as follows: 3G cellular wireless, Bluetooth (802.15.1), Wi-MAX (802.16) and Wi-Fi (extratechnically and commonly recognized as 802.11).

Current WSN links tend to be asymmetric, which means that both ends are often nodes in different categories [12-14]. It is not uncommon for a single endpoint to be designated the "base-station" (BS), which is often stationary but has a high-bandwidth connection to other networks. Since a "client node" is typically transportable, the nodule at the other termination of the association uses its connection to the BS to communicate through additional protuberances.

Since their inception, wireless sensor networks have developed tremendously in scope and are now used in a wide range of fields, including ecological disciplines and medical investigation; communications; updation and improvement; agronomy; investigation; and army facilities. Some researchers have noted how tough and stimulating WSN development may be despite the network's numerous advantages[15-18]. WSNs are now deployed using some programming approaches that focus mostly on low-level (LLB) system concerns. HLB approaches have been established and certain favourable resolutions anticipated for simplifying WSN design and abstracting from technological LLB particulars.

It is anticipated that the information in this chapter, which is based on evidence and metadata, will serve as an introduction for those with an interest in WSNs to the various machineries of WSNs, as well as certain of the most important solicitations and standards of WSNs, and the various developments and challenges they present [19].

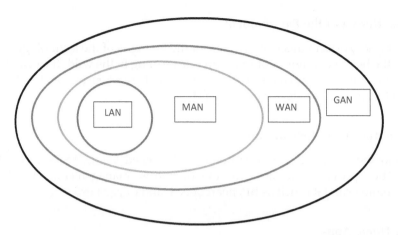

FIGURE 15.1
Types of wireless networking.

15.3 Various Kinds of Mobile Networks

Figure 15.1 shows that there are basically five types of WNs. The following is a brief overview of these.

15.3.1 WPAN

Wirelessly applied personal area networks that use low-power, short-expanse WN technologies such as Bluetooth or wireless USB are known as wireless PANs. It is possible for a wireless PAN to cover an area as large as several kilometers [20].

15.3.2 WLAN

In a limited space such as a workstation lab, home-based organization, or workplace, this wirelessly applied local area network or wireless computer network (WCN) joins two or supplementary expedients over wireless communiqué (WC) to establish a LAN. Moving from one part of the location to another within it while still connected to the WN is possible with this. Using a gateway, a wireless LAN could also afford connection to the internet. Wireless LANs based on IEEE 802.11 specifications and marketed as Wi-Fi products are the norm today. Because of the convenience with which they may be set up and used, WLANs are becoming increasingly popular in the home [21]. Businesses that provide wireless connections to their employees and customers tend to use them as well.

15.3.3 WMAN

In wirelessly applied metropolitan area networks users can communicate and connect to a variety of computer assets in a topographical section to the extent of an urban neighbourhood. The term MAN denotes the connectivity of local area networks (LANs) in a city area into the greater set-up of WMAN [22]. An urban area's several LANs can be interconnected via "point-to-point links" using the term "WMAN."

15.3.4 WWAN

Wireless wide area network is one of many variations of WNS. The larger a WAN is, the more technology has to change to accommodate it. Web sheets and live spilling video are all distributed via WNs of numerous sizes. Data is transferred through mobile networks like 2G, 3G, 4G LTE, and 5G while using wireless wide area networks (WANs), which differ from wireless local area networks (LANs) [23]. Wireless WAN, or mobile broadband, is sometimes referred to as such (MBB). A wireless service provider offers these devices, which can be found locally, regionally, or even globally (WSP). To access the network, check email, or establish a virtual private network (VPN) from within a WN's perimeter, users with CNs and wireless WAN cards can employ wireless WAN connection. It is possible to integrate wireless WAN capabilities with the help of several CNs. A WWAN can also be a sequestered set-up that distances a big region. Nodes on turrets, automobiles, airplanes, and structures, for example, make up a "mesh network" (also called a MANET). A "low power, low-bit rate (LBR) wireless WAN, (LPWAN)," suggested to transfer tiny packs of information between strategies, usually battery-operated sensors, might also be used to transmit data. The wireless WANs typically use "encryption and authentication" measures to protect their connections because RMSs don't propose a materially sheltered joining pathway. As reported, some primary GSM encryption processes are flawed, and safety experts have warned that wireless WAN is not very safe [24].

15.3.5 WGAN

A few systems that are made up of a variety of consistent CNs (WANs) shield a wide geographic area referred to as a wide area network (WAN). In many ways, it is the same as the internet, which is a GAN. GANs, dissimilar LANs and WANs, shield a considerably broader area. The transmission of consumer communications from one LAN to one more is a major difficulty for any wireless GAN because GANs are used to support MTCNs across multiple wireless LANs. A broadband (BB) wireless GAN is one of the most prevalent types of GANs. Global satellite internet network (SIN) with transferable telephone terminals (BB wireless GAN) link LAN CNs to BB internet via BB [25, 26].

15.4 Classification of Wireless Sensor Networks

The wireless sensor network is broadly classified in five categories based on their application, listed below in Table 15.1.

15.5 The IEEE 802.15.4 Expertise

Short-range communication systems, such as "IEEE 802.15.4 wireless technology," are designed to provide solicitations with high output and other prospective needs in WPAN [7]. According to Buratti et al. [7], the "IEEE 802.15.4 wireless technology" has a number of advantages, including low cost, shortened intricacy, squat power intake, and shortened broadcast data degree. WSNs are one of the most common applications of this technology.

TABLE 15.1

Wireless Sensor Network Classification

Type	Terrestrial WSN (TWSN) [27]	Underwater WSN [28, 29]	Underground WSN [30, 31]	Multimedia WSN [32]	Mobile WSN [33]
No. of Nodes	Hundreds to thousands of economical nodes	Many fewer sensors than the TWSN	TWSN has a greater number of nodes.	A number of low-cost sensor nodes equipped with cameras and microphones	A considerable number of units have been deployed
Deployment	Pre-intended or ad hoc method	Underwater deployment of nodes and a vehicle. Deployment of sparse nodes	Used in caves or mines	To provide proper coverage, they are deployed in a pre-planned manner. A well-executed deployment simplifies data retrieval and compression	Possesses the ability to change and adapt to its environment
Cost	A reduced price	Extra costly compared to TWSN	More costly than TWSN in placement of apparatus and upkeep	Highly overpriced	Highly overpriced
Challenges	It is challenging to implement grid, optimum, and 3D placement models in pre-planned TWSN	Nodes are being deployed	Challenging to deploy and to collect information	Demand for bandwidth is high, and energy usage is high. Cross-layer design, compression methods, and QoS provisioning	Sensing area coverage, localization, self-configuration, and deployment

According to Buratti et al. [7], the IEEE 802.15.4 Working Group focuses on standardizing the "ISO/OSI protocol stack" [18]. The industry consortia ZigBee Alliance and 6LowPAN are two options for describing the higher layers. A few standard-defined "physical and MAC layers"-related technical aspects will be briefly addressed [3126]. In addition, various physiologies connected to higher layers will be discussed, with a focus on the "Zigbee and 6LowPan".

15.6 Design Challenges in WSNs

The deployment of sensor networks, a subset of those that are begun in WSN systems, is said to provide a number of difficulties. Over wireless lossy spots there is no substructure for the sensor nodes to communicate with each other.

Applications and Challenges 221

As a further projected difficulty, the sensor nodes' energy supply is a constrained, non-renewable natural resource. Akyildiz et al. [1, 2] have shown that the development of WSNs can only be fully reaped when the methods used to monitor and manage natural resources have been planned from the start (energy source).

A few major design problems for WSNs are as follows.

15.6.1 Scalability

The number of nodes in an SN might range from a few to many. In addition, the deployment density can be adjusted to suit your needs. The node concentration could increase to the point where a node has a large number of neighbours within its transmission range as a result of collecting data at high resolution. They should be capable to sustain and reserve enactment successfully in the SNs they are placed in [27].

15.6.2 Culpability Tolerance

SNs are used in hazardous environments on a regular basis and are vulnerable. Node failures are attributed to hardware issues or response deficiency of somatic supply. As expected, the number of node failures is substantially higher than in WNs that have been reinforced or established with infrastructure. There must be protocols that can identify these failures in the nodes immediately and that are adequately resilient to handle a comparatively great number of node disappointments while preserving and protecting the network system's functionality. For the routing protocol project, this is of special importance because it makes alternative channels for packet redirection available. However, different deployment scenarios necessitate different culpability tolerance requirements [28].

15.6.3 Cost of Production

Only if specialized SNs could be mass produced at a price that rivalled that of existing data collection methods could sensor networks compete with them in terms of information gathering. An SN should have a price point that is extremely cheap [28].

15.6.4 Hardware Limitations

In order for SNs to function, they must contain at least one detecting component (sensor), a treating component, a conducting component, and a power source. In some instances, the nodes may contain in-built devices or other expedients, such as localization prearrangement, that help with location-conscious steering. However, each supplementary feature originates at a price and upsurges the node's power ingesting and corporal scope [29]. As a result, additional capability must be constantly weighed against the need to keep costs and power consumption low.

The network's topology: even though WSNs have progressed in many ways, the networks are still constrained in terms of energy sources, processing supremacy, storing (memory), and transportation capabilities. It's clear that energy is of the utmost importance, as seen by the vast number of procedures that have been established to conserve energy and so enable a network to function. According to certain reports, one of the most important approaches that could help reduce WSN energy consumption rates is the maintenance of the topology.

15.6.5 The Transmission Media

Radio communication over the prominent ISM bands is typically used to implement communication and interaction among nodes. However, optical or infrared communication is used by some sensor networks, with the electromagnetic device having benefits of robust and efficient intrusion removal [30].

15.6.6 Energy

Sensors are crucial to observing systems because they consume a lot of energy. Sensor nodes are small, light, and only have a limited amount of power and battery life. Sensor lifespan is influenced by the amount of energy it consumes. Node operations, for example detecting and statistics gathering, besides system processes such as data transmission via multiple communiqué rules, use energy.

Replacement or recharging is not always possible due to the small size of batteries. Other energy sources exist, but they can't completely replace the requirement for energy management. Scheming and realizing numerous energy-effective hardware and software practices for WSNs is critical in managing the limited battery [31].

15.6.7 Harsh Environment Conditions

WSN sensor nodes are exposed to RF interface, trembling, extremely harsh conditions, great humidity, muck, which decrease their enactment. Because of the severe environment, sensors may malfunction and provide incorrect information .

15.6.8 Self-Management

In a WSN, great numbers of sensor nodes are typically installed in a fixed location. However, the WSN topology changes regularly due to the loss of nodes. It is also possible to reorganize the network by adding extra sensors at any time. As a result, a sensor network system must be flexible enough to adjust to shifting connectivity. The topology of the network is constantly changing while it is being operated [32]. We must ensure that once deployed, the WSN can function autonomously, allowing for network configurations, adaptations, and repairs.

15.6.9 Heterogeneity

Interconnecting nodes with diverse sensing, computation, and processing capacities creates heterogeneity in the network. When two fully different WSNs communicate with one another, heterogeneity is created. New communication and network configuration challenges can arise when dealing with heterogeneity. We need new routing protocols that use sensor node heterogeneity wisely in order to extend the life of WSNs.

15.6.10 Redundant Data

The collection of data from sensor nodes is one of the primary goals of the WSN. Sensed data is processed and sent to the base station or sink by sensor nodes. Data redundancy can occur when multiple nodes pick up on the same piece of information. There is a greater danger of receiving multiple data packets at the sink or base station if sensor nodes are distributed in a dense manner. There are times when nodes in close proximity to one another sense and send identical packets. In order to avoid wasting energy or resources, this data should not be sent to the base or sink node. In addition, it is possible that the most critical data is overlooked [32].

WSN places a high value on data freshness, as well as the accuracy and integrity of the information it transmits. Real-time operations are required by a number of WSN applications, and data must be able to sink within a reasonable amount of time. Previous round data should not be incorporated into the current one [19]. Maintaining the freshness of data is what this means. When a network isn't working properly, data can get stale and be sent to the wrong place.

15.6.11 Event-Driven Challenge

The WSN should be built on an event-driven technological framework. Priority should be given to data that is critical to the success of the project. We need a system that can detect and communicate crucial information in real time.

15.6.12 Quality of Service (QoS)

A sensor network's ability to meet the demands of a particular application is measured by its quality of service (QoS). The quality of service can be characterized in two ways: application-specific and network-specific. Network coverage, the optimal number of sensors that can be activated, the quality and accuracy of sensor data measurement, latency, and delay are some of the QoS application-specific factors. Energy and bandwidth management, as well as satisfying application needs, are addressed in a QoS network viewpoint. QoS for WSNs should be able to handle node addition and removal. It is difficult to manage the QoS settings of a sensor network because the topology of the network is constantly changing and the routing information is inaccurate [33].

15.6.13 Deployment

The term "deployment" refers to the process of placing sensor nodes in the field. Depending on the application and demographics of the area, deployment can be structured or ad hoc. It is possible to drop sensor nodes from helicopters into remote and inaccessible locations. If the number of sensor nodes is too high, the network may get congested due to the number of simultaneous broadcasts. Aside from that, it is possible that all of these transmissions are redundant since they pick up on the same thing. In the event that sensor nodes are deployed in a sparse or insufficient number, the data yield will be poor or insufficient. If nodes are distributed at random, self-configuration is important. In the event of malfunctioning sensors or the death of sensor nodes, the property of fault tolerance states that the WSN should continue to function. Sensor nodes in WSN start failing when their power drops below a specific threshold value, and sometimes they are unable to transmit accurate information to the next node or sink. Failures and attacks on sensor networks are all too common. A faulty sensor is a sensor that gives an incorrect reading, either because of an attack or because of a change in the environment. Even if these sensors are functional or even alive, the data they produce is inaccurate. In such dire circumstances, the network must continue to function and provide the bare minimum of services [33]. Energy is constantly being depleted by WSN sensor nodes. The rate of energy depletion varies from node to node, depending on a variety of factors. Many nodes die early and rest later in the network life cycle as a result, adding insult to injury. Sensor nodes cannot be repaired or maintained after deployment. In the event that certain sensors fail, the network should remain active and connected. This means that WSNs need mechanisms that can handle failures in order to function properly.

15.6.14 Localization

In many real-world circumstances, nodes are distributed at random, making it difficult to discover and manage them in the absence of any supporting infrastructure. Localization is the process of determining the exact position of the nodes that have been deployed. Algorithms for finding a place are used to perform localization. Distributed, accurate, resilient, and scalable algorithms are ideal [34]. There are several situations in which they should allow for the mobility of nodes.

15.6.15 The Consumption of Power

It was originally specified that most of the difficulties faced by WSNs were due to a lack of power supplies. The scope of the nodes bounds the amount of power that can be generated (battery). Because of this, careful consideration must be given to both software and hardware design in terms of energy efficiency. It is possible that data compression can save energy in radio transmission, but it requires additional power for the processing that goes along with it. Depending on the application, some nodes may need to be turned off in order to conserve energy, while others may need to be operational at all times. Sensor networks, according to Puccinelli and Haenggi [28], offer a powerful combination of distributed sensing, processing, and communication. Sensor networks can be used in countless ways, but they also pose a number of issues because of their uniqueness. Sensitive networks affect hardware design (HWD) in four steps, including "power source(s), processor(s), and communication hardware(s)." HWD platforms have been constructed to test and apply the numerous concepts that have been developed by various scholars to suit all areas of learning, notably the technical and scientific elements [29]. As previously stated, in order to simplify the strategy and distribution of WSNs, several HLB events have been predicted, designed, and established for their resolution in the current WSN programming procedures that focus on low-level systems.

Model-driven engineering (MDE) is emerging as a promising option, and these HLB techniques, according to BenSaleh et al. [4], should ease the enterprise and disposition of WSNs while alleviating some of their difficulties.

15.7 Conclusion

A vision for future trends in research and opportunities such as MDE approaches on WSNs is recommended.

15.8 Conclusion and Future Scope

From applications to technological hurdles, this chapter focuses on WSNs' most pressing issues. When constructing a WSN, it is essential to identify the most suitable skill and statement protocol. These decisions are influenced by a variety of factors, most notably the requirements of the application.

This chapter was devoted to describing some of the constraints that must be met by a WSN and the many aspects that must be taken into account while creating a WSN. The goal is to aid in the selection of the most appropriate technology by WSN designers. For the IEEE 802.15.4 standard, there are a variety of performance levels to choose from, as well as a standard.

Acknowledgments

We thank our parents for their support and motivation.

References

[1] Akyildiz IF, Su W, Sankarasubramaniam Y, Cayirci E. A survey on sensor networks. IEEE Communications Magazine. 2002;**40**(8):102–114.
[2] Akyildiz IF, Su W, Sankarasubramaniam Y, Cayirci E. Wireless sensor networks: A survey. Computer Networks. 2002;**38**(4):393–422.
[3] Akkaya K, Younis M. A survey on routing protocols for wireless sensor networks. Elsevier Journal of Ad Hoc Networks. 2005;**3**(3):325–349.
[4] BenSaleh MS, Saida R, Kacem YH, Abid M. Wireless sensor network design methodologies: A survey. Journal of Sensors. 2020;**2020**:9592836.
[5] Bharathidasan A, Anand V, Ponduru S. Sensor Networks: An Overview. Department of Computer Science, University of California, Davis. Technical Report; 2001.
[6] Boukerche A. Algorithms and Protocols for Wireless, Mobile Ad Hoc Networks. Hoboken: John Wiley & Sons;2009.
[7] Buratti C, Conti A, Dardari D, Verdone R. An overview on wireless sensor networks technology and evolution. Sensors. 2009;**9**:6869–6896.
[8] Chen DB, Zhang NL, Zhang MG,Wang ZH, Zhang Y. Study on remote monitoring system of crossing and spanning tangent tower. IOP Conference Series: Materials Science and Engineering. 2017;**199**(1):1–6.
[9] Chong CY, Kumar SP. Sensor networks: Evolution, opportunities, and challenges. IEEE Proceedings. 2003;**91**(8):1247–1254.
[10] Culler D, Estrin D, Srivastava M. Overview of sensor networks. IEEE Computers. 2004;**37**:41–49.
[11] Durisic MP, Tafa Z, Dimic G, Milutinovic V. A survey of military applications of wireless sensor networks. In: Proceedings from the 2012 Mediterranean Conference on Embedded Computing (MECO). Piscataway: IEEE; 2012. pp.196–199.
[12] Furtado H, Trobec R. Applications of wireless sensors in medicine. In: 2011 Proceedings of the 34th International Convention MIPRO. Piscataway: IEEE; 2011. pp. 257–261.
[13] Hac A. Wireless Sensor Network Designs. Etobicoke, Ontario: John Wiley & Sons; 2003. p. 2003.
[14] Haenggi M. Opportunities and challenges in wireless sensor networks. In: Ilyas M, Mahgoub I, editors. Handbook of Sensor Networks: Compact Wireless and Wired Sensing Systems. Vol. 1. BocaRaton, FL: CRC Press; 2004. pp. 1–14

[15] Hao J, Brady J, Guenther B, Burchett J, Shankar M, Feller S. Human tracking with wireless distributed pyroelectric sensors. IEEE Sensors Journal. 2006;**6**:1683–1696.

[16] IEEE 802.15.4 Standard. Part 15.4: Wireless Medium Access Control (MAC) and Physical Layer (PHY) Specifications for Low-Rate Wireless Personal Area Networks (LR-WPANs). Piscataway: IEEE; 2006.

[17] IEEE Standard 802-2002 at the Wayback Machine, IEEE Standard for Local and Metropolitan Area Networks: Overview and Architecture, page 1,section 1.2: "Key Concepts," "BasicTechnologies."

[18] IEEE 802.15.4 Working Group (see also the website: www.ieee802.org/15/pub/TG4.html

[19] Lee DS, Lee YD, Chung WY, Myllyla R. Vital sign monitoring system with life emergency event detection using wireless sensor network. In: Proceedings of IEEE Conference on Sensors. Piscataway: IEEE; 2006.

[20] Matin MA, Islam MM. Overview of wireless sensor network. In: Matin MA,editor. Wireless Sensor Networks – Technology and Protocols. London: IntechOpen; 2012.

[21] Nwankwo W, Olayinka AS, Ukhurebor KE. The urban traffic congestion in Benin City and the search for ICT-improved solution. International Journal of Scientific &Technology Research. 2019;**8**(12):65–72.

[22] Nwankwo W, Olayinka AS, Ukhurebor KE. Nanoinformatics: Why design of projects on nanomedicine development and clinical applications may fail? In: Proceedings of the 2020 International Conference in Mathematics, Computer Engineering and Computer Science (ICMCECS). Piscataway: IEEE; 2020. pp. 1–7.

[23] Nwankwo W, Ukhurebor KE. Anx-ray of connectivity between climate change and particulate pollutions. Journal of Advanced Research in Dynamical and Control Systems. 2019;**11**(8) Special Issue:3002–3011.

[24] Nwankwo W, Ukhurebor KE. Investigating the performance of point to multipoint microwave connectivity across undulating landscape during rainfall. Journal of the Nigerian Society of Physical Sciences. 2019;**1**(3):103–115.

[25] Omojokun G. A survey of Zigbee wireless sensor network technology: Topology, applications and challenges. International Journal of Computer Applications. 2015;**130**(9):47–55.

[26] Ong J, You YZ, Mills-Beale J, Tan EL, Pereles B, Ghee KA. Wireless passive embedded sensor for real-time monitoring of water content in civil engineering materials. IEEE Sensors Journal. 2008;**8**:2053–2058.

[27] Pan J, Hou Y, Cai L, Shi Y, Shen SX. Topology control for wireless sensor networks. In: Proceeding of the 9th ACM International Conference on Mobile Computing and Networking. Piscataway: IEEE; 2003. pp. 286–299.

[28] Puccinelli D, Haenggi M. Wireless sensor networks: Applications and challenges of ubiquitous sensing. IEEE Circuits and Systems Magazine. 2005:19–29.

[29] Rodriguez-Zurrunero R, Utrilla R, Rozas A, Araujo A. Process management in IoT operating systems: Cross influence between processing and communication tasks in end-devices. Sensors. 2019;**19**(4):805.

[30] Sohraby K, Minoli D, Znati T. Wireless Sensor Networks: Technology, Protocols and Applications. Hoboken: John Wiley & Sons; 2007.

[31] The Industrial Consortia ZigBee Alliance (www.zigbee.org) and 6LowPAN.

[32] Tiwari P, Saxena VP, Mishra RG, Bhavsar D. Wireless sensor networks: Introduction, advantages, applications and research challenges. HCTL Open International Journal ofTechnology Innovations and Research. 2015;**14**:1–11.

[33] Raghavendra C, Sivalingam K, Znati T. Wireless Sensor Networks. New York: Springer; 2004.

[34] Rajaravivarma V, Yang Y, Yang T. An overview of wireless sensor network and applications. In: Proceedings of the 35th Southeastern Symposium on System Theory. Vol. 2003. Piscataway: IEEE; 2003. pp. 432–436.

16

A Novel Heuristic for Maximizing Lifetime of Target Coverage in Wireless Sensor Networks

Pooja Chaturvedi, A.K. Daniel, and Vipul Narayan

CONTENTS

16.1 Introduction 227
16.2 Problem Formulation 228
16.3 Proposed Heuristic 229
16.4 Mathematical Validation 230
16.5 Simulation Results and Analysis 230
    16.5.1 Scenario 1 232
    16.5.2 Scenario 2 232
    16.5.3 Scenario 3 232
    16.5.4 Scenario 4 233
    16.5.5 Scenario 5 234
16.6 Conclusion and Future Work 234
References 242

16.1 Introduction

WSN has recently been considered a most important research topic in the scientific fraternity. WSN is a network organized as a collection of a number of tiny, low weight, inexpensive, and highly constrained nodes. The main source of energy for these nodes is usually a battery, which cannot be replaced or recharged frequently and conveniently. The design and deployment of WSN is dependent on various factors such as dynamic network topology, fault tolerance, reliability, hardware and resource constraints, and energy consumption. Approaches which efficiently utilize the network resources are of great significance while designing networks for different applications, which range from medical to battlefield to structural monitoring [1].

Coverage is another challenging issue in the field of sensor networks, which ensures the monitoring of the region of interest for the longest possible duration while satisfying the application QoS requirements. Coverage approaches are broadly classified as area coverage, which aims to provide monitoring of an entire region, and target coverage, which is about ensuring coverage of a fixed set of points [2]. The research shows that the sensor nodes consume least energy in the idle state as compared to the active state. Thus the approaches which keep the least number of nodes in the active state can achieve better energy conservation and hence enhance network lifetime. Several heuristics have been proposed in the literature to address the target coverage problem from the perspective of energy efficiency and connectivity. The scheduling-based approach, which activates a subset of nodes, is considered to be the most energy efficient approach. The set cover is defined as the collection of sensor nodes which can collaborate to monitor a predefined set of targets. The nodes in the set covers are activated according to a schedule determined by the base station. There are various strategies proposed in the literature to determine the set covers which are classified as non-disjoint or disjoint set covers [3–10]. The non-disjoint set covers restrict the node to be included in only one set cover, where as in the disjoint set cover-based approach, the nodes can participate in multiple set covers. However, in this case the number of set covers in which a node can participate is limited by the initial battery of the nodes. The network lifetime in scheduling-based approaches is proportional to the number of set covers obtained.

One such node strategy is proposed in [11], in which the nodes are divided into a number of set covers based on node contribution, trust values, and probabilistic coverage model. The observation probability is used as a metric to determine the status of node to keep as active or put to sleep. All the nodes which are included in the set covers are activated periodically to perform the environment sensing and data transmission tasks.

16.2 Problem Formulation

Let us assume that a set of x sensor nodes and a set of y targets are randomly deployed in the target area, and then the target coverage problem is defined as the objective of maximizing the observation time of each target with the desired quality of service parameter. A target is said to be monitored by a sensor node, if its distance is less than the sensing range (r_s) of the sensor node. A sensor node in the sensor network can exist in either sleep or active mode. The most efficient strategy to solve this problem is to determine the number of set covers which can cover all the targets for the longest possible duration. The energy efficient target coverage can be defined using linear programming as follows:

$$\text{Maximize} \sum_{k=1}^{sc} t_k \tag{1}$$

$$\text{Subject to} \sum_{k=1}^{sc} c_{ij} t_k \leq e_i \text{ for all sensors } x_i$$
$$t_k \geq 0 \text{ for all sensor covers } sc_k \tag{2}$$

Where t_k is the operational time of a set cover, e_i is the initial energy of the nodes, and sc is the number of set covers obtained for the considered network conditions. The operational time of the set cover is limited by the initial energy of the node. c_{ij} is a Boolean variable which represents the coverage constraint and is defined as:

$$c_{ij} = \begin{cases} 1, & \text{if a sensor } s_i \text{ covers a target } y_j \\ 0, & \text{otherwise} \end{cases} \tag{3}$$

Maximizing Lifetime of Target Coverage

Coverage constraint is defined on the basis of observation probability. The observation probability is defined as the product of detection of coverage probability and the trust values.

$$P_{op}(j) = 1 - \prod_{l \in ns}\left(1 - P_{ij}^{obs}\right) \qquad (4)$$

such that $ns_j = \{i, P_{op}(j) > 0 \text{ and } j \text{ is active}\}$ and P_{ij}^{obs} is determined as follows:

$$P_{ij}^{obs} = \text{cov}(i,j) \times T_{ij} \qquad (5)$$

The coverage probability and trust value of the nodes are determined as in [8].

The upper bound on the activation time of the set cover is limited by the minimum energy level of all the nodes which are included in the current set cover. The maximum operational time for target t_j denoted by A_j is determined by the following equation:

$$A_j = \sum_{i=1}^{x} c_{ij} e_i \qquad (6)$$

$$A_{\max} = \min\{A_1, A_2 ... A_{sc}\} \qquad (7)$$

The approach proposed in [8] aims to determine the maximum number of disjoint set covers which can cover all the targets with the desired confidence level and are activated according to a schedule determined by the base station on the basis of the observation probability. The main objective in this approach was to include the minimum number of nodes to keep in the active state in a set cover.

The network lifetime for the target coverage is defined as the duration until all the targets are monitored with the desired confidence level by at least one sensor node. This chapter aims to enhance the network lifetime as an extension to the approach proposed in [8]. The improvement in the approach here is achieved by including the highest priority node in the set cover. The priority of the node is calculated using the equation below:

$$p_i = \sum_{j=1}^{y} \frac{c_{ij}}{\sum_{i=1}^{x} c_{ij}} \qquad (8)$$

The consideration of the priority of nodes in determination of the set cover achieves network lifetime closer to the upper bound, hence increasing the system efficiency.

16.3 Proposed Heuristic

The proposed set cover-based scheduling strategy consists of the following steps:

a. Determine the coverage constraint on the basis of the observation probability.

b. Based on the coverage constraint, identify the critical target. A target is said to be critical if it is covered by the least number of sensor nodes.
c. Identify the critical sensor. A sensor node which covers the critical target is said to be the critical sensor.
d. Determine the upper bound on the network lifetime using the equation defined above.
e. Determine the number of set covers which can cover all the targets on the basis of the priority assigned to the nodes using the conditions:
   i. Select the node which can cover the maximum number of the targets.
   ii. Select the node which has the maximum residual energy.
   iii. Determine the set cover including only one critical sensor.

16.4 Mathematical Validation

To understand the effect of considering the priority of nodes in the determination of set covers, consider a network scenario consisting of four sensor nodes and two targets. The targets are monitored by the nodes as follows:

$$y_1 = \{x_2, x_3, x_4\} \text{ and } y_2 = \{x_1, x_2\}$$

Based on the sensor target coverage constraint, we can determine the critical target as y_2 and critical sensors are determined as: $\{x_1, x_2\}$. Using equation 7, the upper bound on lifetime is achieved as two units. But if we use the heuristic proposed in [8], only one set cover can be generated which will achieve lifetime of one time unit. But if we consider the priority of the nodes as defined in equation 8 and also condition (iii) that only one critical node is selected in a set cover, then we obtain the different set covers as $C_1 = \{x_2\}$, $C_2 = \{x_3, x_4\}$, $C_3 = \{x_1, x_3\}$, and $C_4 = \{x_1, x_4\}$ which can be activated as $\{x_2\}$ followed by the set cover $\{x_3, x_4\}$ or $\{x_2\}$, $\{x_1, x_3\}$ or $\{x2\}$ followed by the set cover $\{x_1, x_4\}$. In both cases, the lifetime achieved is closer to the upper bound, i.e. two time units. The node contribution is defined as the number of targets it can monitor. The different parameters for the considered network are summarized in Table 16.1.

The node contribution and priority of the nodes are shown in Figure 16.1 (a) and (b). The network lifetime obtained for the different operational time of the set covers is shown in Figure 16.1 (c). It can be observed from the results that the network lifetime is maximum (i.e. two time units) and is equal to the upper bound, if we consider the activation time for each set cover as one time unit.

16.5 Simulation Results and Analysis

To evaluate the performance of the proposed heuristic, we have considered four different network scenarios which consist of varying numbers of nodes and targets deployed in a square target region of dimensions 100×100 m². The operational time of activation of the set covers is considered as 0.1, 0.2, 0.5, and 1 time units.

Maximizing Lifetime of Target Coverage 231

TABLE 16.1

Node Contribution and Priority

Node ID	Contribution	Priority
1	1	0.5
2	2	0.833
3	1	0.333
4	1	0.333

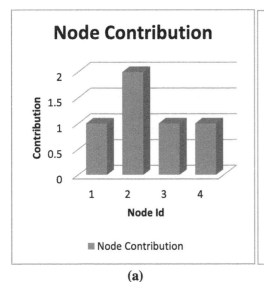

(a)

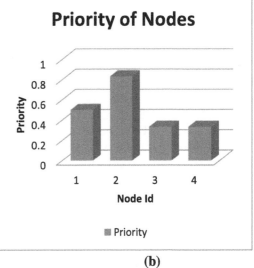

(b)

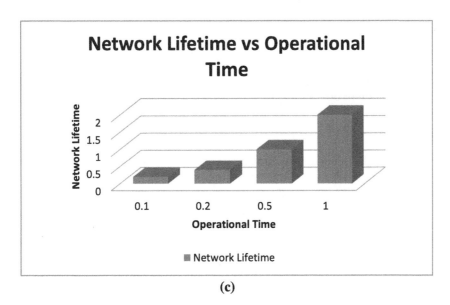

(c)

FIGURE 16.1
(a) Node Contribution, (b) Node priority (c) Network lifetime.

TABLE 16.2

Node Contribution and Priority for Scenario 1

Node ID	Contribution	Priority
1	3	0.5
2	4	0.694444444
3	3	0.444444444
4	4	0.694444444
5	4	0.694444444
6	2	0.277777778
7	2	0.277777778
8	4	0.694444444
9	3	0.44444
10	3	0.47778

16.5.1 Scenario 1

In the first network scenario, we consider a network which consists of 10 nodes and 5 targets randomly deployed in the target region of 100 × 100 m². Based on the sensor target coverage constraint, we can determine the critical target as y_3 and critical sensors are determined as: $\{x_2, x_4, x_5, x_8, x_{10}\}$. Using equation 7, the upper bound on lifetime is achieved as 5 time units.

The different parameters for the considered network are summarized in Table 16.2. The node contribution and priority of the nodes are shown in Figure 16.2 (a) and (b). The network lifetime obtained for the different operational time of the set covers is shown in Figure 16.2 (c). It can be seen in the results that the maximum network lifetime achieved is 5 time units, which is close to the upper bound.

16.5.2 Scenario 2

In the second network scenario, we consider a network which consists of 20 nodes and 10 targets randomly deployed in the target region of 100 × 100 m². Based on the sensor target coverage constraint, we can determine the critical target as y_3 and critical sensors are determined as: $\{x_1, x_9, x_{16}, x_{17}, x_{18}, x_{19}\}$. Using equation 7, the upper bound on lifetime is achieved as 6 units. The different parameters for the considered network are summarized in Table 16.3. The node contribution and priority of the nodes are shown in Figure 16.3 (a) and (b). The network lifetime obtained for the different operational time of the set covers is shown in Figure 16.3 (c).

16.5.3 Scenario 3

In the third network scenario, we consider a network which consists of 30 nodes and 15 targets randomly deployed in the target region of 100 × 100 m². Based on the sensor target coverage constraint, we can determine the critical target as y_2 and y_3 and the critical sensors are determined as: $\{x_1, x_4, x_5, x_6, x_7, x_9, x_{11}, x_{16}, x_{17}, x_{18}, x_{19}, x_{21}, x_{26}, x_{27}, x_{28}, x_{29}\}$. Using equation 7, the upper bound on lifetime is achieved as 10 units. The different parameters for the considered network are summarized in Table 16.4. The node contribution and priority of the nodes are shown in Figure 16.4 (a) and (b). The network lifetime obtained for the different operational time of the set covers is shown in Figure 16.4 (c).

Maximizing Lifetime of Target Coverage

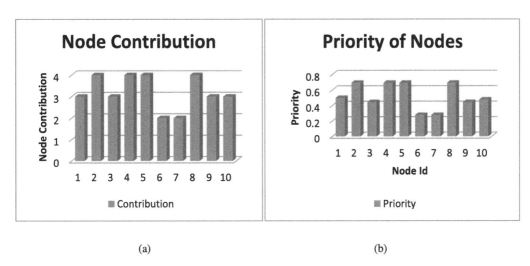

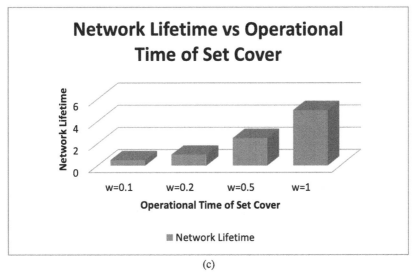

FIGURE 16.2
(a) Node contribution, (b) Node priority, (c) Network lifetime for scenario 1.

16.5.4 Scenario 4

In the fourth network scenario, we consider a network which consists of 40 nodes and 20 targets randomly deployed in the target region of 100 × 100 m². Based on the sensor target coverage constraint, we can determine the critical target as y_3 and critical sensors are determined as: $\{x_1, x_2, x_4, x_6, x_{13}, x_{14}, x_{16}, x_{17}, x_{19}, x_{22}, x_{23}, x_{25}, x_{27}, x_{29}, x_{31}, x_{33}, x_{35}, x_{37}, x_{39}\}$. Using equation 7, the upper bound on lifetime is achieved as 19 units. The different parameters for the considered network are summarized in Table 16.5. The node contribution and priority of the nodes are shown in Figure 16.5 (a) and (b). The network lifetime obtained for the different operational time of the set covers is shown in Figure 16.5 (c).

TABLE 16.3

Node Contribution and Priority for Scenario 2

Node ID	Contribution	Priority
1	7	0.72253
2	0	0
3	2	0.18803
4	5	0.51069
5	5	0.53089
6	7	0.7218
7	8	0.69872
8	7	0.66698
9	5	0.56581
10	4	0.38803
11	2	0.23377
12	3	0.26496
13	6	0.59005
14	6	0.57894
15	6	0.59005
16	8	0.84274
17	7	0.75428
18	4	0.42051
19	6	0.65428
20	1	0.07692

16.5.5 Scenario 5

In the fifth network scenario, we consider a network which consists of 50 nodes and 20 targets randomly deployed in the target region of 100 × 100 m². Based on the sensor target coverage constraint, we can determine the critical target as y_3 and y_{10} and critical sensors are determined as: $\{x_1, x_2, x_4, x_6, x_8, x_{12}, x_{13}, x_{14}, x_{15}, x_{16}, x_{17}, x_{19}, x_{21}, x_{22}, x_{23}, x_{25}, x_{27}, x_{29}, x_{31}, x_{33}, x_{35}, x_{37}, x_{39}, x_{42}, x_{44}, x_{46}\}$. Using equation 7, the upper bound on lifetime is achieved as 24 units. The different parameters for the considered network are summarized in Table 16.6. The node contribution and priority of the nodes are shown in Figure 16.6 (a) and (b). The network lifetime obtained for the different operational time of the set covers is shown in Figure 16.6 (c).

The comparison of the network lifetime achieved using the proposed heuristic with the upper bound is shown in Figure 16.7. It can be inferred from the results that the network lifetime is closer to the upper bound in most of the cases. The network lifetime increases in proportion to the operational time of the set covers and is maximum for the operational time of one time unit.

16.6 Conclusion and Future Work

The chapter presented an efficient heuristic to address the target coverage problem such that the network lifetime is maximized, as an extension to the heuristic proposed in [8]. The set cover determination strategy is improved by selecting the node with the maximum priority to be included in the set cover. The additional condition is incorporated as

Maximizing Lifetime of Target Coverage

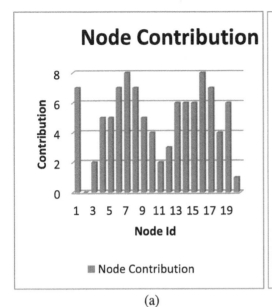

(a)

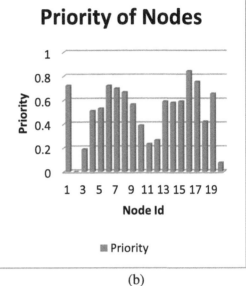

(b)

FIGURE 16.3
(a) Node vontribution, (b) Node priority, (c) Network fifetime for scenario 2.

TABLE 16.4

Node Contribution and Priority for Scenario 3

Node ID	Contribution	Priority
1	11	0.72215
2	3	0.176879
3	4	0.249145
4	9	0.590465
5	7	0.485747
6	12	0.805483
7	7	0.473842
8	9	0.579932
9	9	0.623582
10	8	0.524437
11	2	0.158824
12	3	0.176923
13	10	0.660184
14	10	0.634054
15	10	0.649073
16	12	0.814759
17	10	0.673203
18	5	0.333333
19	8	0.539869
20	2	0.1125
21	4	0.285808
22	5	0.294979
23	10	0.649073
24	10	0.642983
25	10	0.656916
26	11	0.746598
27	10	0.689076
28	5	0.32549
29	9	0.619141
30	2	0.105556

selecting only one critical sensor node in the set cover. The performance of the proposed heuristic is evaluated for different network scenarios of varying number of nodes, targets, and different activation times of each set cover. The results show that the proposed heuristic achieves network lifetime close to the maximum upper bound.

To study the performance of the proposed protocol for heterogeneous network architecture is our future work. The least contributing nodes may be relocated to further improve the coverage requirement. Future work may also study the proposed protocol on the basis of different network parameters such as data transmission and packet delivery rate, etc.

Maximizing Lifetime of Target Coverage 237

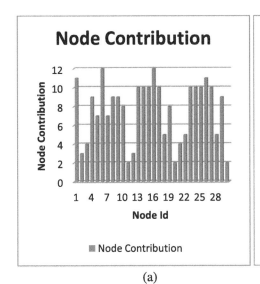

(a)

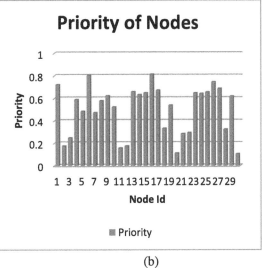

(b)

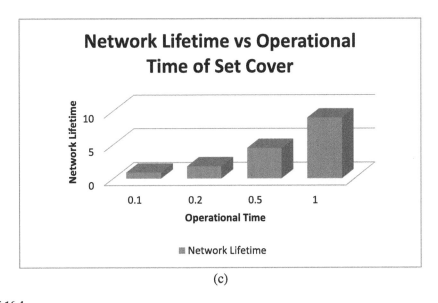

(c)

FIGURE 16.4
(a) Node contribution, (b) Node priority, (c) Network lifetime for scenario 3.

TABLE 16.5

Node Contribution and Priority for Scenario 4

Node ID	Contribution	Priority
1	12	0.499606
2	15	0.638598
3	16	0.646164
4	10	0.40999
5	17	0.692958
6	12	0.495351
7	11	0.440512
8	14	0.552619
9	9	0.347103
10	9	0.373479
11	10	0.399412
12	12	0.482359
13	12	0.510117
14	11	0.451463
15	14	0.586159
16	10	0.422284
17	16	0.651176
18	14	0.567244
19	11	0.461528
20	10	0.401878
21	16	0.661843
22	9	0.372663
23	14	0.581401
24	9	0.368327
25	14	0.582481
26	7	0.28333
27	17	0.697257
28	10	0.396423
29	16	0.667239
30	9	0.349819
31	15	0.631027
32	13	0.521207
33	15	0.62059
34	8	0.322378
35	14	0.583579
36	11	0.430598
37	13	0.55038
38	10	0.407614
39	13	0.540592
40	10	0.401251

Maximizing Lifetime of Target Coverage

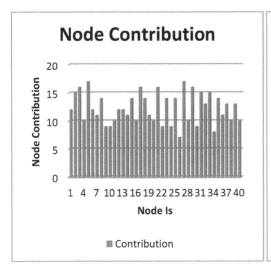

(a)

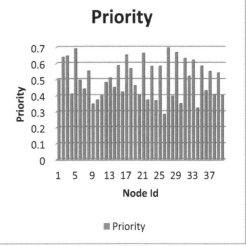

(b)

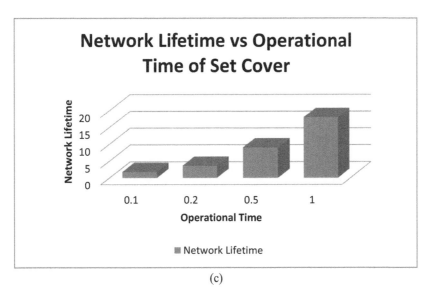

(c)

FIGURE 16.5
(a) Node contribution, (b) Node priority, (c) Network lifetime for scenario 4.

TABLE 16.6

Node Contribution and Priority for Scenario 5

Node ID	Contribution	Priority
1	12	0.40707
2	15	0.511045
3	16	0.519786
4	10	0.328666
5	17	0.557532
6	12	0.397593
7	11	0.353067
8	14	0.445271
9	9	0.278855
10	9	0.300162
11	10	0.319464
12	12	0.385651
13	12	0.410907
14	11	0.36158
15	14	0.468672
16	10	0.335778
17	16	0.526083
18	14	0.454754
19	11	0.370195
20	10	0.32124
21	16	0.534596
22	9	0.296757
23	14	0.467964
24	9	0.294716
25	14	0.468448
26	7	0.227483
27	17	0.559198
28	10	0.316645
29	16	0.534412
30	9	0.278955
31	15	0.505134
32	13	0.415192
33	15	0.497688
34	8	0.25641
35	14	0.466635
36	11	0.345039
37	13	0.440628
38	10	0.324982
39	13	0.429692
40	10	0.325705
41	9	0.278955
42	15	0.505134
43	13	0.415192
44	15	0.497688
45	8	0.25641
46	14	0.466635
47	11	0.345039
48	13	0.440628
49	10	0.324982
50	13	0.429692

Maximizing Lifetime of Target Coverage

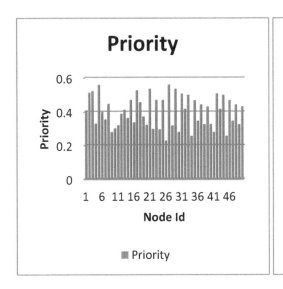

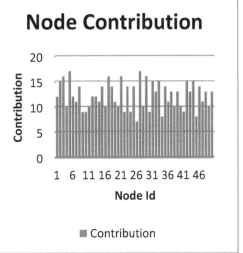

(a)   (b)

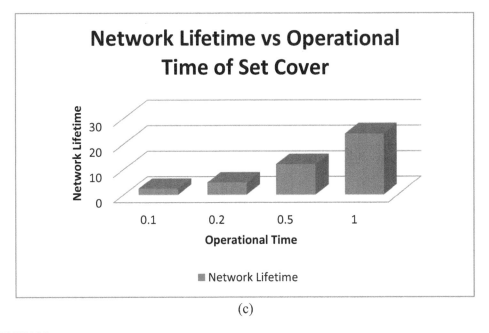

(c)

FIGURE 16.6
(a) Node contribution, (b) Node priority, (c) Network lifetime for scenario 5.

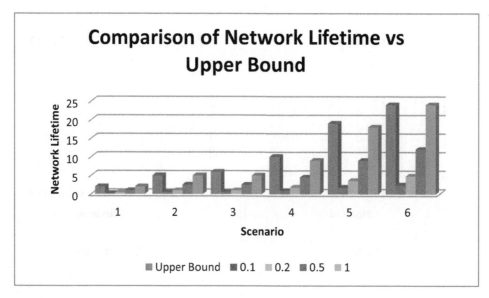

FIGURE 16.7
Comparison of network lifetime vs. upper bound.

References

[1] Akyildiz, I.F., Su, W., Sankarasubramaniam, Y., et al. "A survey on sensor networks," *IEEE Commun. Mag.*, 2002, 40 (8), pp. 102–114.
[2] Slijepcevic, S., Potkonjak, M. "Power efficient organization of wireless sensor networks," *Proc. IEEE Int. Conf. on Communications*. IEEE, 2001, pp. 472–476.
[3] Cardei, M., Thai, M.T., Li, Y., et al. "Energy-efficient target coverage in wireless sensor networks." *Proc. IEEE Infocom*. IEEE, 2005.
[4] Manju, A.K.P. "High-energy-first (HEF) heuristic for energy efficient target coverage problem," *Int. J. Ad Hoc Sens. Ubiquit. Comput.*, 2011, 2 (1), pp. 45–58.
[5] Zorbas, D., Glynos, D., Kotzanikolaou, P., et al. "Solving coverage problems in wireless sensor networks using cover sets," *Ad Hoc Netw.*, 2010, 8, pp. 400–415.
[6] Mini, S., Udgata, S.K., Sabat, S.L. "Sensor deployment and scheduling for target coverage problem in wireless sensor networks," *IEEE Sens. J.*, 2014, 14 (3), pp. 636–644.
[7] Heinzelman, W.R., Chandrakasan, A., Balakrishnan, H. "Energy-efficient communication protocol for wireless micro sensor networks," *HICSS '00: Proceedings of the 33rd Hawaii International Conference on System Sciences*, Volume 8, IEEE, 2000, p. 8020.
[8] Narayan, V., Daniel, A.K., Rai, A.K. "Energy efficient two tier cluster based protocol for wireless sensor network," *2020 International Conference on Electrical and Electronics Engineering (ICE3)*. IEEE, 2020.
[9] Narayan, V., Daniel, A.K. "Multi-tier cluster based smart farming using wireless sensor network," *2020 5th International Conference on Computing, Communication and Security (ICCCS)*. IEEE, 2020.
[10] Narayan, V., Daniel, A.K. "A novel approach for cluster head selection using trust function in WSN," *Scalable Computing: Practice and Experience*, 2021, 22 (1), pp. 1–13.
[11] Chaturvedi, P., Daniel, A.K. "A novel sleep/wake protocol for target coverage based on trust evaluation for a clustered wireless sensor network," *Int. J. Mobile Network Design and Innovation*, 2017, 7 (3/4), pp. 199–209.

17

Network Recovery in Dense and Emergency Areas Using a Temporary Base Station and an Unmanned Aerial Vehicle

Sayanti Ghosh, Sanjay Dhar Roy, and Sumit Kundu

CONTENTS
17.1 Introduction 243
17.2 Problems in the Disaster Area 244
    17.2.1 Classification of the Disaster Area 244
    17.2.2 Communication Network Partially or Fully Damaged 244
17.3 Related Work 244
    17.3.1 Internet of Things (IoT) 244
    17.3.2 Unmanned Aerial Vehicle Assisted Communication 245
    17.3.3 Motivation and Contribution 245
17.4 Proposed System Model 246
    17.4.1 Dense Area 246
    17.4.2 Area Spectral Efficiency (ASE) for Dense Area 249
    17.4.3 Emergency Area 249
    17.4.4 Area Spectral Efficiency for Emergency Area 250
17.5 Simulation Results and Discussions 251
17.6 Conclusion and Future Work 254
Acknowledgment 256
References 256

17.1 Introduction

Almost every year, disaster happens, either natural or man-made, somewhere in the world. In 2020, for example, there were devastating floods in Indonesia, Cyclone Amphan in India, and floods in Bangladesh, India, Japan, and China due to heavy rains, etc. Cyclone Amphan was among the strongest, deadliest, and costliest storms in the history of the Bay of Bengal. Although it is impossible to control a natural or man-made disaster using advanced communication technology, we aim to save people's lives and minimize losses in crisis zones or non-functional areas (NFAs). In some areas, the base stations (BSs) are either moderately or completely damaged. Voice and data communication between responders and victims is an important part of the disaster management system.

17.2 Problems in the Disaster Area

The development of new technology for establishing a connection link in a disaster area or NFA has increased significantly. However, it is not always possible to reach disaster locations. In this chapter, we discuss these problems and try to resolve them.

17.2.1 Classification of the Disaster Area

Disaster areas can be divided into two parts: dense areas and emergency areas. In particular, an ad-hoc network can be set up with a temporary base station (T-BS) in a dense area to provide wireless connection for normal users [1]. The IoT gateway acts as a central hub for IoT users or one of the key elements in this ecosystem [2, 3]. It is used for communication with all sensors and remote connections such as the internet, applications, or users. IoT devices may connect to the IoT gateway using Bluetooth low energy (LE) and Zigbee for short-range wireless communication. Wi-Fi links them to the internet public cloud for long-range wireless communication LTE through ethernet LAN or fiber optics WAN [4]. In an emergency area, a UAV works as an aerial communication platform for serving ground users [5].

17.2.2 Communication Network Partially or Fully Damaged

A communication network has been proposed in dense and emergency areas using a T-BS and a UAV. A T-BS is placed in the dense area, and a UAV is placed in the emergency area after a disaster. The dense area defines the area where BSs are partially damaged. However, the area where BSs are fully damaged is called an emergency area. Cellular users (CUs) communicate to T-BS in a dense area, and smart home devices such as IoT devices communicate to IoT gateways (IoT-GWs). In the emergency area, CUs communicate to the UAV only. An analytical framework has been developed for evaluating the outage probability, data rate of users and ASE for dense and emergency areas.

17.3 Related Work

17.3.1 Internet of Things (IoT)

In 1988, Internet of Things (IoT) was coined by Kevin Ashton [6]. It refers to a worldwide network for communication using standard protocols. In [7], the authors consider a cellular-based uplink IoT network incorporating NOMA transmission. In [8], the authors consider energy-efficient ultra-reliable communications for uplink resources in narrowband IoT (NB-IoT) networks. The link-level coverage performance has been developed for the NB-IoT uplink network based on signal repetition [9]. In a disaster scenario, IoT-based wireless technology has been developed using a gateway selection algorithm [10]. The applications of IoT are early warning systems for fire detection, smart cities, self-driven cars, smart grids, industrial internet, and for managing disasters and reducing tasks.

17.3.2 Unmanned Aerial Vehicle Assisted Communication

In an emergency area, a UAV is the most promising wireless technology for serving ground users [11]. In [12], the authors discuss UAV-based uplink throughput analysis using experimental measurements and simulations over a cellular network. Static and mobile UAVs have been considered for downlink wireless communication in an underlaid device to device (D2D) communication network. In [13], the authors discuss UAV-enabled wireless systems analytical frameworks based on machine learning, optimization theory, stochastic theory, and game theory, etc. In [14], the authors discuss the optimal UAV flight trajectory for the dual-sampling scheme in wireless networks. The flight performance of UAVs includes some characteristics that are studied in [15]. In [16], the authors consider uplink and downlink UAV coverage performance in three-dimensional (3D) system modeling. In [17], the authors discuss the analytical action of the path-loss of the cellular network. In [18, 19], the authors study UAV-assisted wireless communication and cellular-connected UAVs for the fifth generation (5G) cellular network. Performance evaluation of the maximum endurance of a small hovering UAV is discussed in [20, 21]. In [22, 23], the authors study the stochastic geometry of the aggregate interference and coverage probability in radio frequency (RF) networks.

17.3.3 Motivation and Contribution

Earlier researchers have pursued IoT and UAV-based work for a non-calamity framework. This chapter has proposed a UAV-assisted uplink for an IoT-enabled D2D network in a disaster scenario. In this chapter, disaster areas are divided into two categories: dense area and emergency area. An inner circle is defined as where the BSs are partially damaged, and can be covered by small cells. A T-BS is placed in each cell in the inner circle. Hence, CUs and IoT users will communicate with the T-BS and IoT-GW, respectively. The rescue teams can save people from the NFA region with the help of IoT users. An outer circle is defined where the BSs are fully damaged, and here all users are cellular users. CUs will communicate to the UAV only in the outer circle. These CUs are far away from the T-BS, and due to limited battery power, UAV cannot traverse the whole NFA.

This chapter can be summarized thus:

- A system model is proposed for post-disaster scenarios where disaster areas are divided into two categories, i.e., dense and emergency areas.
- T-BSs, CUs, IoT users, and IoT gateways are considered in a dense area. Accordingly, the outage probability has been developed analytically.
- The area spectral efficiency has been evaluated for a cellular user in dense areas and emergency areas.

The system model under consideration and the relevant analysis are discussed in section 4. In section 5, analytical and simulation results are discussed. Section 6 briefly concludes and indicates some future research directions.

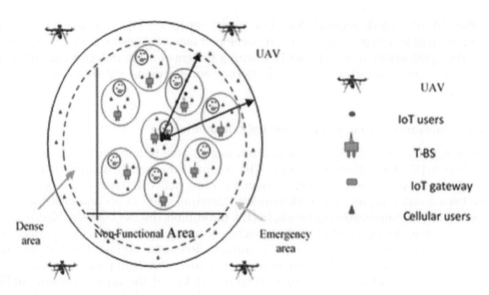

FIGURE 17.1
System model for uplink signal analysis.

17.4 Proposed System Model

In this system model, the BSs are partially or fully damaged in a post-disaster scenario. Figure 17.1 shows the uplink signal analysis and establishment of communication networks in dense and emergency areas. A T-BS can be placed in each cell in a dense disaster area, where all the CUs can communicate with their respective T-BS. Also, an IoT user can communicate to their respective IoT gateway in each cell. The area of each cell in the inner circle is denoted as *a meter*^{2} and this area is also called the dense area. All the T-BS locations follow a homogeneous poisson point process (HPPP) with density λ.

17.4.1 Dense Area

At the T-BS, the signal-to-interference noise ratio (SINR) becomes

$$\gamma = \frac{P_{cu} g_{c,b} R^{-2b}}{\sum_{j \in \Phi_I} P_s g_j |x_j - B_0|^{-2b} + \sum_{z \in Z} P_o g_z D^{-2b} + \sigma_n^2} \quad (1)$$

where, $g_{c,b} = |h_{c,b}|^2$, $g_j = |h_j|^2$, and $g_z = |h_z|^2$. P_{cu} is the signal power emitted by the CU transmitter, b is the amplitude loss exponent as $b = \frac{\alpha}{2}$, α is the propagation constant, R represents the distance between CU and T-BS, and σ_n^2 denotes the noise power. $h_{c,b}, h_j,$ and h_z represents the Rayleigh fading channel which is exponentially distributed of the unit mean.

Now, we consider for the analysis whether the receiver is located at the origin, and it is denoted as B_0. The distance between an interfering IoT user (located at x_j) and B_0 is $\|x_j - B_0\| = X_j$, where j represents the points of interferers' point process Φ_j. g_j and g_z represent channel gains with independent and identical distribution (i.i.d.) of the form (1). Here, the interferer set Φ_I is distributed according to HPPP over the entire plane $\mathbb{R}^2$. Spectrum sharing systems are limited interference, so thermal noise is negligible in the region of our interest. D represents interfering users from other cells. P_s represents the interference power of IoT users, and P_o represents the interference power of other cell users. To make the analysis tractable, the interference is negligible from users of other cells. Hence, the SINR at the T-BS becomes

$$\gamma = \frac{P_{cu} g_{c,b} R^{-2b}}{\sum_{j \in \Phi_I} P_s g_j |x_j - B_0|^{-2b} + \sigma_n^2} \tag{2}$$

where, $I = \sum_{j=1}^{L} I_j$. I_j is the interference caused by the j-th interferer. The MGF (moment generating function) is defined in [21]. The MGF $M_I^{j(s)} = E\left[e^{-sI_j}\right]$. $E[.]$ denotes the expectation, and s denotes Laplace variable.

The MGF of the aggregate interference for the j-th interferer can be given as in [22]:

$$M_I^{j(s)} = E_{X_j, g_j}\left[e^{-sI_j}\right] \tag{3}$$

The PDF (probability density function) of X_j can be denoted as $f_{X_j}(x_j) = \frac{2\pi X_j}{A_I}$ for $\mathbb{X}_G < X_j < \mathbb{X}_E$, and $f_{X_j}(x_j) = 0$ for others. $A_I = \pi(\mathbb{X}_E^2 - \mathbb{X}_G^2)$, where A_I is the total area surrounding the interferer nodes.

By averaging e^{-sI_j} using the PDF of interferer distance, we get [22]:

$$M_{I/g_j}^{j(s)} = \int_{\mathbb{X}_G}^{\mathbb{X}_E} e^{-sP_s X^{-\alpha}} g_j \left(\frac{2\pi X}{A_I}\right) dX \tag{4}$$

$$M_{I/g_j}^{j(s)} = \frac{2\pi}{A_I} \int_{\mathbb{X}_G}^{\mathbb{X}_E} e^{-(sP_s g_j)X^{-\alpha}} X dX \tag{5}$$

$$M_{I/g_j}^{j(s)} = \frac{2\pi}{\alpha A_I}\left(\mathbb{X}_E^2 E_{\frac{2+\alpha}{\alpha}}\left(\frac{P_s s g_j}{\mathbb{X}_E^\alpha}\right) - \mathbb{X}_G^2 E_{\frac{2+\alpha}{\alpha}}\left(\frac{P_s s g_j}{\mathbb{X}_G^\alpha}\right)\right) \tag{6}$$

$E_\mathbb{N}(.)$ is the generalized exponential integral. The MGF of the interference from the j-th interferer can be defined as:

$$M_{I/g_j}^{j(s)} = \frac{\pi (P_s s)^{\frac{2}{\alpha}-2} \mathbb{B}^{-\frac{4}{\alpha}-2}}{(-1)^{\lambda+\frac{2}{\alpha}} 4\alpha^2 (1+\alpha)\Gamma(\lambda) A_I}(\mathbb{Q}(\mathbb{X}_E) - \mathbb{Q}(\mathbb{X}_G)) \tag{7}$$

where, $\mathbb{Q}(\mathbb{X})$ is defined as in [21]. Each interferer is assumed to be independent, and the MGF of I given N can be denoted as:

$$M_{I/N}(s) = \prod_{j=1}^{N} M_I^{j(s)} = \left(M_I^{j(s)}\right)^N \tag{8}$$

Finally, we get [23]:

$$M_{I/N}(s) = e^{\rho A_I \left(M_I^{j(s)} - 1\right)} \tag{9}$$

where, ρ is interferer density.

The exact outage probability expression for the dense area can be expressed as:

$$\mathbb{P}_{out} = P\left(\frac{P_{cu} g_{c,b} R^{-\alpha}}{\sum_{j \in \Phi_I} P_s g_j |x_j - B_0|^{-\alpha} + \sigma_n^2} \leq \gamma_{th}\right) \tag{10}$$

$$\mathbb{P}_{out} = P\left(h_{c,b} \leq \frac{\gamma_{th}\left(\sum_{j \in \Phi_I} P_s g_j |x_j - B_0|^{-\alpha} + \sigma_n^2\right)}{P_{cu} R^{-\alpha}}\right) \tag{11}$$

$$\mathbb{P}_{out} = 1 - exp\left(-\gamma_{th}\left(\frac{\sum_{j \in \Phi_I} P_s g_j |x_j - B_0|^{-\alpha} + \sigma_n^2}{P_{cu} R^{-\alpha}}\right)\right) \tag{12}$$

Averaging concerning I, we get:

$$\mathbb{P}_{out} = 1 - exp\left(-\frac{\gamma_{th} \sigma_n^2}{P_{cu} R^{-\alpha}}\right) E_I\left[exp\left(-I\left(\frac{\gamma_{th}}{P_{cu} R^{-\alpha}}\right)\right)\right] \tag{13}$$

$$\mathbb{P}_{out} = 1 - exp\left(-\frac{\gamma_{th} \sigma_n^2}{P_{cu} R^{-\alpha}}\right) M_I\left(\frac{\gamma_{th}}{P_{cu} R^{-\alpha}}\right) \tag{14}$$

However, the SINR at the IoT-GW can be expressed as:

$$\gamma_{GW} = \frac{P_s g_j |x_j - B_0|^{-2b}}{\sum_{j' \neq j} P_{s'} g_{j'} |x_{j'} - B_0|^{-2b} + \sigma_{wn}^2} \tag{15}$$

In equation (15), P_s represents the power of IoT users, and $P_{s'}$ represents the interference power of IoT users. $g_{j'}$ denotes interfering channel gain, and $|x_{j'} - B_0|$ indicates the distance between B_0 and $x_{j'}$.

The outage probability of the desired user is:

$$\mathbb{P}_{out,GW} = P(\gamma_{GW} \leq \gamma_{th}) \tag{16}$$

17.4.2 Area Spectral Efficiency (ASE) for Dense Area

The ASE is defined as the sum of the maximum average data rates/Hz/unit area.

$$\text{ASE} = \frac{1}{|a|}\sum_{m=1}^{M} E\left(\log_2\left(1+\gamma_m\right)\right) \times P\left(Cellular\,user\right) + \frac{1}{|a|}\sum_{l=1}^{L} E\left(\log_2\left(1+\gamma_l\right)\right) \times P\left(IoT\,user\right) \quad (17)$$

$P(Cellular\,user)$: Probability that a cellular user is associated with the T-BS in the dense area.

$P(IoT\,user)$: Probability that an IoT user is communicating to the T-BS via IoT-GW in the dense area.

$E\left(\log_2\left(1+\gamma_m\right)\right)$ and $E\left(\log_2\left(1+\gamma_l\right)\right)$ are the ergodic rates of the mth cellular user and l^{th} IoT user.

17.4.3 Emergency Area

In the emergency area, all CUs are assumed to be uniformly distributed in A with density $\kappa = \frac{N}{A}$ users/m^2. It is assumed that the UAV provides adjustable half power beam widths with $\theta \in \left(0, \frac{\pi}{2}\right)$ as shown in Figure 17.2. Further, the antenna gain is approximately modeled in the direction (ϖ, ψ) as in [18]:

$$G = \begin{cases} \dfrac{G_0}{\varpi^2} & -\theta \leq \varpi \leq \theta, -\theta \leq \psi \leq \theta \\ g \approx 0 & otherwise, \end{cases} \quad (18)$$

where, $G_0 \approx 2.2846$; ϖ and ψ denote the azimuth and elevation angles, respectively [18]. For simplicity, $g = 0$, but g satisfies $0 < g < \frac{G_0}{\varpi^2}$. The UAV covers the region in an NFA with a radius $\bar{r} = H\tan\theta$, where H is the height of the UAV. The sequence used to consider the UAV's location is $\{c[k], k \in \{1,\ldots,K,K+1\}\}$, where $c[k]$ denotes the UAV's location at time slot k. The distance between the UAV and CU$n \in \{1,2,\ldots N\}$ at time slot $k \in \{1,\ldots K\}$ is given by [18]:

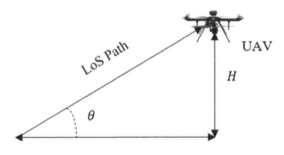

FIGURE 17.2
Uplink signal analysis in an emergency area.

$$r(k) = \sqrt{\|c[k] - c_n\|^2} \qquad (19)$$

As the UAV is located in a fringe area, thus the channel power gain between the UAV and a CU is given by

$$h_u(r) = \frac{\mathcal{B}_0}{H^2 + (r(k))^2} \qquad (20)$$

where, $\mathcal{B}_0$ is the channel power gain at the reference distance 1 meter. The propulsion power consumption for a rotary-wing UAV with speed is represented as in [18]:

$$F(V) = \Delta_0 \left(1 + \frac{3V^2}{\mathcal{U}_{tip}^2}\right) + \Delta_i \left(\sqrt{1 + \frac{V^4}{4v_0^4} - \frac{V^2}{2v_0^2}}\right)^{1/2} + \frac{1}{2}\chi_0 \kappa s \mathcal{A} V^3 \qquad (21)$$

where, Δ_0 and Δ_i denote the blade profile power and induced power, rotor disc area $\mathcal{A}, \mathcal{U}_{tip}$ represents the tip speed, v_0 denotes mean rotor induced velocity, χ_0 and s are the fuselage drag ratio and rotor solidity, respectively. At time T_i seconds(s), the rotary-wing UAV serves the CUs. The UAV flying time is less than the sum hovering time, i.e., $\sum_{i=1}^{K+1} T_i \gg \frac{\mathcal{L}}{V}$ where $\mathcal{L}$ denotes the total traveling distance. The mission completion time can be represented as $T_{completion} = \sum_{i=1}^{K+1} T_i + \frac{\mathcal{L}}{V} \approx \sum_{i=1}^{K+1} T_i$. The altitude range of UAV is $H \in [H_{min}, H_{max}]$, where $H_{min} > 0$ and $H_{max} > H_{min}$.

At the UAV, the received signal to noise ratio (SNR) can be represented as in [18]:

$$\gamma_u(r) = \frac{P_u h_u(r) G}{N_0} = \frac{\zeta H^2 \tan^2 \theta}{\theta^2 (H^2 + r^2)} \qquad (22)$$

where, $\zeta = \frac{P_u \mathcal{B}_0 G_0 \rho \pi}{N_0}$, P_u denotes the CU transmit power and N_0 denotes the noise power. The outage probability for the desired user in the emergency area can be expressed as:

$$\mathcal{P}_{out} = \mathcal{P}(\gamma_u(r) \leq \beta_{th}) \qquad (23)$$

where, β_{th} denotes target threshold. Hence, the data rate is $\mathbb{R}_u(r) = \log_2(1 + \gamma_u(r))$.

17.4.4 Area Spectral Efficiency for Emergency Area

In the emergency area, the ASE of the cellular users under the coverage of a UAV can be represented as:

$$\text{ASE} = \frac{1}{|A|} \sum_{n=1}^{N} E(\log_2(1 + \gamma_{u,n}(r))) \qquad (24)$$

TABLE 17.1

Simulation Parameters and Their Values

Simulation parameters	Values
a	200 meter (m)
A	700 meter
$\mathcal{A}$	0.79 meter
$\mathcal{U}_{tip}$	200 meter/s
v_0	7.2
H_{min}	10 meter
s	0.05
H_{max}	50 meter
α	2

where, $E\left(\log_2\left(1+\gamma_{u,n}(r)\right)\right)$ is the ergodic rate of the nth cellular user.

17.5 Simulation Results and Discussions

The results here verify the performance of the proposed IoT-enabled D2D network for the T-BS and rotary-wing UAV communication in the cellular uplink scenario. The numerical results are verified using extensive Monte Carlo simulation in MATLAB. Some network parameters are considered as $P_{cu} = 20$ dBm, $M = 3$, $L = 4$, and $N = 4$. In Table 17.1, some important parameters are shown.

Figure 17.3 shows the impact of γ_{th} on outage probability of a CU concerning R. We consider $R = 5$ m, 10 m, and 15 m for this plot. For a particular γ_{th}, the outage increases if the distance, R, increases. The probability of outage increases as the γ_{th} increases for a fixed R.

Figure 17.4 illustrates that the outage probability of a CU increases if the γ_{th} increases. Moreover, if the P_{cu} increases, the outage probability increases. For example, at $\gamma_{th} = 4$ dB, if the P_{cu} increases from 15 dBm to 25 dBm, the outage probability decreases by 18 percent. Also, the analytical and simulation results are presented in Figure 17.4.

In Figure 17.5, the area spectral efficiency vs. area with respect to P_{cu} is shown. For a fixed P_{cu}, the ASE decreases as the a increases for a fixed P_{cu}. If the P_{cu} increases, then the SINR increases. Therefore, the ergodic rate increases; as a result, area spectral efficiency increases.

Figure 17.6 shows the outage probability vs. γ_{th} in case of the emergency area with respect to P_u. The outage probability increases as the γ_{th} increases for a fixed P_u. However, if the P_u increases, then the outage probability decreases. For example, at $\gamma_{th} = 4$ dB, if the P_u increases from 20 dBm to 25 dBm, the outage probability decreases by 15 percent.

Figure 17.7 shows the outage probability vs. H in terms of θ. Moreover, if the H increases, then the channel power gain decreases. If the channel power gain decreases, then the SNR at the UAV decreases as a resulting outage probability increases. However, if the θ increases, then the outage probability increases. For example, at $H = 32$ m, if the θ increases from 20^0 to 30^0, then outage probability rises by 78.87 percent.

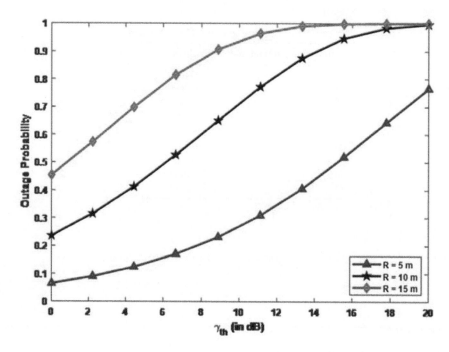

FIGURE 17.3
Impact of γ_{th} on the outage in terms of R (in dense area) at $a = 200$m.

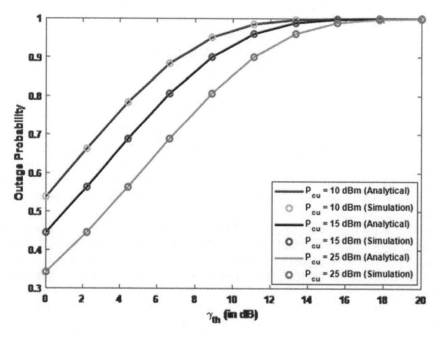

FIGURE 17.4
Outage probability of several values of γ_{th} at $N = 4$.

Network Recovery

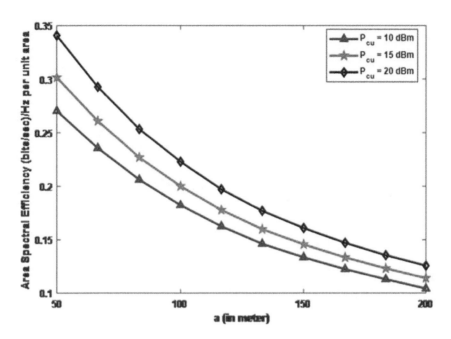

FIGURE 17.5
Area spectral efficiency vs. area (in dense area).

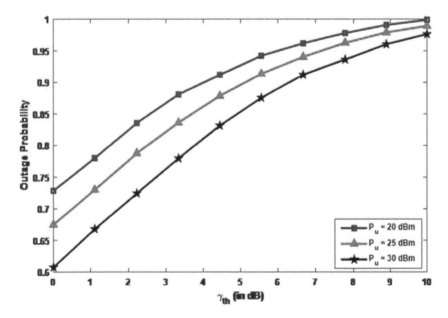

FIGURE 17.6
Outage probability vs. γ_{th} (in emergency area) at $A = 700$m.

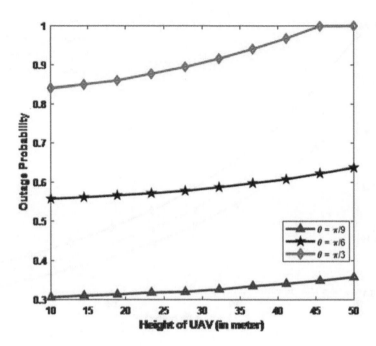

FIGURE 17.7
Outage probability vs. height of UAV in emergency area.

Figure 17.8 illustrates the impact of H on data rate in terms of P_u. It is shown that as the H increases, then the SNR decreases, so the data rate decreases. Also, it is observed that if the P_u increases, then the SNR increases; as a result, the data rate increases.

Figure 17.9 shows the area spectral efficiency vs. UAV for different values of A. The ASE decreases as the H increases for a particular A. It follows as the H increases, then the SNR decreases. If the SNR decreases, then the ergodic rate decreases; as a result, area spectral efficiency decreases; however, if the area in an emergency area decreases, then the area spectral efficiency increases.

17.6 Conclusion and Future Work

AT-BS and UAV-based networks have been presented in the uplink scenario. In a disaster-affected region, a T-BS and UAV provide assistance for establishing the connection. A T-BS has been placed in each cell where the BSs are partially damaged (dense area). It provides a communication network for cellular users and links between IoT gateway and IoT devices. The randomly located IoT devices deliver information to the IoT gateway and assist the rescue teams in providing necessary information about the disaster. However, a UAV provides a communication network for cellular users in an emergency area where BSs are fully damaged. The outage probability has been estimated for the proposed network in dense and emergency areas. Also, the area spectral efficiency for dense and emergency regions has been observed. Accordingly, the effect of several parameters such as UAV

Network Recovery 255

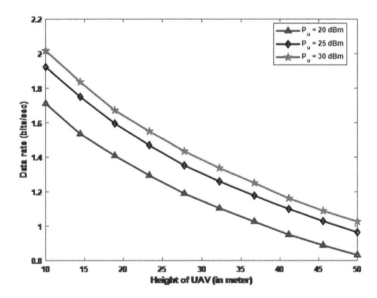

FIGURE 17.8
Impact of H on the data rate in emergency area.

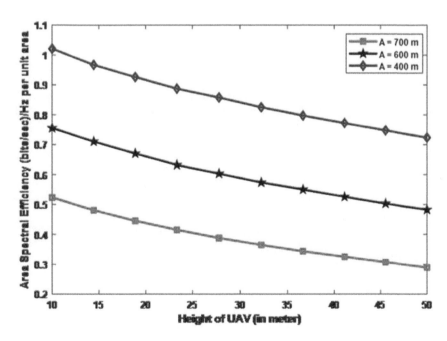

FIGURE 17.9
Area spectral efficiency vs. height of UAV (in emergency area).

speed, the height of UAV, etc., has been shown. Our analysis is helpful for applications in rural areas and urban areas. In the future, we can improve the performance of our proposed work by considering the downlink scenario. Also, the rotating UAV and its different movement, and scheduling approach, will be considered.

Acknowledgment

This research is funded by the TCS-RSP with Cycle 15.

References

[1] Partha Pratim Ray, Mithun Mukherjee, and Lei Shu. "Internet of Things for disaster management: State-of-the-art and prospects." *IEEE access* 5 (2017): 18818–18835.

[2] Akram Al-Hourani, Sithamparanathan Kandeepan, and Abbas Jamalipour. "Stochastic geometry study on device-to-device communication as a disaster relief solution." *IEEE Transactions on Vehicular Technology* 65, no. 5 (2015): 3005–3017.

[3] Vinod Kumar Tiwari and Vijay Singh. "Study of Internet of Things (IoT): A Vision, Architectural Elements, and Future Directions." *International Journal of Advanced Research in Computer Science* 7, no. 7 (2016).

[4] Amir H. Alavi, Pengcheng Jiao, William G. Buttlar, and Nizar Lajnef. "Internet of Things-enabled smart cities: State-of-the-art and future trends." *Measurement* 129 (2018): 589–606.

[5] Abdu Saif, Kaharudin Dimyati, Kamarul Ariffin Noordin, Nor Shahida Mohd Shah, Qazwan Abdullah, Mahathir Mohamad, Mahmod Abd Hakim Mohamad, and Ahmed M. Al-Saman. "Unmanned Aerial Vehicle and Optimal Relay for Extending Coverage in Post-Disaster Scenarios." *arXiv preprint arXiv:2104.06037* (2021).

[6] Akash Sinha, Prabhat Kumar, Nripendra P. Rana, Rubina Islam, and Yogesh K. Dwivedi. "Impact of Internet of Things (IoT) in disaster management: A task-technology fit perspective." *Annals of Operations Research* 283, no. 1 (2019): 759–794.

[7] Zhou Ni, Ziru Chen, Qinbo Zhang, and Chi Zhou. "Analysis of RF energy harvesting in uplink-NOMA IoT-based network." In *IEEE 90th Vehicular Technology Conference (VTC2019-Fall)* (2019): 1–5.

[8] Jia-Ming Liang, Kun-Ru Wu, Jen-Jee Chen, Pei-Yi Liu, and Yu-Chee Tseng. "Energy-efficient uplink resource units scheduling for ultra-reliable communications in NB-IoT networks." In *IEEE 29th Annual International Symposium on Personal, Indoor and Mobile Radio Communications (PIMRC)* (2018).

[9] Md Sadek Ali, Yu Li, Song Chen, and Fujiang Lin. "Narrowband Internet of Things: Repetition-based Coverage Performance Analysis of Uplink Systems." *J. Commun.* 13, no. 6 (2018): 293–302.

[10] Kaljot Sharma, Darpan Anand, Munish Sabharwal, Pradeep Kumar Tiwari, Omar Cheikhrouhou, and Tarek Frikha. "A disaster management framework using Internet of Things-based interconnected devices." *Mathematical Problems in Engineering* (2021).

[11] Partha Pratim Ray, Mithun Mukherjee, and Lei Shu. "Internet of Things for disaster management: State-of-the-art and prospects." *IEEE Access* 5 (2017): 18818–18835.

[12] Basem M. ElHalawany, Rukhsana Ruby, and Kaishun Wu. "D2D communication for enabling Internet-of-Things: Outage probability analysis." *IEEE Transactions on Vehicular Technology* 68, no. 3 (2019): 2332–2345.

[13] Joaquim Porte, Alan Briones, Josep Maria Maso, Carlota Pares, Agustin Zaballos, and Joan Lluis Pijoan. "Heterogeneous wireless IoT architecture for natural disaster monitorization." *EURASIP Journal on Wireless Communications and Networking* 2020, no. 1 (2020): 1–27.

[14] Hesham G. Moussa and Weihua Zhuang. "Access point association in uplink two-hop cellular IoT networks with data aggregators." *IEEE Internet of Things Journal* 7, no. 6 (2020): 5386–5400.

[15] Xingqin Lin, Vijaya Yajnanarayana, Siva D. Muruganathan, Shiwei Gao, Henrik Asplund, Helka-Liina Maattanen, Mattias Bergstrom, Sebastian Euler, and Y.-P. Eric Wang. "The sky is not the limit: LTE for unmanned aerial vehicles." *IEEE Communications Magazine* 56, no. 4 (2018): 204–210.

[16] Tomasz Izydorczyk, Gilberto Berardinelli, Preben Mogensen, Michel Massanet Ginard, Jeroen Wigard, and Istvan Z. Kovacs. "Achieving high UAV uplink throughput by using beamforming on board." *IEEE Access* 8 (2020): 82528–82538.

[17] Mohammad Mozaffari, Walid Saad, Mehdi Bennis, and Mérouane Debbah. "Unmanned aerial vehicle with underlaid device-to-device communications: Performance and tradeoffs." *IEEE Transactions on Wireless Communications* 15, no. 6 (2016): 3949–3963.

[18] Xiaonan Liu, Zan Li, Nan Zhao, Weixiao Meng, Guan Gui, Yunfei Chen, and Fumiyuki Adachi. "Transceiver design and multihop D2D for UAV IoT coverage in disasters." *IEEE Internet of Things Journal* 6, no. 2 (2018): 1803–1815.

[19] Haiyun He, Shuowen Zhang, Yong Zeng, and Rui Zhang. "Joint altitude and beamwidth optimization for UAV-enabled multiuser communications." *IEEE Communications Letters* 22, no. 2 (2017): 344–347.

[20] Liang Liu, Shuowen Zhang, and Rui Zhang. "Multi-beam UAV communication in cellular uplink: Cooperative interference cancellation and sum-rate maximization." *IEEE Transactions on Wireless Communications* 18, no. 10 (2019): 4679–4691.

[21] Mohammad Mozaffari, Walid Saad, Mehdi Bennis, Young-Han Nam, and Mérouane Debbah. "A tutorial on UAVs for wireless networks: Applications, challenges, and open problems." *IEEE Communications Surveys & Tutorials* 21, no. 3 (2019): 2334–2360.

[22] Fenyu Jiang and Chris Phillips. "High Throughput Data Relay in UAV Wireless Networks." *Future Internet* 12, no. 11 (2020): 193.

[23] Rui-lin Liu, Zhong-jie Zhang, Yu-fei Jiao, Chun-hao Yang, and Wen-jian Zhang. "Study on flight performance of propeller-driven UAV." *International Journal of Aerospace Engineering* (2019).

18

Wireless Sensor Networks with the Internet of Things

Ashish Bagwari, Geetam Singh Tomar, Jyotshana Bagwari, Jorge Luis Victória Barbosa, K.S. Sastry, and Manish Dixit

CONTENTS

18.1 Introduction 260
18.2 WSN Methodology 260
18.3 Architecture of WSN with IoT 261
18.4 Applications of WSN with IoT 263
      18.4.1 Military Application 263
      18.4.2 Environmental Application 263
      18.4.3 Agricultural Application 264
      18.4.4 Health Application 264
      18.4.5 Infrastructure Monitoring Application 264
      18.4.6 WSNs for Power Engineering Systems 264
18.5 Advantages of WSN with IoT 266
      18.5.1 Effective in Harsh Environments 266
      18.5.2 Data Collection Process in WSN 266
      18.5.3 Long-distance Communication 266
      18.5.4 Protecting Hardware and Data Assets 266
18.6 Challenges of WSN with IoT 266
      18.6.1 Real-time Monitoring 266
      18.6.2 Security and Safety 267
      18.6.3 Quality of Service 267
      18.6.4 Configuration 267
      18.6.5 Availability 267
      18.6.6 Data Integrity 268
      18.6.7 Scalability 268
      18.6.8 Power Consumption 268
      18.6.9 Communication 269
18.7 Conclusion and Future Work 269
Acknowledgments 270
References 270

DOI: 10.1201/9781003326205-22

18.1 Introduction

With the development of wireless networking technologies, every aspect of our daily lives has altered substantially. One of the technologies that is developing the quickest is the Internet of Things. With the addition of IoT, several devices may be connected in the real world, effectively altering our daily lives. Since this is the case, the need for constant and consistent communication is quickly expanding, especially in highly dynamic fields [1]. The IoT has been defined as the interconnection and exchange of data amongst autonomous systems (things). New technologies and applications may flourish thanks to the supremacy of IoT. Transceivers, microcontrollers, and protocols for the transmission and control of sensor data are often included in these sensors and actuators (including home appliances, security cameras, and environmental monitoring sensors) [2]. Sensors and other real-time components are linked together to deliver collected data to consolidated databases, where it is aggregated and made accessible to authorized users.

The characteristics of IoT using wireless technologies are considerably different from those of regular wired or wireless networking systems since there are so many more communication devices [3]. However, since each IoT device detects and transmits some data to a specific IoT server, data produced by a large number of objects may have some effect on the performance of the network as a whole. Therefore, IoT networks will operate reliably and sustainably for a very long time with little to no human intervention. The cornerstone of the IoT-based systems all around us is heterogeneous WSN, which connects a variety of intelligent sensors and will soon bring considerable improvements [4]. These products' quick technological advancement has led to energy consumption issues [5]. The rapid increase in communication and information exchange has led to uncontrollable increases in energy use and carbon emissions.

18.2 WSN Methodology

However, for the sensor nodes to be truly useful in most scenarios, they need to be able to operate reliably over the course of years rather than months or even weeks. This is especially true for tasks like environmental control and protection, agricultural monitoring, border surveillance and protection, and so on, due to various application criteria [6]. The energy used by the sensors determines how long an application will last, and dead nodes might have an impact on data dependability, device compatibility, and accuracy. A sensor node typically consists of four main components:

- processing subsystem
- sensing subsystem
- communication subsystem
- power supply subsystem

These four points are shown in Figure 18.1, with individual process:

The sensor nodes might cause a 500 to 1,000 percent rise in total energy usage (CPU, radio, etc.) during data receiving. Otherwise, the sensor nodes require a lot less effort

WSNs with the Internet of Things 261

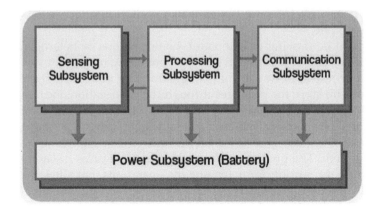

FIGURE 18.1
A process of WSN-based sensor nodes.

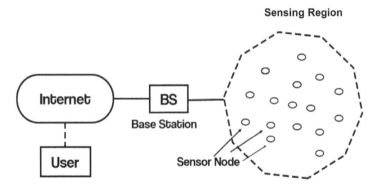

FIGURE 18.2
WSN architecture with IoT.

and resources. The sensor module analyzes the surrounding environmental factors and converts ambient energy into electric signals. The processor module processes the information to acquire information on events occurring nearby to the sensor, and the data is then communicated via radio transmitter to a destination node. The sensor nodes are powered by a battery, which is often not rechargeable or replaceable, particularly when the sensor nodes are anticipated to function for a longer period of time without human interaction throughout the application [7]. As a result, the IoT community is increasingly driven to provide clean, efficient energy solutions.

18.3 Architecture of WSN with IoT

In a sensor field, the sensor nodes are commonly scattered. Each of these dispersed sensor nodes is able to collect data and transmit it back to the sink and end users. As illustrated in Figure 18.2, data are routed back to the end user through a multi-hop design without infrastructure via the sink. Internet allows the user/sink to interface with the sensor node/task management node.

The nodes might be moving or stationary, homogeneous or heterogeneous, aware of their position or not, and could contain a power generating unit or not. On the other hand, for the sake of our research, we are going to assume that nodes are (i) preprocessing, (ii) analyze the user, (iii) store the packages through the base station and, (iv) mine the data through the sensor nodes (scalability of large data). In response to the occurrence that is being monitored, the sensor generates analogue signals [8]. The analog-to-digital converter (ADC) takes these signals and transforms them into digital data, which is then sent to the processing unit. The processing unit, which in most cases has a limited amount of onboard storage, is responsible for managing the operations that allow the sensor node to interact with other nodes in order to perform the defined sensing functions. In general, the on-board storage is of a modest quantity.

The node, along with the other nodes and the base station, are linked together through a transceiver unit to create the wireless sensor network (WSN). When the nodes are deployed in a specific location, they have the ability to spontaneously build a communication network in that location. As can be seen in the diagram, sensor nodes are often installed in a topology that is not regular, and they are densely crowded (Figure 18.3). The phenomenon is detected by the remote sensors, which are also known as motes. This information is then preprocessed and sent once the perceived amount has been converted into data. The observed phenomena are ultimately analyzed in a centralized site known as the base station (BS), which is equipped with necessary processing capabilities [9]. Establishing a route between the motes and the base station is necessary for the data transmission process. This route often requires more than one hop. Some operations let the sensor node

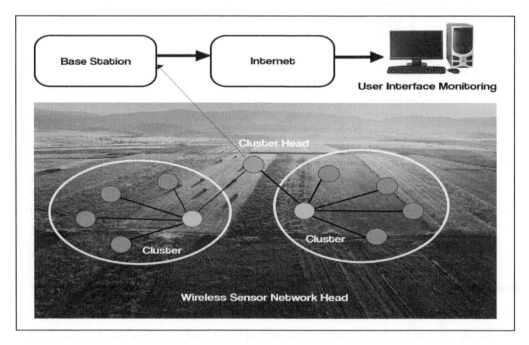

FIGURE 18.3
Structure of wireless sensor network (WSN) in the agriculture field.

interface with other nodes in order to carry out the specified sensing duties; however, this will depend on the routing algorithm.

18.4 Applications of WSN with IoT

Today, sensors capable of monitoring a variety of physical events have been produced successfully. The expansion of micro-electro-mechanical systems (MEMS) technology has made intelligent sensors possible. In addition to environmental elements such as temperature, pressure, etc., sensors can now detect a myriad of additional phenomena. Research [10] addressed the design and implementation of a solar-powered precision agricultural (PA) network with a wireless sensor network (WSN) by applying IoT architecture to satisfy the need for identifying highly effective methods for a smart agriculture management system. Through IoT real-time data exchanges, this system provides farmers with crucial information on saltwater incursions, soil moisture, water level, wet conditions, temperature, and the general state of the land in a user-friendly and simply accessible format. Besides PA networks, IoT-based WSNs find interesting applications in different fields such as military, weather monitoring, infrastructure management, health management, etc. The sections below briefly explain some of those possible applications.

18.4.1 Military Application

Military command, control, communications, computing, intelligence, surveillance, reconnaissance, and targeting (C4ISRT) systems might include wireless sensor networks [11]. During the Cold War, acoustic networks were used for submarine monitoring. A few of these sensors continue to detect earthquakes in the area. C4ISRT sensing qualities include rapid deployment, self-organization, and error tolerance. All of these characteristics make WSNs a suitable C4ISRT sensing technology. Sensor networks are densely distributed, have low deployment costs, and sensor failure has little effect on performance. Consequently, sensor networks are suitable for battlefields. Such applications need sensor security. Monitoring friendly troops, equipment, and ammunition; battlefield surveillance; reconnaissance of opposing forces and terrain; targeting; combat damage assessment; and nuclear, biological, and chemical (NBC) attack detection and reconnaissance are among the military applications of sensor networks [12].

18.4.2 Environmental Application

Environmental concerns are increasing. Environmental scientists are studying the influence of industrial and other activities on the environment, especially climate change. Environmentalists also monitor wildlife, aqualife, and bird migration. Remote sensing satellites assist in anticipating weather, rainfall, thunderstorms, and forest fires. Good agricultural outputs need soil and moisture monitoring. Wireless sensors are widely used [13]. Sensors for environmental monitoring need temporal, spatial, and spectral integration. Further, environmental sensors must be put in distant air, surface, and undersea environments. Long-term unattended operation is required. Sensors may need power-scavenging technologies like solar cells. Communication systems must work with varied media and barriers. As radio signals attenuate depending on the weather, communication must be ensured under all situations.

18.4.3 Agricultural Application

IoT-based intelligent agriculture will be very helpful for farmers. Intelligent agriculture employs GPS-controlled robotics, intelligent irrigation, and intelligent warehouse management. All of these methods are performed by sensors controlled by a remote smart device or Internet-connected computer. In Figure 18.3, the blue-colored cluster node is a remote-controlled, GPS-enabled robot that performs duties such as weeding, spraying, sensing moisture, frightening birds and animals, etc. Sensors measure temperature and humidity for the intelligent warehouse management system cluster node, which is in pink [14]. A smart irrigation cluster head mode adjusts the water pump level after wetting the soil. AVR Microcontroller Mega 16/32 utilizes the Zigbee module, ultrasonic sensor, Raspberry Pi, and Temperature Sensor LM35 for operations. AVR studio v4, Proteus 8 Simulator, Dip Trace, SinaProg, and the Raspbian operating system are software and moisture sensors. The IoT improves agricultural production and overall productivity.

18.4.4 Health Application

WSNs are frequently used in healthcare. Wireless medical sensor networks (WMSN) are a cutting-edge aspect of the healthcare industry that enhances patient care without sacrificing comfort. The memory, processing power, battery life, and bandwidth of WMSN devices are restricted [15]. These medical sensors (e.g., ECG electrodes, pulse oximeter, blood pressure, and temperature sensors) are implanted in a patient's body to gather physiological data and transmit it over a wireless channel to the hand-held devices of healthcare providers, as seen in Figure 18.4. Healthcare applications such as continuous patient monitoring, mass-casualty disaster monitoring, large-scale in-field medical monitoring, and emergency response have profited from wireless medical sensor technologies. WMSNs also provide novel methods for analyzing acute disorders (such as Parkinson's).

18.4.5 Infrastructure Monitoring Application

Every society's progress depends on its infrastructure. Building roads, bridges, dams, pipelines, airports, etc. is expensive and takes a long time. Economic development depends on monitoring the use and health of such systems. With terror threats mounting, it is important to defend national infrastructure [16]. Even during construction, design faults/defects and human mistakes create accidents. Such incidents involve human deaths, thus early warning signs are crucial. In 2009, a bridge under construction in Kota, Rajasthan, collapsed, killing 48. WSN technology may provide early warnings of imminent incidents.

18.4.6 WSNs for Power Engineering Systems

WSNs have a significant role to play in power engineering systems, especially in operation and management of networks, addressing grievances of consumers, and planning future expansion. Due to recent technological advances in renewable energies and deregulation policies, several smart and micro grids are established around the world [17]. Though this

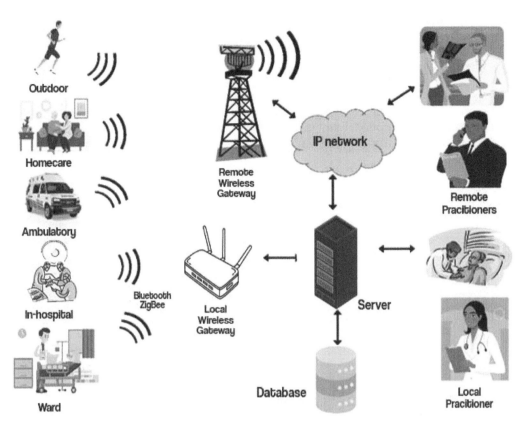

FIGURE 18.4
Typical architecture of medical application with WSN IoT.

trend has resulted in several benefits, monitoring such grids is becoming a challenge. Some of the reasons for this include, but are not limited to: enterprise class supervisory and control systems are very expensive; real-time monitoring of power apparatus requires a lot of skilled and qualified human resources, which again can be expensive; and portions of networks may be owned by different individuals privately. In modern smart grids, data typically is generated at every load point for every 30-minute interval [17, 18] and is sent to central servers for further processing, such as measurement and billing. With appropriate use of WSNs, such tasks can be taken up in edge and fog computing layers without having to burden the central servers. Demand estimation is an important task in future planning [18] and WSNs can help in aggregating the data collected and then provide the summarized information. In the case of commercial and industrial consumers, it is possible to deploy WSNs for large appliances that consume higher amounts of energy, since overall costs in deploying WSNs and their maintenance can be far lower when compared to conventional monitoring systems [17, 18]. Apart from the above applications, several other benefits – such as load forecast or dedicated energy dashboards for large-scale power consumers – can be realized with the same WSN infrastructure, as the collected information can be processed as needed. Further research needs to be undertaken on how the collected raw data can be processed and how the information can be channeled, and on design and development of newer, yet smarter WSN architectures for smart grids.

18.5 Advantages of WSN with IoT

18.5.1 Effective in Harsh Environments

Initially, wireless sensor networks were created for battlefield surveillance in military operations where a cable network was impractical. In these settings, a wireless sensor network might be installed with low danger to monitor combat conditions. Wireless sensor nodes may be dropped from an aircraft into any desired region to transmit data for processing.

18.5.2 Data Collection Process in WSN

When it would be impossible or dangerous for a human to physically collect sensor data, WSNs are advantageous. It would be safer and less expensive to have a sensor node at the bottom of a reservoir as opposed to physically monitoring it.

18.5.3 Long-distance Communication

Their network architecture makes WSNs scalable for environmental monitoring. If you have deployed sensor nodes to monitor a certain area and want to expand, you may configure more nodes to communicate on the same network and deploy them in the expanded area. Adding sensors to the network does not require changing old ones.

18.5.4 Protecting Hardware and Data Assets

Overheated servers may cause data loss and unscheduled downtime in data centers. Data centers are crammed with computers, and adding connected sensors may add to the cable congestion. Wireless sensor nodes make real-time temperature sensing in data centers simpler without requiring unnecessary wiring.

18.6 Challenges of WSN with IoT

The complexity of the IoT is achieved via the presentation and communication of a variety of heterogeneous artifacts in a variety of circumstances, which further complicates the deployment of security systems. The majority of existing WSN security research focuses on finding answers to nebulous problems without taking into account the influence of the IoT's guiding principles and distinguishing characteristics, as detailed in this chapter.

18.6.1 Real-time Monitoring

This is a challenging topic for sensor networks that govern their available resources. In this instance, the IoT system requires efficient service gateway architecture to limit the quantity of data that must be delivered by continuously analyzing user data. In addition, data-driven intelligent middleware architecture is necessary to convey real-time data only when the reading exceeds the threshold [19].

18.6.2 Security and Safety

In applications that are used in the real world, safety, trust, and privacy are other challenges that are vital to consider. The path to achieving varying degrees of safety may be challenging as well as straightforward. Some security solutions are appropriate for M2M deployments in which the device and the server already have an established trust relationship with one another. [20] With this "IP to the field" approach, sensor nodes have extra obligations in addition to the standard sensor functions they provide. As a result of this added obligation, the sensor nodes will be faced with new responsibilities or difficulties. The topics of network setup, service quality (QoS), and security will be explored as possible future work. The following issues are taken into consideration.

Depending on the level of sophistication of the application, WSNs may be able to provide data with confidentiality, verification, fairness, and usability even in the absence of Internet access. For an attacker to be successful in adding malicious nodes to the network or in blocking or catching nodes, they need to engage in physical activities in close proximity to the WSN [21]. The connection of WSNs to the Internet, on the other hand, makes it possible for malicious actors from all over the globe to carry out their operations.

18.6.3 Quality of Service

In terms of the intelligence that is provided to the sensor nodes, each disparate device that is connected to the IoT is required to contribute to the overall quality of service. These heterogeneous devices enable a task to be distributed among several nodes, each of which has access to a certain set of resources [22]. Because of changeable network configurations and connection properties, the currently available quality of service (QoS) techniques on the Internet still need to be improved.

18.6.4 Configuration

In addition to the management of quality of service and security, sensor nodes need to manage a variety of tasks. These tasks include networking for a new node that is joining the network [23], ensuring self-healing by locating and removing faulty nodes, and addressing management for the construction of scalable networks, amongst other tasks. However, self-configuring the most recent node on the Internet does not happen as a normal function. For this reason, the user is required to install the proper software and take enough precautions to avoid device failures in order for network configuration to function without difficulty.

18.6.5 Availability

WSNs may be exploited if there are hacked nodes in the network [24]. Additional funds would need to be budgeted in order to pay for the implementation of an encryption scheme for WSN security. In spite of this, researchers have developed major approaches, some of which include modifying the code in order to reuse it, while others involve using supplemental communications in order to achieve their aims. In addition to this, several methods have been developed in order to access the data. Availability is very necessary in order to maintain the operating services of WSNs. Additionally, it contributes to the maintenance of the whole network up to the point when it is shut off.

18.6.6 Data Integrity

When a rogue node joins the network and injects erroneous data or when a vacillating wireless channel corrupts the original data, WSN may be infiltrated [25]. In addition, during the decryption of aggregated data, the base station is able to categorize the encrypted and aggregated data based on the encryption keys. Nevertheless, a malfunctioning network may be to blame for the loss of data or the change of data. As a result, it is essential that the integrity of the data be preserved all the way through the transmission of the data packets.

18.6.7 Scalability

The needs of an application will determine the total number of nodes that will be deployed inside that application. Depending on the use case, the density of the network might be anything from a few nodes to thousands of nodes. To function well in environments with different densities, communication and data aggregation methods are necessary [26]. In networks with a low population density, only a small number of nodes are able to transmit data, but in networks with a high population density, several nodes are able to send data, which may result in multiple data values being produced for the same phenomena at the same time. The algorithms that aggregate the data need to be able to detect instances of data being collected more than once. Additionally, the batteries are depleted since several sensors are transmitting the same data readings.

18.6.8 Power Consumption

Sensor networks have their own special set of issues when it comes to networking. Only if the application continues to collect and transmit data for longer than the required life will the deployment of WSNs for that application be considered justifiable. According to the findings of recent research, the most significant user of energy in WSNs is radio communication. According to the authors [26] on the cost of computation compared to the cost of communication in future platforms, which were presented by [27], it is possible to carry out 3,000 instructions for the same cost as the transmission of one bit over a distance of 100 meters. In general, the amount of power that is lost is negatively proportional to the square of the distance. When it comes to WSNs, on the other hand, the power loss is often of the fourth order. This finding lends credence to the idea that communication protocols should be developed in such a way as to utilize the least amount of energy feasible.

Two different strategies are used in such networks to save energy and reduce the amount of money spent on communication costs. The first occurs at the media access control (MAC) and networking levels, when nodes shut off their radios (using adaptive duty cycling) when they are not needed for communication. The second method is termed data reduction via in-network processing, and it is also known as data aggregation. In this method, correlations in the data are used in order to minimize the quantity of the data and, as a result, the cost of transmission. According to [27], the total radio power consumption P_c is as follows:

$$P_c = N_T[P_T(T_{on}+T_{st}) + P_{out}(T_{on})] + N_R[P_R(R_{on}+R_{st})] \tag{1}$$

where P_T/P_R is the power used by the transmitter/receiver, P_{out} is the power output by the transmitter, T_{on}/R_{on} is the time the transmitter/receiver is turned on, T_{st}/R_st is the time the transmitter/receiver is started up, and N_T/N_R is the number of times the transmitter/

receiver is switched on per unit time, where both the task at hand and the MAC scheme in use influence this value.

18.6.9 Communication

Consumption of electricity is the primary focus of attention in WSNs. The nodes are powered by batteries. The power output of the battery is limited, both financially and in terms of its size and capacity. After the battery has been completely depleted, it is not feasible to recharge or replace it in the vast majority of situations. In the case that the battery dies, the node ceases functioning and is said to be dead. An increase in the number of dead nodes has the effect of reducing the amount of data collected and sent to the BS. Because of this, the network cannot support the application in its current state. It is advisable that while developing WSNs or any component of them, attention should be given to spending as little energy as possible. This is true for both the WSNs themselves and any component of them.

Each node in a multi-hop ad hoc sensor network fulfills two distinct but complementary functions: the role of data originator and the function of data router. Each node, in its capacity as the data originator, is obligated to collect data from the on-board sensors, analyze that data, and then transfer it to the next node, which will then send it to the BS. Each sensor node is responsible for both the generation of data as well as the relaying of information that is sent by the nodes in its immediate vicinity. The communication range of a node in a WSN is restricted due to the use of low-power communication protocols. It is necessary to have multi-hop communication in a big network so that nodes may transmit the data that has been sent to them by their neighbors to the data collector, also known as the sink.

18.7 Conclusion and Future Work

A few fields in which WSNs may be put to use were dissected in this article. There was also discussion of the difficulties involved in the design of sensor nodes and the construction of wireless communication infrastructure. In WSNs, routing needs to overcome obstacles related to the life duration of the network, which are caused by energy limits. We also highlighted some of the research issues that arise while constructing WSNs for current applications. In sensor networks, one of the primary challenges is determining how essential data may be handled in a manner that uses less power. As a result, several different techniques for data aggregation used in order to reduce the amount of power consumption were discussed in this work. In this study, a review of prior works defining the function of IoT in WSN was provided, and then different data aggregation methodologies offered in earlier works were discussed. The strategies for data aggregation put an emphasis on the network's quality of service (QoS), energy conservation, increased longevity, and increased levels of network security.

Mobile base stations are thought to increase a network's longevity by ensuring that its stored energy is used consistently. This is in contrast to stationary base stations, which present a significant limitation. In integration, WSNs and IoT must have interoperable layered functions. WSN nodes must be upgraded to withstand Internet security assaults. WSNs must enhance their capacity and IoT must adapt its layered operation to work with WSNs.

Acknowledgments

We thank our parents for their support and motivation.

References

[1] M. A. Matin (ed.), *"Wireless Sensor Networks – Technology and Applications*. Intech, 2012.
[2] M. Murthy, *Ad Hoc Wireless Networks: Architectures & Protocols*. Pearson Education, 2007.
[3] I.F. Akyildiz, "Wireless sensor networks: A survey," *Computer Networks*, 38(4) (2002), 393–422.
[4] S. Kazem, M. Daniel, Z. Taieb, *Wireless Sensor Networks: Technology, Protocols, and Applications*. John Wiley & Sons, 2007.
[5] K. Pardeep, G.L. Sang, J.L. Hoon, "E-SAP: Efficient-strong authentication protocol for healthcare applications using wireless medical sensor networks," *Sensors*, 12(2) (2012), 1625–1647.
[6] M. Al Ameen, J. Liu, K. Kwak, "Security and privacy issues in wireless sensor networks for healthcare applications," *Journal of Medical Systems*, 36 (2012), 93–101.
[7] L. Konrad, "Mercury: A wearable sensor network platform for high-fidelity motion analysis," in *Proceedings of the 7th ACM Conference on Embedded Networked Sensor Systems*, pp. 183–196. ACM, 2014.
[8] I. Jawhar, N. Mohamed, K. Shuaib, "A framework for pipeline infrastructure monitoring using wireless sensor networks," in *Wireless Telecommunications Symposium*, pp. 1–7. IEEE, 2007.
[9] A. Mainwaring, J. Polastre, R. Szewczyk, D. Culler, J. Anderson, "Wireless sensor networks for habitat monitoring", in *Proceedings of the ACM International Workshop on Wireless Sensor Networks and Applications*, pp. 399–423. ACM, 2002.
[10] F. Flammini, A. Gaglione, F. Ottello, A. Pappalardo, C. Pragliola, A. Tedesco, "Towards wireless sensor networks for railway infrastructure monitoring," *Electrical Systems for Aircraft, Rail-way and Ship Propulsion*, 19(21) (2010), 1–6.
[11] C. Kameswari et al., "BriMon: A sensor network system for railway bridge monitoring," in *Proceedings of the 6th International Conference on Mobile Systems, Applications, and Services*, pp. 2–14. ACM, 2015.
[12] K. Flouri, "A versatile software architecture for civil structure monitoring with wireless sensor networks," *Smart Structures and Systems*, 10(3) (2012), 209–228.
[13] Y. Cho, M. Kim, S. Woo, "Energy efficient IoT based on wireless sensor networks for healthcare," in *20th International Conference on Advanced Communication Technology (ICACT)*, pp. 294–299. IEEE, 2018.
[14] H.W. Kim, D. Kyue, "Technology and security of IoT," *Journal of the Korea Institute of Information Security and Cryptology*, 22 (2012), 7–13.
[15] S.A.H. Antar, N.M. Abdul-Qaw, S. Almurisi, S. Tadisetty, "Classification of energy saving techniques for IoT-based heterogeneous wireless nodes," *Procedia Computer Science*, 171 (2020), 2590–2599.
[16] N. Kaur, S.K. Sood, "An energy-efficient architecture for the Internet of Things (IoT)," *IEEE Systems Journal*, 11(2) (2017), 796–805.
[17] K.S.S. Musti, H. Iileka, F. Shidhika, "Industry 4.0 based enterprise information system for demand-side management and energy efficiency," in N. Prakash and D. Prakash (eds.), *Novel Approaches to Information Systems Design*, pp. 1570–1591. IGI-Global, 2020.

[18] K.S.S. Musti, "Quantification of demand response in smart grids," in *IEEE India Council International Subsections Conference (INDISCON)*, pp. 278–282. IEEE, 2020.

[19] A.S. Abdul-Qawy, P.P.J.E. Magesh, T. Srinivasulu, "The Internet of Things (IoT): An overview," *International Journal of Engineering Research and Applications*, 5 (2015), 71–82.

[20] X. Zhou, "Green communication protocols for mobile wireless networks," Ph.D. thesis, University of Ottawa, 2017.

[21] M. Healy, T. Newe, E. Lewis, "Wireless sensor node hardware: A review," in Proceedings of *IEEE Sensors*, pp. 621–624. IEEE, 2008.

[22] H.M. Fahmy, *Concepts, Applications, Experimentation and Analysis of Wireless Sensor Networks*. Springer, 2016.

[23] K.I. Kim, "Clustering scheme for (m, k)-firm streams in wireless sensor networks," *Journal of Information and Communication Convergence Engineering*, 14(2) (2016), 84–88.

[24] Y. Cho, S. Lee, S. Woo, "An adaptive clustering algorithm of wireless sensor networks for energy efficiency," *Journal of the Institute of Internet, Broadcasting and Communication*, 1 (2017), 99–106.

[25] M.S. Islam, G.K. Dey, "Precision agriculture: Renewable energy based smart crop field monitoring and management system using WSN via IoT,"in *International Conference on Sustainable Technologies for Industry 4.0 (STI)*, pp. 1–6. IFIP, 2019.

[26] K. Begum, S. Dixit, "Industrial WSN using IoT: A survey," in *International Conference on Electrical, Electronics, and Optimization Techniques (ICEEOT)*, pp. 499–504. IEEE, 2016.

[27] S. Sarkar, K.U. Rao, J. Bhargav, S. Sheshaprasad, A. Sharma, " IoT based wireless sensor network (WSN) for condition monitoring of low power rooftop PV panels," in IEEE 4th International Conference on Condition Assessment Techniques in Electrical Systems (CATCON), pp. 1–5. IEEE, 2019.

19

Wireless Sensor Networks for Energy, E-Health, Building Maintenance and Agriculture Areas, and Simulation Results

Nakka Marline Joys, N. Thirupathi Rao, and Debnath Bhattacharyya

CONTENTS

19.1 Introduction 273
19.2 Related Works 274
    19.2.1 Various Attacks on Wireless Sensor Networks 275
    19.2.2 Solutions to Overcome Attacks on WSNs 276
19.3 Materials and Methods 278
    19.3.1 Proposed Model 278
        19.3.1.1 Security Technologies Applied in the Model 278
        19.3.1.2 Wireless Sensor Network in Agriculture Protection Model Proposed 279
        19.3.1.3 Preparation for Model Implementation 281
        19.3.1.4 Protection of the Sensor Node Message in the Process of Transmission 281
        19.3.1.5 Verification of the Content of the Protected Message in the Process of Reception 283
19.4 Characteristics of the Model 283
19.5 Proposed Model Simulation Results and Discussions 286
19.6 Results and Discussions 288
19.7 Conclusion 290
Acknowledgements 290
References 290

19.1 Introduction

The goal of this study is to come up with a model for a security sensor network that is independent of communications infrastructure and offers point-to-point security, as well as to figure out how to adapt it to sensor networks in other human activities that need data security for management reasons. The food and agricultural industries are among the biggest and most essential parts of the economy. The complexity of this biological production system is contributed to by a wide variety of factors, including people, machines, nature, chemistry, biology, the weather, and climate.

In a system as complicated as this one, decisions have to be made based on a diverse variety of available data. The ability to collect agricultural monitoring data is impacted by a number of factors, including geographic location, weather and climate, proximity to markets, transportation and storage options, and, of course, the fact that farming practices and activities are as diverse as the people who participate in them [1].

Real-time sensor data may be collected using wireless sensor networks (WSNs) in the agricultural sector. Due to the dynamic nature of the information being monitored, the system is able to perform agri-parameter monitoring in real time. When you have to make a choice in a short amount of time, it may be daunting to consider a lot of information coming from a variety of sources. If you want to work with large amounts of data, you are going to need to be a good analyst [2, 3].

Recent progress in agriculture is based not just on the amount and quality of data obtained, but also on the safety of that data. To put this another way, security risks like eavesdropping, interruption, or even physical attacks have the potential to change the data as well as the structure of the network. A security model that allows for the processing and storage of data from the source node (origin) is required both by the communication network and the application in order to guarantee that sensor data remains correct, undamaged, and encrypted all the way through the transmission process to the destination. A data security architecture is developed with the use of encryption, digital envelopes, digital signatures, and a public key infrastructure in order to protect data from sensors, wireless networks, and data processing applications. This is done so that the data may be kept secure (PKI). Since the security of the model is not reliant on the communications infrastructure, sensor networks and other human activities that need data security for management objectives may also benefit from the model's security. The requirements for the security of a distributed system are satisfied. These requirements include authentication and authorization, in addition to data integrity, trustworthiness, and availability.

The primary objective of this research is to develop a point-to-point security sensor network model that is independent of communications infrastructure, and then figure out how to apply that model to sensor networks in other human activities that require data security for management reasons. The remaining components are ordered in the following manner: in the third part, we will discuss the various materials and methods that may be used. The fourth part of this chapter contains many models that may be used to secure wireless sensor networks. The fifth part provides an evaluation of the performance of the model. In the sixth portion, we provide an updated model, and in the seventh section, we bring the structure of the work to a close.

19.2 Related Works

Sensor networks, also known as SNets, are a kind of widely distributed system that made up of several sensor nodes that are placed in the field and have the ability to interact with each other using wireless networks [4]. This system intends to identify, with the assistance of a distributed network of sensors, which data most accurately reflects what it has observed.

Wireless sensor networks, often known as WSNs, are networks of sensors that are spread out over the world and do not have any central support. These sensors are used to monitor a wide range of physical and environmental factors in order to provide accurate data. WSNs are finding more and more applications in the agricultural sector. Examples of monitoring technologies include precision agriculture equipment and process control and monitoring systems.

The following sets of security requirements must be met by sensor networks in order for them to be considered secure: authentication, authorization, data integrity, trustworthiness, availability, non-repudiation, trust, and privacy [5].

Authentication refers to the process of determining whether or not a member of a sensor network really is who it claims to be. Authorization is the process of determining whether or not someone is authorized to participate in a certain activity. It is essential to take precautions to prevent data from being corrupted or deleted in an unauthorized way. If the integrity of the data is preserved, then sensor data will not be changed without the user's permission while it is being provided, processed, or stored. The inaccessibility of data by unauthorized persons is an essential component of "reliability." There are precautions that are taken to ensure the safety of the transmission, as well as monitoring of the transmission methods that are used in sensor networks. The fulfillment of these requirements guarantees that the system is in working order and that all authorized users have access to the service. In accordance with the provisions of section X.509, "In general, a person or organization may show confidence in another person or organization by expressing trust in that other person or organization" (International Telecommunication Union Telecommunication Standardization Sector). In order for a transaction to be considered legal, it must be non-repudiated, which means that all parties involved must recognize it and it cannot be questioned. Trust is a vital factor to take into account; nevertheless, securing user information from unauthorized access is a difficult task due to the fact that it may apply to the whole dataset, not just a portion of the data. To protect the confidentiality of data, one must make every effort to restrict access to just those persons who have a legitimate need for the information.

19.2.1 Various Attacks on Wireless Sensor Networks

There are two different kinds of assaults that may be made against the security of a WSN: passive and active. Active attacks are made with the goal of compromising either the data's privacy or its integrity.

The communication architecture of a WSN network is built up using layers. As a result of the layered construction, it is possible to carry out a wide range of attacks, including the following.

WSNs are susceptible to physical layer assaults, which may be carried out by an adversary either by seizing control of a node or by blocking its usage of a radio channel. A signal with a high energy level is sent in order to prevent sensor nodes from communicating with one another.

At the link-layer level, protocols are used to guarantee that neighboring nodes may use the same wireless channels. This ensures that links can be abstracted. The functionality of the protocol may be altered by the attacker in many different ways: by sending a collision packet, by adding and probing messages to learn about the communication template, or by stifling communication. There is a large number of strategies that may be used in order to interfere with the capability of the WSN to communicate. In a traditional hole attack, an adversary may try to take control of the network by redirecting all traffic to a compromised node that seems to be a genuine participant in the network but is really under their command. Nodes that have been compromised by an attack or are known to be malicious are unable to route messages. An adversary may tamper with, change, or invent routing information in order to obstruct the flow of data across a network. Attackers can send as many requests for connections as they see fit, as long as they do so before using all of the resources necessary for each connection or reaching the maximum allowed number of assaults on the transport layer. Because of this attack, valid nodes have a difficult time performing their normal functions.

Threats at the application layer include things like data corruption, overload, repudiation, and malicious code. This kind of assault, in addition to depleting the network's resources, also results in the nodes' use of a considerable quantity of energy. Attacks may be classified into one of five categories: eavesdropping, traffic analysis, interruption, hijacking, and physical assaults. Information may be sent by the sensor network to an eavesdropper. Listening in on talks allows eavesdroppers to directly threaten nodes since they are able to hear what they are saying. There are two approaches one might use while eavesdropping on conversations. Eavesdropping is defined as listening in on someone else's conversation without actively participating in the conversation. The sensor or aggregation point is approached by a person who is attentively listening to what is going on around them.

Simply by analyzing the traffic on the network, a potential attacker might figure out how the WSN is organized. Attackers could utilize traffic analysis to discover additional information about the nodes' owners as well as the vulnerabilities of the nodes. An effort may be made by adversaries to compromise a sensor application. The attacker is now gathering information from various sensor nodes in order to make the attack as successful as is practically possible. In order to carry out the attack, two different strategies are used. By adding erroneous information, modifying existing data, or simply adjusting values, data may be corrupted, made useless, or even completely removed from existence. When the environment of the sensor is changed, such as when heat, humidity, or other elements are added to the sensors, data from the sensors may cease.

Once the hijacker succeeds, they have successfully seized control of the sensor. The attacker will choose sensors that have the capability to not only listen in but also disrupt the network. This kind of attack is the most challenging to defend against because it comes from nodes that have already been validated. The attacker eavesdrops on a conversation between two nodes and participates in the conversation. In order to deceive the nodes, the attacker will send them misleading information in order to deactivate the hardware of the sensor and compromise its functionality. Error-tolerant systems ensure that the functioning of the network will continue normally despite the failure of individual nodes. Theft, attacks on software and communication protocols, attacks on key management protocols, attacks on sensor schemes, and attacks on protocols that synchronize time are all potential threats to wireless sensor networks.

In order to devise effective security measures, it is essential to have a solid understanding of the many types of attacks. There are several different ways in which wireless networks may be categorized. Taxonomies may be used to classify WSN assaults according to their purpose, behavior, and objectives in order to get a better understanding of these attacks and to develop more effective defence solutions.

19.2.2 Solutions to Overcome Attacks on WSNs

The authors of [6] examine authentication in order to find a way to stop unauthorized access to agricultural WSNs. WSNs are used for farm monitoring since they give remote user identity in addition to key agreement. Utilization in real time is a possibility, but it is not an easy process.

The procedures for establishing the system, enrolling users and agricultural experts, signing in, authenticating and agreeing on a session key, changing the password, and adding a dynamic node are much improved by the strategy presented in [7]. An experienced user, a base station, a sensor node, and an interconnecting device were the four components of the system that Ali and his colleagues constructed instead. The core entities in this architecture are in charge of handling key agreement and authentication respectively (base stations). If there are no base stations, people will not be able to develop trusting relationships with one another.

In article [8], the author makes the case that authenticating sensors and resources is the most effective way to stop people from exploiting them at times when they are most required. Using BAN logic (Burrows–Abady–Needham), this study takes an in-depth look at security and authentication for coal mines in wireless sensor networks. In order for BAN logic to function, both parties need to be able to authenticate one another and come to an agreement on a session key. The findings of the official risk assessment indicate that it is competent to provide protection against a wide range of threats to security. The suggested authentication method offers strong authentication, which contributes to an increased level of safety.

Article [9] established a unique technique for recognizing users and coming to an agreement on keys in heterogeneous ad hoc wireless sensor networks. Using the recommended method's lightweight key agreement protocol, it is feasible to protect a session key by using a generic sensing node as the security mechanism. Even in the case where the user is never connected to the GWN, the user, the sensor node, and the gateway node may all share the same authentication thanks to the method that has been recommended (GWN). Because the architecture of the WSN makes very little use of resources, the only calculations that are required are hash and XOR (exclusive disjunction).

The authors of [10] and [11] put out key agreement mechanisms for the WSN environment in their respective publications. This served as the foundation for a proposed user authentication and key agreement. The method proposed by [12] overcomes the security vulnerabilities described in the previous paragraph. The proposed protocol supports mutual authentication and session key agreement among all of the participants by using BAN logic. This ensures that the protocol will be successful. The proposed protocol not only resolves each and every one of the aforementioned security problems, but it also satisfies each and every one of the requirements for security, such as requiring a low amount of power consumption, preserving user anonymity while utilizing mutual authentication, and providing an easy-to-use phase for changing passwords.

It is possible to overcome all of the difficulties that were present in prior systems and guarantee that security is maintained both within and outside of the network by using the new protocol developed by [13]. Because ECC is used, the authentication process for WSNs is far more fool-proof than that of other protocols.

Dos Santos developed a model for agricultural forecasting using data from a study that was published in 2019 [14]. He is able to keep an eye on the harvest thanks to WSN, a prediction model that utilizes the ARIMA (Autoregressive Integrated Moving Average) model, and LoRaWAN (Long Range Wide Area Network) technology (short and medium area coverage). A need exists for farmers to have data on the timing of when they should grow arugula, and a WSN was established in order to gather this information. According to the findings of this study, the LoRa technology might be utilized to monitor agricultural land by using a prediction model that sends notifications to farmers in advance of odd events. The inability of the model to do real-time monitoring is a problem that has to be addressed. In the future, this should be done via a mobile application instead.

During the course of their investigation, [15] came up with an original strategy for protecting LoRaWAN. This article walks through the process of using proxy nodes to improve the performance of the LoRaWAN gateway. Proxy nodes are analogous to a gate in that they check to see whether there is sufficient consistency in the flow of data from sensor nodes to the end node. It is possible for attackers to take advantage of the fact that the proxy node is the site with the highest level of trust. Two of the most significant obstacles to overcome in terms of security in LoRaWAN are the generation of session keys and the updating of keys.

19.3 Materials and Methods

At the sensor level, we are looking for a method to protect wireless sensor networks used in agriculture. This paradigm assures the security of the data that is obtained as well as the capacity of sensors to connect with one another in a secure manner. As a consequence of our research into the topic of WSN security, we were able to determine the threats that are most likely to affect WSNs and come up with a set of security guidelines. During the process of developing the model, it was ensured that both symmetric and asymmetric cryptography, in addition to the public key infrastructure, were adhered to in a stringent manner.

The following databases were used in order to carry out the research: ABI/INFORM Global, Academic Search Premiere, ACM Digital Library, IEEE Xplore, ScienceDirect, and Google Scholar.

When we looked at WSN implementation in agriculture as a distributed system, we analyzed attacks on WSNs making use of many levels of architectural flaws in network communication to target security needs such as authentication and authorization in addition to data integrity and trustworthiness. Cryptographic security solutions were required since the model's WSN security requirements mandated their use.

Building a WSN to monitor farms was accomplished with the assistance of the OMNeT++ software's Flora architecture. In order to evaluate how effectively the proposed model functions, we assess the amount of energy it uses at the receiver, the amount of energy it uses per bit, and the amount of energy it uses at the transmitter node.

19.3.1 Proposed Model

Specifically, it is made to meet the requirements for the secure operation of sensor networks, as discussed above. If all of the players in the model are able to safely communicate and receive data, as well as identify themselves, then the fundamental aim of the model has been achieved. This guarantees that those responsible for making decisions have access to reliable data in order to guarantee that the messages cannot be questioned. The use of data encryption, digital envelopes, digital signatures, and public key infrastructure (PKI) are all components of the approach that is being implemented to secure sensor networks.

19.3.1.1 Security Technologies Applied in the Model

Data is transformed into a form that can only be deciphered by the one who is supposed to receive it after it has been encrypted. The generation of secret keys as well as the encryption of communications both make use of symmetric cryptography. With the use of asymmetric cryptography and digital certificates issued by PKI, digital communications may be authenticated, and secret keys can be kept secure.

Encryption of communications and secret keys is handled by the model through digital envelope technology. In the current circumstance, the information included in the transmission will only be known to its intended receiver. The first layer of protection for communications is provided by a kind of encryption known as symmetric (e.g., DES, 3-DES, RC2, AES, or a private algorithm). After that, information is encrypted by utilizing the public key of the recipient's digital certificate in conjunction with an asymmetric encryption method such as RSA. The message is first decoded using the secret key, which is obtained via the process of decryption utilizing the private key. It is encrypted using the signer's private key using an asymmetric cryptographic technique, such as the RSA algorithm, so that only the person who signed the message can decode it. This guarantees that no one else may access the message's contents. A digital signature is an additional step that must be taken before the communication can be considered complete. The receiver is responsible for authenticating the digital signature. Certificates are the means through which the private key of an identifiable (notified) entity is linked to its corresponding public key. The private key belonging to the certification authority is used in order to sign the certificate, and anybody in possession of that key is afterwards able to validate the certificate.

Encryption and public key infrastructure allow for the safe transmission of data as well as management messages that are sent between two members in a wireless sensor network. A secret key is generated by the sender, and this key is then used in an asymmetric manner in order to encrypt the message. During the time that the message is in transit, asymmetric cryptography is used to ward off any attempts at eavesdropping on the confidential keys. The receiver makes use of the secret key that was extracted via the process of asymmetric cryptography in order to decode the message. Examining the sender's digital signature allows the recipient to determine whether or not the communication is genuine. The public key infrastructure software makes it easy to swap out the private key and establish a one-of-a-kind link between the user and the public key. As can be seen in Figure 19.1, the wireless sensor network at the farm was developed with the intention of protecting it.

19.3.1.2 Wireless Sensor Network in Agriculture Protection Model Proposed

Establishing a sensor network requires a number of steps, including installation and configuration. A risk assessment is carried out as part of the initial configuration of the sensor network in order to choose the symmetric approach and secret key that will be used for the purpose of message security. Users have access to public symmetric algorithms, or they may pay to have a bespoke algorithm developed for them. It is also required to choose a certifying authority in order to get digital certificates for sensor networks. These digital certificates will be used to digitally sign and distribute secret network keys. PKIs may be developed by users for their own individual usage in conjunction with certifying agencies. Sensor nodes, data processing and aggregation nodes, and the application for administering and processing the sensor network all need to be outfitted with crypto-algorithms, crypto-keys, and digital certificates, respectively. The figure 19.3 shows the Digital Envelope in Agriculture Sector.

Following the completion of the data collection process, the sensor node will generate messages. When a message is digested, it is encoded using MD5, SHA-1 (Secure Hash Algorithm), and SHA-256 as the message digest techniques. This message digest is

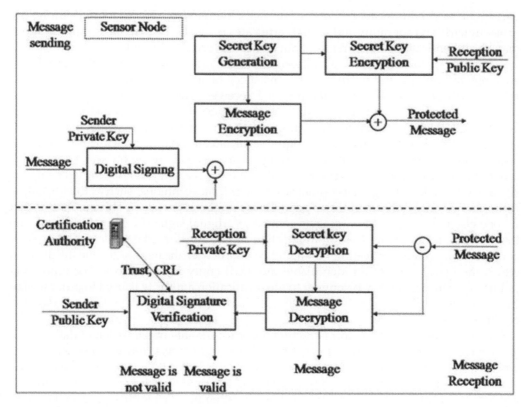

FIGURE 19.1
The data collection process from various sensor nodes.

encrypted using an asymmetric cryptographic algorithm, and the encryption is done using the private key, also known as the PrKs, of the person who signed the message. A digital signature for the communication is provided in the form of encrypted message digests, which are also included in the message. Following this step, the message and the digital signature are both encrypted using symmetric cryptography together with a private key. After that, PuKr's public key SKey is decrypted, and in the same fashion, it is appended to the previously encrypted message E (M + DS). In order to digitally sign a message, the sender must first verify the authenticity of his or her own digital certificate, as well as the trust chain and current status of the recipient's digital certificate. Only then may the message be digitally signed. When you check the trust chain of a certificate, you may determine whether or not it was issued by a renowned certifying body and is, as a result, trustworthy. As a result of a lack of confidence in the receiver, the process of delivering information may be suspended. Either by querying the OCSP (Online Certificate State Protocol) or by utilizing a list of revoked certificates, it is feasible to determine whether or not digital certificates have been compromised. When a digital certificate is taken from a sensor or app, the device loses its ability to provide users with a safe environment. It is not reliable to use sensor nodes that have been compromised by hackers or are known to produce inaccurate data. In light of the current circumstances, further contact is not possible. The data collected by the sensor nodes are considered to be invalid and out of date the moment the digital certificate is revoked from those nodes. If a hacking attempt is made on the receiver (server or data processing program), as a result, its certificate may be revoked or suspended. In contrast, the digital certificate of the recipient is checked, but the sender's secret key is left safe. This indicates that the receiver is no longer available to be used. As a direct consequence of this, the sensor node is unable to do any more data analysis in this predicament.

19.3.1.3 Preparation for Model Implementation

During the receiving step of the transaction, the receiver is responsible for decrypting the secret key and ensuring that the digital signature is accurate. Decryption of the secret key is accomplished with the help of the recipient's personal private key, known as PrKr. You will need the secret key, which is composed of M and DSS, in order to interpret message E. An asymmetric technique is employed to decode the message digest, and the public key of the sender of the message is utilized to verify the digital signature. The receiver of a message must apply the same message digest algorithm as the message it received, which is denoted by the letter M, in order to validate the message. In this scenario, the check is considered successful if the digest of the message (MD1) matches the digest of the encrypted message (MD). In the event that this is not the case, the examination will be failed. In this scenario, the receiver is aware that the sender is genuine, that the message has not been tampered with, and that the sender will not be able to back out of sending any further messages in the future. In this stage of the process, the digital certificates of both the sender and the receiver are checked to confirm that the trust chain has not been broken. The trust chain is used in order to validate the sender's certificate. If the sensor node of the attacker is unable to relay the information because the sender's certificate chain cannot be trusted, the message will be erased. Because the digital certificate of the sender's sensor node on the receiver's end has been revoked or placed on hold, it is not possible to trust the data that was received. This is analogous to a scenario in which both the public and private keys of the receiver have been revoked or suspended. As a direct consequence of this, there is considerable uncertainty over the safety of the secret key.

19.3.1.4 Protection of the Sensor Node Message in the Process of Transmission

If the criteria for the security of a distributed system are satisfied, it is feasible to conduct a test to determine whether or not a digital signature is secure. A credible authority issues a digital certificate to each and every sensor node in the network. If the same key that was used to authenticate the message's signature was also used to authenticate the message, then the sender's public key may be used to verify an authenticated message. Once the sender's digital signature has been verified, you will no longer be able to stop an email from being sent.

This concept is referred to as non-repudiation. Data integrity can only be protected by using symmetric cryptography, but your private key may be shielded from prying eyes using asymmetric encryption. The certifying authority gains the community's confidence when it distributes digital certificates to each and every user of the sensor network. Only legitimate, trustworthy data with a high level of integrity that has been successfully sent may be processed after transmission. The issuance of digital certificates to persons, sensors, aggregation nodes, and applications once they have been enrolled in the PKI application protects their privacy. What should be done with the data collected by a wireless sensor network? When using the paradigm, the information that is acquired by sensors may be managed with more ease. If the digital certificate of the sensor node has been revoked or

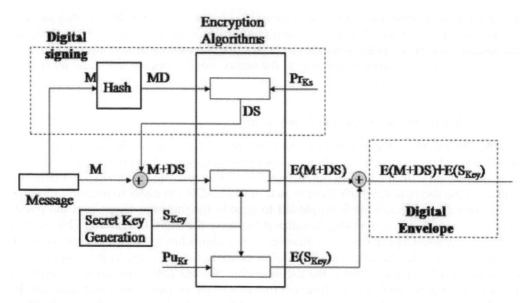

FIGURE 19.2
The proposed encryption algorithm.

suspended for whatever reason, the certificate that is associated with that sensor node will be changed. In the event that either the sender's or the receiver's digital certificate is revoked or suspended while the verification process is taking place, the message will not be able to be sent. In this scenario, the data may be sent to an aggregate node or an application in order to undergo data processing. At that location, it will be investigated in the context of the factors that led to the revocation or suspension of the digital certificate(s). Using this approach (Figure 19.2), you are able to handle data in a variety of different ways depending on the cause for the status change of a certificate. In a wireless sensor network, the transfer of the secret key has to be done in a secure manner.

This approach makes it possible to convey the secret key to each sensor node in a safe manner. It is possible to utilize this option during a data protection procedure in order to reduce the amount of work that the CPU of the node is required to do while it is sending a message. If there is a security breach or if the key has to be updated on a regular basis, the model may be modified such that at launch, all nodes utilize the same secret key, and a new secret key can be sent to all nodes in the event that there is a need to do so. When you send a message, you may have peace of mind knowing that it will make it to its destination without incident. No matter how many nodes the communications have to pass through on their route to their final destination, this function ensures that they remain private and safe from the point where the data was generated all the way to their final location. When moving from one node to the next in a network, it is difficult to read a message that has been encrypted. This verifies not only that the privacy of the message was not compromised while it was being sent, but also that the data came from the same source as the person who was authorized to send it (the source node). When data is merged and sent out from the aggregation node, it is of the utmost importance to ensure that the digital message signature is maintained. In this scenario, it is feasible to confirm that the original data has not been modified in any way and that it did, in fact, come from the source node. This indicates that it might be applied in a wide range of sensor network topologies in the agricultural sector, such as designs based on sensor node mobility (stationary, mobile, and hybrid architectures) and architectures based on hierarchy (single-tier architecture, multi-tier architecture). In addition, we have included data security in every stage of the design process, from the sensor node to the application that analyzes the data, as well as from the management application to the sensors. This suggests that the model application will not be affected by the movement, flow, or rest of the sensor node, nor will it be affected by its position in the hierarchy-based design. Zigbee, Wi-Fi, Bluetooth, GPRS/3G/4G (General Packet Radio Service), and WiMAX (Worldwide Interoperability for Microwave Access) are not among the communication technologies used by the suggested model. This method does not need any particular wireless transmission technology since the operations of the model are carried out both before and after the message is sent. The complexity of the model is increased when there is an increase in the number of nodes that need specific hardware and software. Because of this, we will need to reevaluate the process through which we construct wireless networks for agricultural sensor nodes. Middleware and sensors that are connected to the soil, environment, and plants all need the installation of crypto-modules.

19.3.1.5 Verification of the Content of the Protected Message in the Process of Reception

An irrigation system in a field is composed of sensors that monitor temperature, soil moisture, and nutrients; gateway nodes that collect and combine data; irrigation system management units with two sensors that regulate water flow and nutrient intake; and a control center that monitors and manages the entire system. In order to figure out how to provide plants with the ideal conditions for growth, we analyze the temperature, the amount of moisture in the air, and a number of other environmental elements. The testing will take place in the authorized region. During the experiment, the competitors intend to eavesdrop on the wireless sensor network in order to get data without authority or to modify data in order to cause the experiment to provide inaccurate findings. The optimal model that was recommended and sensors equipped with crypto-modules will be employed in order to guarantee that the process of data collection will not hinder the capacity of the wireless sensor network to receive management messages. The agricultural data is collected by the sensor network and then sent to the sink node by means of a gateway sensor (SkN). Data is sent from the sink node to the experimental control center internal network through an internet connection so that a determination can be made about the amount of watering and nutrients to be applied. Before the sensor network is formed, the PKI-based security model is implemented as shown in Figure 19.4. To begin with, it was found that it was possible to read messages, manipulate data, and replace sensor nodes with nodes that provided fraudulent data. This was the first stage in the process. Not only does the transmission of data through wireless sensor networks provide a security concern, but it does so whenever data is sent over the internet.

19.4 Characteristics of the Model

The second step of the procedure involves the addition of node private keys as well as sensor node private keys. In addition to digital signatures and verification software, computers come equipped with a variety of digital certificates and cryptographic keys that may be used for the purpose of gathering information and making decisions.

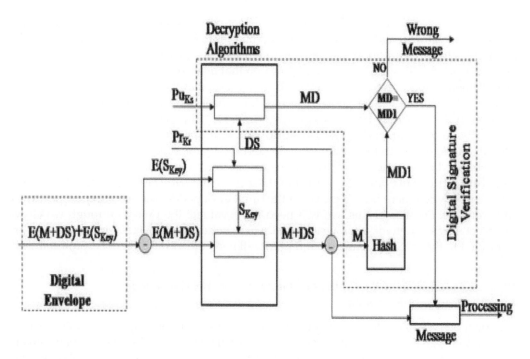

FIGURE 19.3
The digital envelope in the agriculture sector.

The control center (Figure 19.4 will send a message to each of the sensors in order to instruct the sensor nodes to send in the data that was collected by an application that is specifically designed for that purpose. The internet and other wireless sensor networks are used in order to transmit a message that gathers data for further analysis. It is possible for the communication in the wireless sensor network to be set up in such a way that, prior to the data being delivered to the application that collects the data, the sink node does an analysis of the data and organizes it. It is possible for an application that gathers data to utilize the sink node in place of the sink for a sensor node. If the SkN does not analyze data and instead only transports it from the experimental field to the data processing application, as shown in Figure 19.5, the SNA node digitally signs the message (1). At long last, the message and its associated digital signature are merged (2). Messages and digital signatures are both symmetrically encrypted and digitally secured by the secret key of the SNA node. Both the private key of the SNA node and the public key of the data-collecting application node are encrypted with the use of asymmetric cryptography. The transmission includes private keys that have been encrypted as well as digital certificates. The nodes that make up a sink network are responsible for ensuring the safety of the internet connections that data processing applications use. The data collected by the sensor nodes are encrypted before being communicated over the internet, and the identification of the SkN nodes can be verified with complete safety. At the control center, the connection is first fragmented, and then the digital certificate of the SNA node is checked for authenticity. In order to decode the secret key, the RSA method as well as the application's private key are used.

The AES algorithm as well as a private key are essential components for the purpose of decoding the message. In order to decipher the digital signature, it is necessary to make

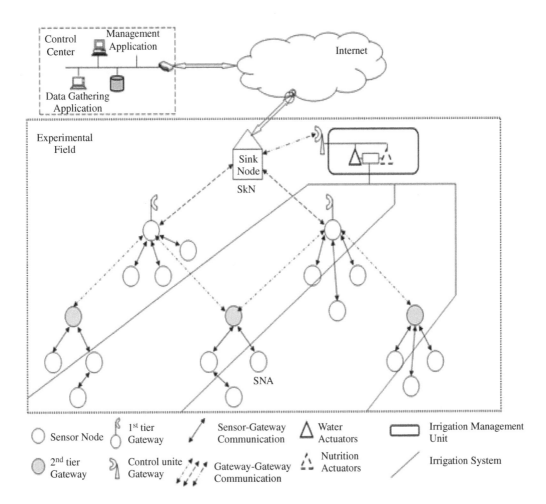

FIGURE 19.4
The control center on the agriculture field.

use of both the public key of the SNA node and the digital certificate that was contained in the message. Before it is delivered, the message is subjected to a hash alteration, much like the SNA node that was responsible for transmitting the message itself. After that, it is contrasted in order to determine whether or not the message has still kept its integrity, authenticity, and the ability to be challenged. It is conceivable that the message was changed or that it was transmitted by someone other than the intended recipient.

The list of revoked certificates that have been granted by the certifying authority may be reviewed by the recipient in order for them to determine the trust route (whether or not the sender's certificate can be trusted), the validity of the certificate, and the length of the certificate. If there is an error in the verification procedure, the data from sensors whose permissions have been revoked or who aren't allowed to collect data will be disregarded. In the same way as before, actuators A1 (irrigation) and A2 (nutrition) receive the management message from the management application. However, this time the management application is the one to broadcast it, and the actuators are the ones to receive it.

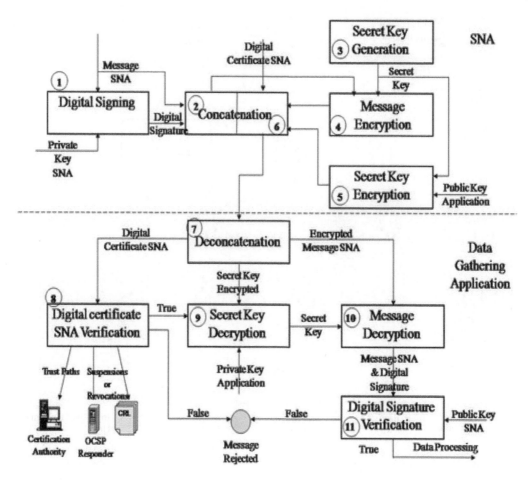

FIGURE 19.5
The flow of the proposed model.

19.5 Proposed Model Simulation Results and Discussions

The proposed model for the security key is evaluated using LoRa and TelosB-based wireless sensor networks (WSNs) for short- and long-range situations, respectively.

LoRa is a low-power long-range wireless technology that enables the successful transport of smaller amounts of data over greater distances. It is not necessary to get a license in order to make use of the ISM band, which is shorthand for the industrial, scientific, and medical purposes for which it is used. Using LoRa technology, even modest sensors and actuators with few resources are able to send data across distances of tens of kilometers and for many years.

As was said before, cryptography is what generates the private key and secret key that are associated with the sender of the packet (message). [16] demonstrated that the Lora WAN key provisioning technique has a significant flaw due to the fact that the network server is the one responsible for generating both session keys. It is possible that using this method will result in a conflict between the network server and the application server on which one should be given priority. The majority of the time, end nodes are required to continue without upgrading their use of secret keys for the duration of their lifespan. To put it another way, an attacker may take all of the target node's data if they are able to get their hands on the key to the node [17]. On the other hand, if you use our method, you will always be able to ensure that your keys are current. According to the authors of this study, the server-side decryption secret key may potentially be safeguarded by using encryption in conjunction with a public key. As part of the OMNeT++ software development effort, a sensor node and a LoRa gateway application layer that are based on the network protection model described in Section 4 were developed. It was ensured that the key for the message data application was verified by configuring the LoRa gateway's application layer (controller).

The network architecture of the Flora simulation module of OMNeT++ makes use of a modular design and the following elements of a LoRa network: three sensor nodes, one attack packet transmitter, a LoRa gateway, and a LoRa server. This is all done in order to simulate a Flora ecosystem. When sending data to the server, which is referred to as an "attack packet," NetAttack makes advantage of the LoRa gateway.

When code is removed from the application layer of the gateway, data pertaining to AttackPck-0 is sent to a server. Every 20 minutes, the scenario manager will simulate an update to the key by making adjustments to the node application modules and the packet messages (8 bits word).

Chirp spread spectrum is the method that is used for the modulation of LoRa technology. Each LoRa signal is composed of two spreading factor chirps, and the abbreviation SF is used to designate the spreading factor that corresponds to each of these chirps [18]. There are six different data rates (DRs), each of which is provided by six orthogonal SFs. The LoRa message is made more resistant to shifts in frequency by using a low data rate optimization method to reduce the amount of drift caused by the crystal reference oscillator (connected to packet delivery). If our SFs are between 11 and 12, we should be able to achieve this. According to the research, if you want to transfer data over a distance of 5 kilometers in an urban region that has modest obstructions, you will need a signal-to-noise ratio of 12 and 10 dBm to accomplish this. SF = 12 was chosen for this hypothetical circumstance since it works well in rural situations. For the purpose of developing the energy consumption model, the Flora energy model was applied. A radio may be in a variety of states, each of which uses a different amount of power. These states include idle, busy, switching, sleeping, broadcasting, and receiving. It is anticipated that 13 dB of transmission power would be enough for a LoRa network built with a coding rate (CR) value of 6. The power consumption of the transmitter is fixed at 92.5 milliwatts (mW) since the energy model for the wireless network does not include a security model. Approximately 200 joules of energy are used up by each bit that makes up a simulated LoRa packet. The sensor node, which monitors the conditions of the agricultural space by delivering data at intervals of 25 seconds, transmits two data packets per hour. The energy consumption of the sensor node was 1.05 mJ for every 92.5 mW of transmitter power. Because of the higher bit rate, it takes more energy for sensor and gateway node models to communicate with one another on how to avoid being attacked. In addition, each time a new session starts, the secret key as well as the server's transmission acknowledgement need to be changed. Because our idea is implemented as a separate module inside the software that runs on the gateway node, the server acknowledgment transfer is hidden from view. The "receiving state" use of the transmitter was enhanced by 30 percent so that the consumption of the sensor node could be determined.

Over the course of the 25-second transmission time, the sensor node used an additional 0.7 mJ of electricity. The amount of energy required for one transmitting phase is 1.75 mJ. An experiment was carried out in order to have a better understanding of the way in which the delivery package moves. The PLR for SF = 10 was 11/15 while all four nodes, including the NetAttack nodes and the three sensor nodes, were sending out packets. The glitch was caused because the LoRa gateway had to wait longer for bigger data streams than usual, which caused the problem. During the simulated monitoring that took place for two hours, three sensor nodes together sent 15 packets to the server. If further research is done on the subject, it is possible that LoRa gateway delays, also known as waiting times, may be reduced. The attack avoidance mechanism used by this model might stand to benefit from a reduction in the overall size of the SF as well as an acceleration in the rate at which packets are sent over a wireless network.

19.6 Results and Discussions

Every time a message is sent, the node generates a brand-new secret key in order to ensure that the message has not been tampered with in any way. When using this paradigm, there is a greater demand placed on the CPU, memory, and energy used by nodes. The method in which resources are allotted to the wireless sensor network is one that is determined in some way by the parameters that were chosen in advance of the model's actual implementation. This is very important in terms of key lengths and algorithms, risk assessments, the frequency of data transfers, and interactions between management and employees. In order to make the model better, the authors suggest making adjustments to the hardware and software design of the sensor node, in addition to a number of other organizational techniques (organizational aspect).

Modifying the layout of individual nodes, both hardware and software, is a component of model optimization. In order for the suggested paradigm to be functional, it is necessary to augment the existing hardware architecture of the sensor node with crypto-modules and middleware. The fact that the crypto-module provides a secure area for the storage of keys is an additional advantage offered by it. It is possible for the crypto-module to produce not one, but two different sets of keys, one of which is private and the other of which is public. Additionally, it is feasible for it to construct a symmetric key by using an asymmetric method (a secret key). In addition to that, it can encrypt and decrypt messages, perform digital signing, securely store cryptographic keys onboard, manage keys, and encrypt and decrypt secret keys. The TNODE5 microcontroller has been investigated. Texas Instruments is the manufacturer of one of the chips of the TNODE5 system, which is a 16-bit microcontroller. When compared to software implementations on commonly used sensor node systems that are based on the same kind of microcontroller, its hardware implementations for both symmetric and asymmetric cryptography boost power efficiency by orders of magnitude. Additionally, it contains implementations for both symmetric and asymmetric cryptography. The cryptographic portions of this particular microcontroller, on the other hand, do not have the capability to safely store and handle cryptographic keys. The term "middleware" refers to software that provides a secure connection between crypto-modules and the operating system of the node. The planning phase, the preparations for implementation, and the deployment of wireless sensor networks in agriculture are all organizational components that are included in the model that has been provided.

In addition to making changes to the sensor node's hardware and software design, optimizing sensor nodes requires a set of accepted standards for how the model should be used in the sensor network. These standards must be established before optimizing sensor nodes.

During the application of the model, the following organizational measures should be implemented:

Crypto-parameters are now in the process of being developed. The security risk evaluation of the wireless sensor network decides the cryptography methods to be used, as well as the hash functions and key lengths. When crypto-algorithms and key lengths are shortened, the central processing unit (CPU) reacts more quickly and uses less energy overall. It is recommended that data aggregation nodes make use of longer keys because of the greater hardware they possess; however, this also makes them more susceptible to assaults. A technique of encryption and decryption that generates a fresh secret key for each transmission is one of the ideas that has been offered. By employing the secret key, it is possible to have faith in both the sensor node and the message receiver. It calls for a greater amount of processing and energy than the use of either a long-term or a short-term secret code. Up until the time that it is set to expire, all of the nodes will use this key, which will be modified on a regular basis. This will keep the communications secure. As a result of this innovative technology, you no longer have to create a new key for each transfer.

A sensor node has the capability of collecting a wide variety of data types [19–20]. When just one form of data is being sent, the level of encryption and the number of resources used to encode it are both larger than when sending and encrypting numerous kinds of data at the same time. It is important to have a plan in place for the collecting of data that takes into consideration the fact that encryption uses up all of the resources that are available.

As a component of the approach, validation and inspection of certificates will take place. When producing a digital signature as well as a digital envelope, it is essential to check the authenticity of the certificate at many points along the process. This is the action that has to be taken if any of the certificates in the trust chain have had their validity revoked or suspended in any way. In order to carry out this verification, the node must either be able to send the certifying authority a request to check the status of the certificate or have access to a list of certificates that have had their validity revoked. In either case, there is an excessive strain placed on the network's transmission capacities and nodes. In light of recent developments in communications and technology, the authors suggest performing the verification of the digital certificate at the location of the receiver (aggregate node, processing application) [21].

It is possible that your data will be transmitted from one sensor node to another when you are using a sensor network. When messages are sent between two nodes that are geographically close to one another using the model, a different group of nodes bears the burden of carrying those messages. Establishing encrypted communication between the transmitting node and the receiving site is essential to lowering the level of stress experienced by the nodes [22]. It is possible that this scenario involves an end application or a node that collects data based on a set of predetermined criteria. Because of this, communication within the sensor network needs to be established between the location of the transmitter and the location of the receiver.

19.7 Conclusion

WSNs are finding more and more applications in the agricultural sector. The standard of the information that is acquired and analyzed on a farm has a direct bearing on the quantity and quality of the output produced by that farm. To obtain an edge over you, competitors may, for example, tamper with measured data, stop production, or conduct tests in order to steal your thunder. We have developed a system of information security architecture that ensures the confidentiality of sensitive data as it is sent from sensors to management structures and back again. In a simulation, it was discovered that the approach required a significant amount of energy to implement but was successful in preventing communication-based attacks. The greatest amount of energy is used while the hidden key is being frequently changed. In addition to this, both the transmitter and the receiver will need a greater quantity of electricity. The new data protection model for WSNs makes use of technological and organizational solutions to hasten the process of protecting data while simultaneously increasing sensor usage and producing the best possible use of the data.

Acknowledgements

We thank our parents for their support and motivation.

References

[1] Ali, O., Ishak, M.K., Ahmed, A.B., Salleh, M.F.M., Ooi, C.A., Khan, M.F.A.J., & Khan, I. (2022). On-line WSN SoC estimation using Gaussian process regression: An adaptive machine learning approach. *Alexandria Engineering Journal*, 61(12), 9831–9848. doi:10.1016/j.aej.2022.02.067.

[2] Radosavljević, G., Smetana, W., Marić, A., Živanov, L., Unger, M., & Stojanović, G. (2010). Micro force sensor fabricated in the LTCC technology. In *Proceedings of the 27th International Conference on Microelectronics, Nis, Serbia, 16–19 May 2010*, pp. 221–224.

[3] Kwon, K. (2022). Exact solutions for source localization problem with minimal squared distance error. *Applied Mathematics and Computation*, 427. doi:10.1016/j.amc.2022.127187.

[4] Radovanović, M., Vasiljević, D., Krstić, D., Antić, I., Korzhyk, O., Stojanović, G., & Škrbić, B. (2019). Flexible sensors platform for determination of cadmium concentration in soil samples. *Computers and Electronics in Agriculture*, 166, 105001.

[5] Anupong, W., Azhagumurugan, R., Sahay, K.B., Dhabliya, D., Kumar, R., &Vijendra Babu, D. (2022). Towards a high precision in AMI-based smart meters and new technologies in the smart grid. *Sustainable Computing: Informatics and Systems*, 35. doi:10.1016/j.suscom.2022.100690.

[6] Ng, H.S., Sim, M.L., & Tan, C.M. (2006). Security issues of wireless sensor networks in healthcare applications. *BT Technology Journal*, 24, 138–144.

[7] Barclay, I., Simpkin, C., Bent, G., et al. (2022). Trustable service discovery for highly dynamic decentralized workflows. *Future Generation Computer Systems*, 134, 236–246. doi:10.1016/j.future.2022.03.035.

[8] Xu, W., Trappe, W., & Zhang, Y. (2007). Channel surfing: Defending wireless sensor networks from interference. In *Proceedings of the 6th International Conference on Information Processing in Sensor Networks*, pp. 499–508.

[9] Binh, N.T.M., Ngoc, N.H., Binh, H.T.T., Van, N.K., & Yu, S. (2022). A family system based evolutionary algorithm for obstacle-evasion minimal exposure path problem in Internet of Things. *Expert Systems with Applications*, 200, 116943. doi:10.1016/j.eswa.2022.116943.

[10] Chen, H., & Chen, Z. (2022). Energy-efficient power scheduling and allocation scheme for wireless sensor networks. *Energy Reports*, 8, 283–290. doi:10.1016/j.egyr.2022.03.046.

[11] Zahmati, A.S., Hussain, S., Fernando, X., & Grami, A. (2009). Cognitive wireless sensor networks: Emerging topics and recent challenges. In *Proceedings of the 2009 IEEE Toronto International Conference on Science and Technology for Humanity (TIC-STH)*, pp. 593–596.

[12] Deepan, N., & Rebekka, B. (2022). Performance analysis of wireless powered sensor network with opportunistic scheduling over generalized κ–μ shadowed fading channels. *Physical Communication*, 53, 101727. doi:10.1016/j.phycom.2022.101727.

[13] Derdar, A., Bensiali, N., Adjabi, M., et al. (2022). Photovoltaic energy generation systems monitoring and performance optimization using wireless sensors network and metaheuristics. *Sustainable Computing: Informatics and Systems*, 35, 100684. doi:10.1016/j.suscom.2022.100684.

[14] Watro, R., Kong, D., Cuti, S.-F., et al. (2004). TinyPK: Securing sensor networks with public key technology. In *Proceedings of the 2nd ACM Workshop on Security of Ad Hoc and Sensor Networks*, pp. 59–64.

[15] Haque, M.E., Hossain, T., Sarker, M.R., et al. (2022). A hybrid approach to enhance the lifespan of WSNs in nuclear power plant monitoring system. *Scientific Reports*, 12(1), 4381. doi:10.1038/s41598-022-08075-6.

[16] Hidalgo-Leon, R., Urquizo, J., Silva, C.E., Silva-Leon, J., Wu, J., Singh, P., & Soriano, G. (2022). Powering nodes of wireless sensor networks with energy harvesters for intelligent buildings: A review. *Energy Reports*, 8, 3809–3826. doi:10.1016/j.egyr.2022.02.280.

[17] Iqbal, S., & Sujatha, B.R. (2022). Secure key management scheme for hierarchical network using combinatorial design. *Journal of Information Systems and Telecommunication*, 10(37), 20–27. doi:10.52547/jist.15691.10.37.20.

[18] Jebi, R.C., & Baulkani, S. (2022). Mitigation of coverage and connectivity issues in wireless sensor network by multi-objective randomized grasshopper optimization based selective activation scheme. *Sustainable Computing: Informatics and Systems*, 35 doi:10.1016/j.suscom.2022.100728.

[19] Panić, G., Stecklina, O., & Stamenković, Z. (2016). An embedded sensor node microcontroller with crypto-processors. *Sensors*, 16, 607.

[20] Kim, Y., Hyon, Y.K., Lee, S., Woo, S., Ha, T., & Chung, C. (2022). The coming era of a new auscultation system for analyzing respiratory sounds. *BMC Pulmonary Medicine*, 22(1). doi:10.1186/s12890-022-01896-1.

[21] Kooij, L., Peters, G.M., Doggen, C.J.M., & van Harten, W.H. (2022). Remote continuous monitoring with wireless wearable sensors in clinical practice, nurses perspectives on factors affecting implementation: A qualitative study. *BMC Nursing*, 21(1). doi:10.1186/s12912-022-00832-2.

[22] Amutha, J., Sharma, S., & Sharma, S.K. (2022). An energy efficient cluster based hybrid optimization algorithm with static sink and mobile sink node for wireless sensor networks. *Expert Systems with Applications*, 203(C). doi:10.1016/j.eswa.2022.117334.

20

A Survey on Opportunities and Challenges for Next Generation Wireless Sensor Networks

T. Perarasi, M. Leeban Moses, and K. Shoukath Ali

CONTENTS

20.1	Introduction	294
20.2	Growth of Wireless Sensor Networks	294
	20.2.1 Progress	295
	20.2.2 Architectural View	295
	20.2.2.1 Layered Network Architecture	295
	20.2.2.2 Clustered Network Architecture	296
20.3	Next Generation Networks: IoT	297
	20.3.1 Architecture	300
	20.3.2 Challenges	301
	20.3.2.1 Security	302
	20.3.2.2 Platform	302
	20.3.2.3 Interoperability and Standardization	302
	20.3.2.4 Data Storage and Analytics	302
	20.3.2.5 IoT Sensors and Devices	303
20.4	Next Generation Networks: Smart Grid	303
	20.4.1 Remote System Monitoring	303
	20.4.1.1 What Is a Smart Grid?	303
	20.4.1.2 What Makes Up a Smart Grid?	303
	20.4.1.3 The Current Smart Grid Market	304
	20.4.1.4 Why Do We Need Smart Grids?	304
	20.4.1.5 Reaping Rewards	304
	20.4.1.6 The Future Is Now	305
20.5	Green Communication	305
	20.5.1 Energy Monitoring	305
	20.5.2 Algorithmic View	306
	20.5.2.1 MAC Protocols for EH-WSNs	306
	20.5.2.2 Transmission Schemes Classification	306
	20.5.2.3 Routing Protocols for EH-WSNs	307
	20.5.2.4 Schemes Based on Optimization of Battery Operation	307
	20.5.2.5 Link Quality Measurements	307
20.6	Interference Measurements	308
20.7	Consummation	311
Acknowledgments		311
References		311

DOI: 10.1201/9781003326205-24

20.1 Introduction

When it comes to applications like disaster management and border protection, WSN sensors are expected to be deployed at large scales and function autonomously in an inaccessible area where a phenomenon is being observed. One or more nodes having wireless sensing and data networking capabilities make up a wireless sensor network. There are sensor networks in the field that employ wireless sensor nodes with limited battery capacity to gather relevant data. In order to keep the sensor network running for an extended length of time, it is essential to gather data in an energy-efficient manner. For this reason, various protocols for routing, power management, and data transmission have been created specifically for WSNs. This type of WSN has several difficulties, including a high demand for bandwidth, a high energy consumption, the supply of quality of service (QoS), data processing and compression methods, and cross-layer architecture. The physical environment is to be addressed in the future generation of the network.

Applications with a variety of needs may benefit from technology that enables the Internet of Things (IoT). Agricultural, smart city, and smart environment applications, as well as inventory management systems, may all benefit from LoRaWAN's long-range transmission of nodes. One of the most critical IoT KPIs is network life length, which assesses how long it takes for a network to meet its objectives before it has to be replaced. While using a nonrenewable battery to power sensor networks in real-world applications, the network lifetime is often determined by the death of critical nodes that are required to carry out meaningful network operations, which is marked by either a complete loss of power or an accumulation of wasted energy. Due to security, privacy, computational, and energy constraints, as well as issues with reliability, WSN routing is one of the most difficult elements. To be effective, WSN routing systems must handle issues related to data integrity, confidentiality, and availability [1].

20.2 Growth of Wireless Sensor Networks

It is possible to gather information about the surrounding environment using a network of sensor nodes that interact wirelessly. Because of its low power consumption and decentralized nature, most nodes are deployed on-demand. The placement of sensors in a network brings them closer to one another, reducing the distance over which data packets must be sent. As a result of these changes, sensors' battery life and WSN coverage will increase. Installing and moving WSNs is easier than wired systems. Using energy harvesting devices, they may provide real-time capabilities and energy independence, which is incredibly scalable and dependable. As the name suggests, a sensor network collects data from a number of different locations and transmits it to a central place where it may be accessed in real time, and used for various purposes. A computer system's power, memory, computational speed, and bandwidth are all constrained by its size and cost. There are many different ways to structure a wireless sensor network, including a simple star network or an advanced multi-hop mesh network. Routing or flooding can be used to distribute a message. Wireless sensor networks are becoming increasingly important in a wide range of applications, and researchers are taking note.

Opportunities and Challenges for WSNs 295

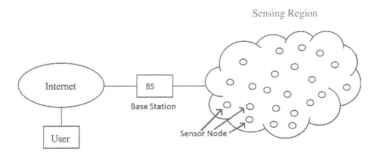

FIGURE 20.1
WSN infrastructure.

The primary role of a WSN is to analyze and transmit data retrieved from faraway sites. The monitoring region may be covered by a vast number of sensors.

20.2.1 Progress

Nodes in WSN are monitored by a processor that is connected to the base station by means of a WSN processing unit and are also shared by internet. An infra structure less network deployed in ad hoc to monitor environmental conditions are wireless Sensor Network (WSN) Figure.20.1. Nodes in WSN are monitored by processor which are connected to base station that are WSN processing unit, and are also shared by internet.

WSNs do not only monitor the sensed parameter. There are various applications like Internet of Things (IoT), surveillance and monitoring for security purposes, threat detection, environmental parameter measurement, and so on. There are many challenges faced by these network, like security, quality of service, performance, optimization in cross-layer, and their scalability. A few components play a vital element in WSNs: (a) Sensors to acquire signals or variables. This sensed information is converted into electrical signals. (b) Radio nodes that help in retrieving data from the sensor and transfer the same to an access point which is the next component [2]. This point receives data by internet and is wireless. An evaluation software is used to present a report to users.

20.2.2 Architectural View

Wireless sensor nodes and base station nodes (sinks) are two distinct types of nodes in a WSN, as shown in Figure 20.2. As a first step, sensors, which are nodes that monitor the physical environment, produce a packet of data and transmit it to the BS (sink). When it comes to managing disasters, security, and crises in a variety of settings including schools and hospitals as well as roadways, sensor network architecture is an ideal solution. WSN employs layered network architecture and clustered architecture as its two primary architectural approaches. These are outlined in detail in the next subsections.

20.2.2.1 Layered Network Architecture

Layered Network Architecture makes use of a single strong base station and many hundreds of sensor nodes. Centrifugal layers are used to arrange nodes in a network.

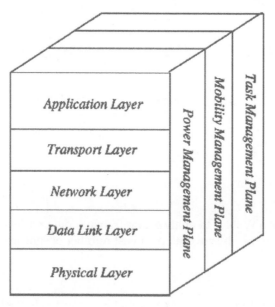

FIGURE 20.2
Layered network architecture.

Layers include application, transport, networking, data linking, and physical (sometimes known as the physical layer). In terms of cross-layers, there are the following:

- power management plane
- mobility management plane
- task management plane

When a sensor network architecture is implemented using layered network architecture, each node participates only in transmissions to surrounding nodes, which reduces power consumption. Scalability and greater fault tolerance make it a better option.

20.2.2.2 Clustered Network Architecture

In a clustered network architecture, clusters are formed by sensor nodes. Using clusters is a nod to the Leach protocol. The Leach protocol's full name is low energy adaptive clustering hierarchy (LEACH).

The properties of Leach protocol are:

- Two-tier clustering design is used.
- Using an algorithm, clusters are formed by pairing nodes together.
- TDMA schedules are created by the cluster head nodes in each of the autonomously formed clusters.
- To save energy, it employs a technique known as "data fusion."

Figure 20.3 demonstrates how a clustered network architecture's data fusion capabilities make it a particularly effective sensor network. In a cluster, each node communicates with

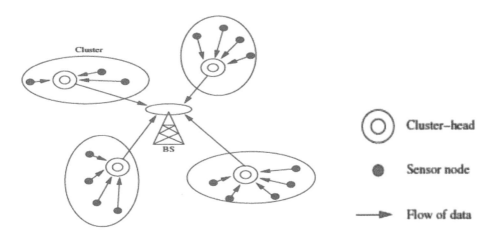

FIGURE 20.3
Clustered network architecture.

the cluster head to acquire data. Grouped data is delivered back to the base station via clusters. The process of creating clusters and selecting a cluster head within each cluster is decentralized.

20.3 Next Generation Networks: IoT

Smart solutions with intelligence are handled by IoT that helps in enhancing connectivity, edge device processing, and real-time data analysis. The term "next generation network" or NGN has been coined to designate the developments in telecommunication core and access networks to be deployed in future decades. A packet-based network is a next generation network (NGN) that provides services to the telecom sector by using broadband technique and also by transporting topologies.

NGNs are characterized by fundamental aspects like transfer of data on packed basis; segregating control functions in terms of users; and also distribution of service from network and interface.

It is predicted that by 2025, many more devices will be connected with the internet and demand will rise. CISCO stated that by 2030, around 500 billion devices will be connected to the Internet of Things. Researchers will benefit from industrial opportunities to create novel strategies. IoT is assisting in converting the world into a smart world. As shown in Figure 20.4, IoT finds many applications; a few of them are real-time multimedia, healthcare systems, smart industries, and smart agriculture. Security requirements and spectrum scarcity are major issues during resource allocation. AI-based usage is a preferred technology.

Consequently, many researchers have contributed to future generation IoT-based applications for different areas, including IoT-based smart healthcare, smart cities, smart agriculture, data analytics, industrial IoT, IoT-based multimedia, and spectrum sharing techniques. A few future WSNs are framed here:

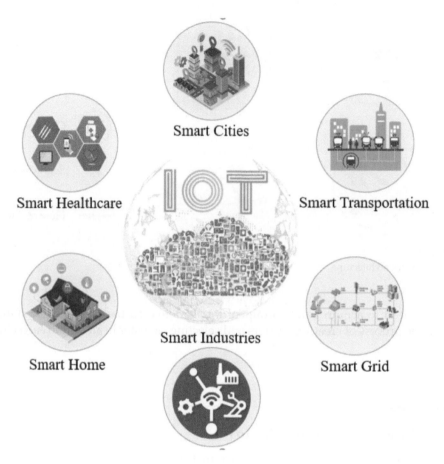

FIGURE 20.4
Next generation Internet of Things (IoT)-based applications.

(a) IIoT (Industrial IoT): Recently, industry has been trying to move to a smart industry structure. The standard named Industry 4.0 will transform existing smart industry into new-perspective cyber-physical technologies that enhance performance, management, control, and fairness. Sensing of information should be performed accurately according to industry standards. There is much research on industrial performance and facility utility. Use of sensors in industries generates the smart industry revolution. Though there are many variations among existing smart methods, smart-with-sensors will rule the industry in the future.

(b) Healthcare IoT: Certain diseases, like coronaviruses and other viral infections, are airborne and spread rapidly. These should be properly monitored and responded to. Due to industrial development and growth, pollution is increasing, worsening lung diseases like chronic obstructive pulmonary disease (COPD) and tuberculosis. Currently, COPD is the most threatening of all diseases, followed by TB. Development in IoT enables patient monitoring and disease control in an easy way. Machine learning (ML) methods are employed in processing images from CT/MRI. This approach provides early detection of disease as well as better treatment.

(c) Fog computing: Telemedical frameworks are created using fog computing technology (FCT) that provides access to edges of the network. FCT reduces latency and reliability. Conventional methodologies may be compared with FCT-based simulation. [1] deals with data transfer with FCT edge nodes with coordination overcoming capacity wastage for controlled IoT devices. ML-based approaches are tested and developed using Q-learning approaches.
(d) 5G technologies: RFID used currently are expensive and do not satisfy the need of 5G with next generation IoT and are not suitable for long-distance communication. To avoid this, I-RFID is proposed, that uses meandering angle technique (MAT) and an inbuilt low-cost printed antenna operating at 860–960 MHz. With the help of this, tag size is reduced by 23 to 33 percent and read range increased by 51 percent. Usage of public key frameworks in 5G vehicular method remains a threat to the privacy and security of networks. Malfunction of approved users or nodes in a network is not detailed by many researchers. Trust-enhanced on-demand routing (TER) mechanism and the Trust Walker (TW) algorithm help make secure vehicular node transmission easier. Precision is validated by throughput, delay, and reliability of the network.
(e) Network-on-a-chip (NoC): Due to scalability and efficiency for heterogeneous networks, network-on-a-chip (NoC) technologies are familiar not only for communication frameworks. Flaws that arise in the router of NoC degrade the network performance. Data transmission takes place by virtual channel (VC) methodologies. Network reliability exists in the proposed mechanism.
(f) Next generation IoT-based smart healthcare: During the pandemic, IoT-based healthcare frameworks were used by many individuals and doctors to get diagnosed with the disease. Several diseases are detected and diagnosed via IoT. Future research in health care will definitely help many people across world. Studies on Alzheimer's are taking place to help patients determine their geospatial location. Sensors as a node in a WSN will be used to track the person. Minor advances in healthcare include diagnosis of epileptic seizures and stroke, tele-healthcare using AR/VR, compact healthcare records, and remote interactive medical training.
(g) Next generation IoT-based smart cities: Next generation IoT enters into smart cities to address many challenges and issues. A challenge that arises in smart cities is that vehicular sensing is not reliable when a device fails. Sensors that access mobility nodes fail. IoT frameworks are required to disseminate many devices.
(h) Next generation IoT-based smart agriculture: With smart agro, monitoring of soil quality and soil erosion is possible, and it helps to save water. This method also improves biodiversity of land, promoting a green environment. In future, this technology will assists in securing food supplies and saving biological systems. It plays a vital role in protecting landscapes and preventing gas emissions and biodiversity wastage. It is also applied to cultivation, preserving natural resources without compromising their quality.
(i) Next generation IoT-based data analytics: Advancements in IoT-based applications increases the number of devices which will continuously gather information from and conduct handshakes with the environment. Smart devices and improvement in data analytical techniques make applications smarter, intelligent, and more useful. As all devices are interrelated in the network, tremendous volumes of information get collected, from which required data are used for further analysis. There is a possibility that all the devices in a node and network will generate the same data and information, which remains as a challenge. This can be easily met by incorporating

machine learning techniques, neural networks, and deep learning algorithms that extensively use various resources.

(j) Next generation IoT-based multimedia applications: The primary goal of IoTs based on multimedia applications is to maximize QoE and QoS in terms of reliability, successful transfer rate, and lifetime of network. These parameters are attained at effective information aggregation, and intelligent MAC layer resource allocation. A limitation is that IoT devices must be small, and therefore have less energy and resources. Redefined architecture is required to offer network access to devices interacting with many different devices, overcoming problems of spectrum scarcity and resource demand.

(k) Next generation IoT-based spectrum sharing techniques: Extension of IoT leads to change in network parameters and performance. A mobile device needs an additional device to handle the information gathered by sensors, which is not managed by existing mobile standards. Some devices at edges require additional support to cope with uplink and downlink frequencies. Scarcity of resources prevents mobiles using an additional device in the existing scheme.

(l) Next generation IoT-based cross-layer protocols: Network clustering methods help IoT-based protocols to achieve effectiveness by proper selecting of routing clusterheads. AI-based IoT devices address many challenges such as low power, high density of users, communication of information over long distances, higher latency, and lossage of data. Cross-layer routing protocols are usually required for energy proficiency, lower latency, and lower information transmission delay. This can be achieved with the help of many learning techniques, like reinforcement learning.

Recently, IoT has entered into all sectors that exist in the world. IoT applications are widely different in purpose, such as smart homes, healthcare, industry, real-time multimedia, and smart agriculture. It is very important to deal with demand, application, and security issues.

20.3.1 Architecture

The Internet of Things (IoT) has a wide range of applications and the use of IoT is expanding quickly. The Internet of Things operates in a variety of ways depending on the specific uses for which it was intended or created. However, there is no widely adhered-to standard architecture for how it works. When it comes to the Internet of Things, its design is dependent on how it is used in various industries. Still, there is a fundamental process flow that underpins the Internet of Things (IoT).

There are four layers (Figure 20.5): the sensing, network, data processing, and application layers; each has a distinct role to play.

These are outlined in detail as follows.

(1) Sensing layer: This layer contains all of the devices that sense, act, and communicate with one another. Data is received, processed, and then sent through a network by these sensors or actuators (physical/environmental factors).

(2) Internet and network gateways, as well as data acquisition systems (DAS), are all part of the network layer. Data-gathering and conversion are two of DAS's primary responsibilities (collecting and aggregating data, then converting analog data to digital data, etc.). Additionally, in addition to providing a link between sensor networks and the internet, advanced gateways handle numerous fundamental gateway tasks, such as virus prevention and filtering.

Opportunities and Challenges for WSNs 301

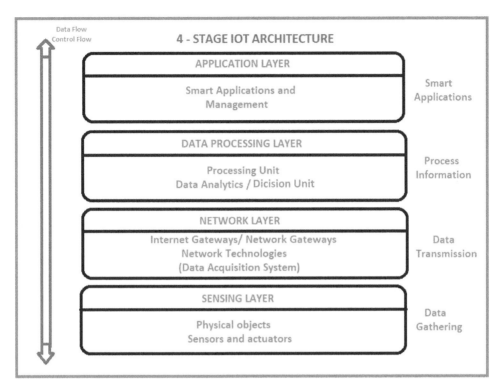

FIGURE 20.5
Four-stage Internet of Things (IoT)-based architecture.

(3) Because of this, data processing is a critical component of IoT systems. Prior to being delivered to the data center, where software programs typically known as business applications monitor and regulate data, data is assessed and preprocessed in this region. Here "Edge IT" or "Edge analytics" comes into play.

(4) In the IoT architecture, there are four layers: physical, network, data, and application. An end-user application such as a farming operation or a healthcare facility may store data in a data center or the cloud for later use by other applications such as defense or aerospace.

20.3.2 Challenges

The five most difficult technical issues you will face while creating your own IoT network are:

- Multiple IoT security flaws affect hundreds of thousands of gadgets.
- The Internet of Things continues to be a confusing jungle of protocols.
- Using big data to make sense of the interconnected world.

The complexity of IoT networks cannot be underestimated by system and network architects. There has been a lot of hoopla around the Internet of Things, from the large number of devices and connections to the possibility for greater efficiency and new sources of money. We can only imagine the scope of the Internet of Things. An IoT project's infrastructure is not easy to build, as many IT teams will tell you. The acquisition of IoT devices necessitates a rigorous bidding procedure, as well as ensuring that they are secure and can interact with one another across a range of protocols (Wi-Fi, radio, or fiber). Monitoring the gateway/network and deciding where to store, manage, and protect data are the next steps. When it comes to data analysis, there's also the matter of where it should be done and by whom. An IoT network's technological issues are laid out in detail in Internet of Business.

20.3.2.1 Security

All of the components of the Internet of Things (IoT) have their own distinct degrees of security, including the device, the gateway, the network, and cloud storage. Endpoint or software attacks are typically enough to break security. No matter where or how the data is stored, end-to-end encryption is critical. There are nearly infinite threats: Some computer platforms (which often have limited memory and computational power resources) may not support advanced security approaches because of poor encryption or few CPU cycles in IoT devices, which are tiny, cheap devices with little or no physical protection at all. The devices themselves may be tampered with or stolen, and if their software is not maintained up to date, they become more vulnerable to cyber-attack. Some argue that a stronger emphasis on identification and the strengthening of network-centric measures such as DNSSEC and the DHCP are needed to prevent attacks on IoT. According to experts, network management and data recording are essential to safeguarding an IoT device's data packets from being intercepted and tampered with.

20.3.2.2 Platform

Platforms for the Internet of Things, whether built in-house or outsourced, have to be scalable and robust. Data security and privacy are top priorities for businesses, as is making sure the platform can manage large volumes of data and integrate with existing legacy systems. In order to properly integrate these platforms into your existing IT infrastructure, you must carefully assess which vendors you want to collaborate with, the associated expenses, and how these platforms will interact. With IoT, network architects have a huge challenge in making sure their network is future-proofed, and that they are making the right decisions when it comes to sensor hardware and radio technologies, as well as cloud platforms.

20.3.2.3 Interoperability and Standardization

Network architects want to avoid being tied to a single vendor and being compelled to utilize an old piece of hardware or software when something better comes to market. Jan Maciejewski, an industry expert and contributing editor at the Internet of Business, believes that there are just too many possibilities. Now that everyone claims to have a platform, the discussion is over which one is the best and how to ensure backwards compatibility. The platform's scalability and capacity to integrate with existing systems are critical considerations when selecting the proper one.

20.3.2.4 Data Storage and Analytics

There is an enormous amount of data generated by IoT devices, and businesses must decide how to store, analyze, and get insight from this data. In other words, how much data should be saved locally and how much in the cloud so that a PaaS or data analytics provider can keep track of it? Furthermore, how can this data be transported and stored while adhering to industry and international regulations? Businesses can take action based on data-driven, proactive choices as a consequence of our investigation. A network management system with a set of predictive analytics tools can provide accurate performance indicators and reporting tools. These include prioritizing traffic, looking at how new services and apps can affect current network capacity, and setting aside time for data-intensive tasks.

20.3.2.5 IoT Sensors and Devices

Distributed sensors are part of a network of wireless sensor nodes that monitor physical or environmental elements such as temperature, sound, and pressure. Sensors communicate with each other over the network. The network engineers will be in charge of monitoring and maintaining these, and the IT department will be in charge of purchasing. Data management, security, and accessibility are all prerequisites for the tendering process. For network architects, Geerdink stresses the need to keep up with the Internet of Things. An architect will have to deal with many more endpoints in the future due to the proliferation of the Internet of Things.

20.4 Next Generation Networks: Smart Grid

Next generation networks optimize efficiency in a real-time duplex between consumers and providers. Many services are provided through building a smart grid, which monitors energy and reduces waste. Open systems, such as smart grids, make renewable energy and electric vehicles more viable. Climate change, energy security, and economic development are just a few of the pressing concerns that smart grids aim to alleviate. By 2030, Korea plans to implement a low-carbon smart grid to create a green-growth economy.

20.4.1 Remote System Monitoring

The existing method of distributing and transmitting power has been shown to be ineffective and unreliable. Grid technology has not evolved much since it was first built, which is why this is the case. In order to solve the drawbacks of the traditional grid, researchers are currently experimenting with smart grid technologies. With a reduction in greenhouse gas emissions of up to 211 million metric tons, this system is more dependable than a conventional grid. Investors are putting their money into this new technology because of this.

20.4.1.1 What Is a Smart Grid?

It is possible to handle the increasing complexity of energy needs in the twenty-first century by automating and controlling a smart grid. This can provide clients with real-time information about how much energy they use in order to reduce utility interruptions and encourage renewable energy sources like solar, wind, and hydro.

20.4.1.2 What Makes Up a Smart Grid?

Smart grids are made up of a number of interconnected moving parts, as is the case with regular grids. Smart grids, as compared to traditional ones, include a more efficient design and functioning. It is possible for intelligent equipment like refrigerators and washing machines to make power utilization decisions based on human preferences. Non-critical operational data, such as power factor, breaker, battery, and transformer health, may be monitored by smart substations.

It is vital to have a smart power meter with two-way communication between the consumer and the power provider. The outcome is that power outages can be identified more quickly, consumers can be charged, and data can be gathered and sent to repair workers. There are several instances of "smart distribution," including long-distance superconducting cables, self-healing and self-optimizing cables, self-balancing systems, and automated monitoring and analysis.

A smart grid's adaptive generation is a part of this system. To optimize energy production and meet voltage, frequency, and power factor criteria, a system may "learn" the unique behavior of power-generating resources through feedback from many grid sites. A broad range of environmentally friendly low-carbon power generating and storage options thus become accessible and affordable.

20.4.1.3 The Current Smart Grid Market

In the United States, the smart grid technology business is expanding. R&D initiatives have been supported by the federal government to the tune of billions of dollars. Approximately $2.5 billion was spent in 2014 on smart grid technology.

Around 150 smart grid technology businesses are thought to exist today, with the United States accounting for 77.4 percent of those firms. In all, the top 25 smart grid suppliers have a market valuation of roughly $2.3 trillion. There has been a total of $60-$300 million invested in General Electric, Honeywell, Itron, and Trilliant Networks so far.

20.4.1.4 Why Do We Need Smart Grids?

As well as being completely in line with today's expectations, smart grids are expected to have major and long-lasting consequences in the near future. When it comes to repurposing old equipment, the technology will do just that. Reduced blackout and burnout risk will be a benefit of this measure. Both energy usage and production costs will be reduced as a result of this new technology. Increased energy demand may be met while making renewable sources of energy more viable by fully using smart grids. In addition, the system will allow for large-scale charging of electric vehicles and near real-time control of energy expenditures for customers.

20.4.1.5 Reaping Rewards

Instead of merely improving power management and adopting greener technologies, a smart grid is all about giving customers an economic advantage. This technique can save the typical home about $600 per year in immediate savings. When customers can see their energy usage in real time, it will drive them to cut their consumption by 5 to 10 percent. It has been demonstrated in studies that people are more willing to cut their energy usage when they have a good idea of how much they use. A year after the introduction of smart grid technology, it is predicted that $42 billion will be saved in energy costs. $48 billion in savings will be made per year in the next five years. Those savings will soar to $65 billion in 15 years and to $102 billion in 30 years. The savings could power Las Vegas for 199 million years, or provide cooling for 378 million houses.

20.4.1.6 The Future Is Now

$3.5 billion in smart grid technology innovation funding was proposed by the Department of Energy from 2016 to 2026. Plug-and-play, self-healing grids and grid automation will be the primary focus of research in the next years.

20.5 Green Communication

In order to reduce the amount of resources required in communication, "green communication" refers to the use of energy-efficient communication and networking equipment and products. Green communication is a cutting-edge field of study that aims to enhance radio networking. Both energy and resource efficiency have a trade-off with user experience (QoS). Reducing energy use saves money on running costs. Reduce carbon emissions is more environmentally friendly. In order to lower the amount of energy needed to run wireless access networks, new algorithms must be developed.

Sensor networks are the foundation of wireless communication today. WSNs may be used in a wide range of applications, including industrial, security, monitoring, tracking, and home automation systems. One of the most pressing challenges for WSNs is the limited supply of energy. Radio optimization, energy harvesting, data reduction, cross-layer optimization, sleep/wake-up policies, load balancing, optimization of power demand, communication mechanism, battery operations, and energy balancing schemes are some of the aspects of energy optimization in EH-WSNs.

20.5.1 Energy Monitoring

Energy collecting methods play an important part in WSNs. Using energy harvesting technologies has had a significant influence on WSN efficiency. This section concludes with a list of important advantages. An alternative energy source for sensor nodes is provided by harvesting energy from the environment, which reduces the need for batteries. For sensor nodes to operate when WSNs are deployed in risky regions where regular access to sensor nodes is difficult or impossible after deployment, a continuous power supply is necessary; this is only attainable by adding energy harvesting technology to WSNs. Environmental energy harvesting permits sensor nodes to continue to operate forever. WSNs are deployed for long-term surveillance of a particular event, and this is the main aim of every WSN that has been deployed. When a source of collected energy is readily available, sensor nodes may continue to perform their basic functions for decades. Traditional WSNs have an upkeep challenge due to their high cost. Batteries should be replaced as soon as they run out, and the sites should be inspected on a regular basis to check on the batteries' health. Using energy collecting methods, the cost of maintenance may be significantly reduced. Because EH-WSNs do not need to be serviced on a regular basis and because they do not utilize batteries, the cost of maintaining them is significantly reduced. WSNs with energy harvesting mechanisms have lower installation costs than ordinary WSN; however the harvester circuits and other hardware components must be included in the overall installation cost. There is also a lower installation overhead with EH-WSNs.

20.5.2 Algorithmic View

MAC protocols for EH-WSNs, classification of transmission schemes, and EH-WSN routing protocols are three basic kinds of schemes based on optimization of communication mechanisms. Here, we will take a short look at each of these methods in detail.

20.5.2.1 MAC Protocols for EH-WSNs

This subset of asynchronous protocols is further divided into three types: those begun by the transmitter, those initiated by the receiver, and those launched by the sink, as well as other variations on these themes. We provide the following breakdown of each category:

(1) Asynchronous protocols initiated by the sender. The EL-MAC, RF-MAC, and Deep Sleep MAC protocols are all covered.
(2) Asynchronous protocols that are initiated by the end user. It is explained how to use ODMAC, EHMAC, QAEE MAC, ERI MAC, and LEB MAC as well as the ED-CR and ERI MAC protocols.
(3) Asynchronous protocols for the sink. The PP-MAC, MTPP-MAC, RF-AASP, and AH-MAC protocols are all detailed in this chapter.

20.5.2.2 Transmission Schemes Classification

There are three types of transmission schemes: fixed transmission, variable transmission, and transmission based on a probability model. Schemes like this are outlined in a succinct manner here.

(1) Static transmission diagrams. Power management in EH-WSNs has been efficiently framed. For controlling communication power efficiency challenges and improving the value of sensor nodes that capture energy, they have presented a framework based on an energy-neutral approach. In the design of this framework, fixed power in circuits, battery efficiency, and storage capacity are all considered.
(2) Use of a wide range of transmission methods. The energy harvesting method is commonly accepted as the most efficient means of powering WSNs. Rechargeable battery storage limitations and the mechanism used to replenish the energy are significant considerations for designing the optimum transmission rules for EH-WSNs. The current framework was designed to discover the optimal data transmission mechanism for a particular time period. The optimal transmission rules for rechargeable battery storage or energy refreshment techniques were identified by means of the proposed framework. Packetized energy is accounted for as it is reloaded in the simulation. In addition, two major issues beset this energy collecting technology: instability and unpredictability in the collected energy. As a result, sensor nodes are dying earlier than expected, decreasing the overall quality and performance of the system.
(3) Transmission schemes based on probabilistic models. It has been demonstrated that the best joint transmission method can be determined in the case of two sensor nodes, one powered by batteries and the other solely by gathered energy. In addition, the base station receives a common message from the network. The goal is to discover the combined transmission method that maximizes throughput while keeping the deadline in mind. For rechargeable batteries, it is nearly impossible or prohibitively expensive to accurately estimate the level of charge.

20.5.2.3 Routing Protocols for EH-WSNs

Routing protocols are a constant problem for both classic wireless sensor networks and EH-WSNs. Traditional wireless sensor networks typically struggle to provide optimal performance after a predetermined amount of time, and the network's performance begins to degrade. In EH-WSNs, the necessity for effective harvesting and management of gathered energy arises. Consequently, both regular WSN and EH-WSN routing protocols take the energy issue into account as a fundamental consideration. Routes based on cost, passive features, geography, and clustering make up the majority of EH-WSN routing protocols. These four subcategories are further subdivided. The following is a succinct description of each.

(1) Protocols for routing based on the cost of routes. The sensor network's lifespan can be extended by reducing the routing cost. The amount of network energy wasted while tracking acceptable routes has been given particular attention in the prior research. The fundamental innovation in this framework addresses the overcharging of limited-capacity batteries in the network, which is the primary source of energy waste. This framework thus reduces energy use, resulting in larger residual energy levels.

(2) Passive properties are used to design routing protocols. Rather than taking into account the present state of the network, routing systems depend on previous data to construct their database of possible routes. Two factors are taken into consideration for the building of the routing tables in this protocol: the rates of solar energy harvesting and information about prior journeys. However, these protocols have an advantage in that they limit energy consumption in route construction overhead, but on the other side, these protocols cannot recover a sensor node that has died, as these protocols are passive in nature.

(3) Geographical routing protocols. If you are looking for a route that does not need any previous knowledge of the area, this is the one for you. Instead of sending out control messages, as is common in geographic routing, this framework's procedure begins with a data packet, a change from prior versions of geographic routing. Neighbors for communication are only selected for communication if the data packet was successfully received at the time of the initial transmission. The quality of the connection, residual energy, location information, and also the rate of energy harvesting are all taken into consideration by this routing architecture when selecting the best possible routes. In order to increase the life of a network, spatial routing selects reliable paths. We also look into opportunistic routing in depth, utilizing the concept of geographic characteristics.

20.5.2.4 Schemes Based on Optimization of Battery Operation

Managing sensor node operations is the major topic of this section. Sensor nodes regularly sense the field and then create data packets, which are then kept in a queue, in EH-WSNs, taking into account the currently available energy profile for the data packet transmission. Furthermore, these methods are referred to as the most efficient in terms of throughput and average latency. An energy-neutral network operation is the goal of the suggested plans.

20.5.2.5 Link Quality Measurements

There are two types of mobility-based schemes: those that use mobile relays and those that use sinks. An overview of these initiatives is given below. Higher latency is a problem for mobile WSNs since they cannot collect data as quickly as fixed WSNs. Slower movement of base stations leads to higher latency since they are gathering data at a slower pace. Mobility-enabled WSNs suffer greatly as a result of this problem. Resolving this problem is found to be an effective means of increasing energy efficiency. With this in mind, the authors of this chapter have come up with an efficient solution to the problem at hand: to gather data, the base station uses a small number of sensors to buffer and aggregate it.

In addition, data transmission to the base station must be limited to the duration of the specific application's lifespan. It is difficult to meet this criteria in a power-limited setting. In order to reduce the energy consumption of WSNs with high data density, an approach was devised that utilized the notion of mobile relays. Energy gaps in WSNs may be addressed utilizing the sink mobility idea. As a result, the energy hole is effectively addressed by the sink mobility method. Sensor nodes use less energy when using the sink mobility method since all data collection is done in a single hop by the mobile sink. Sink mobility is only a serious concern in time-bound emergency circumstances, as sensed data begins to lose relevance over the bound course of time; this delicate problem needs to be treated with caution. Sensor nodes additionally need stringent sleep/wake policies and a forwarding path.

For some applications, the unpredictability of the mobile sink's mobility results in a decrease in performance due to the source sensors' efforts to identify the mobile sink before providing data. Using the smallest number of hops possible to convey data to a mobile sink is the goal, and the rationale is obvious: reducing the number of hops saves energy.

20.6 Interference Measurements

Optimizing EH-WSNs is a difficult task because of a variety of challenges. Multiple, currently available, and very efficient optimization methods may be used to address these multiobjective problem sets. A thorough breakdown of the numerous issues faced by green WSNs is provided. For optimization issues in EH-WSNs, we are attempting to demonstrate some of the major obstacles, as well as some of the techniques, that arise. Aggregate utility and network longevity are included in the performance evaluation of the system. In addition, a multiobjective stochastic technique is used to boost network longevity while also improving aggregate utility. Optimization models and algorithms with multiobjective computing capabilities and stochastic nature are needed to solve bigger real-life optimization issues in several sectors.

To make matters worse, because the processing power of a single sensor node is restricted in WSNs, a distributed method is the only viable option for dealing with a given problem that arises. Communication overhead in the distributed approach is significantly lower than in the centralized approach, as all nodes in the distributed system communicate with each other directly rather than through intermediaries. The framework for multiobjective routing may be used to weigh the benefits of longer network lifespan against QoS. Using this approach, the goal is to minimize battery costs while increasing the amount of energy remaining in the forwarding set. The heuristic technique was applied in both of the above-mentioned works. To solve multiobjective optimization problems, a heuristic strategy is commonly used. Heuristic methods were used to solve the problem quickly, and they relied on the trial-and-error method. The heuristic technique largely delivers results that are quite accurate.

Furthermore, locating sensor nodes in WSNs is a critical challenge that must be addressed. In [2], a multiobjective optimization framework is given for locating sensor nodes properly so that the data may be assessed for its geographical significance. WSN performance evaluation also takes into account the importance of coverage efficiency. Again, a slew of articles using multiobjective frameworks has already appeared in the literature. Increasing coverage is the primary goal, but they also take other desirable outcomes into account. In a multiobjective optimization framework, the optimum solution is selected from a pool of optimal options by taking into account the desired outcome. [3] employed the evolution-based technique. Multiobjective evolutionary optimization employs a population-based strategy. Each iteration generates a new collection of possible solutions, which are then combined in the next iteration. Evolution-based optimization frameworks do not require derivative knowledge, thus their implementation is straightforward. Many optimization problems, including those with multiple objectives, may be solved using evolutionary frameworks because of their broad application.

Since the tactics used in deployment directly affect the effectiveness of sensor nodes, deployment is also a major barrier for WSNs, which is why it is necessary to address this problem. Appropriate deployment options can increase a WSN's life expectancy as well as network performance by reducing energy consumption. WSN deployment problems are addressed using metaheuristic algorithms based on a variety of methodologies in [4]. There is a preference for the metaheuristic method over the heuristics approach. In the metaheuristic framework [5], randomization and optimum solution are the two most important components. The randomization components help prevent solutions from being trapped in local optimums. Making use of the best solution selection stage ensures the convergence of the solutions to optimality.

When it comes to wireless sensor networks, there are two competing goals that must be addressed simultaneously: efficient connection and network longevity [6]. Using the same energy-saving profile, this framework aims to provide greater connection to other schemes. A single solution must be able to optimize all goals concurrently, but standard multiobjective formulations are unable to do so since better connection and a longer lifespan are two important characteristics that have a significant impact on performance. With a well-designed multiobjective optimization formulation, there are several possibilities. The positions of the alternative solutions are also close to or on the Pareto optimum front. It is possible to get a large number of Pareto optimum solutions using nondominated sorting algorithm II simply in a single run.

In WSNs, effective data transfer via energy-efficient or energy-aware routing techniques is also necessary. Network features such as data availability, network longevity, and communication overhead are all regulated by energy-efficient routing algorithms. In order to increase system performance, energy-efficient or energy-aware routing frameworks are used. The use of bio-inspired algorithms in optimization and computational intelligence is on the rise nowadays. Systems based on biological principles frequently employ reactive and proactive strategies in tandem. Bio-inspired approaches, adaptive routing, and enhanced load balancing may all be used to learn the network architecture.

In WSNs, the bit error rate should be kept to a minimum, but the signal-to-noise ratio should be kept at its highest possible level, which can only be achieved by increasing the transmission power. The increase in transmission power has a significant influence on a number of essential network properties, including interference reduction, energy efficiency, and network longevity. As a result, trade-offs between competing goals may be achieved via multiobjective optimization frameworks. When designing effective power allocation strategies and a spectrum sensing module for [7], a multiobjective memetic technique reduces interference and maximizes throughput. The Pareto optimal solution set is found using the trial-and-error method of memetic algorithms, which contain computational intelligence components.

WSNs' dependability and latency are also crucial performance indicators. The multiobjective optimization technique is used in [8] to propose an effective routing architecture. As a result, the goal of implementing quality of service clashes with other goals, such as network cost and network longevity. Fuzzy random variables were used in the routing optimization model to describe the objectives' and constraints' unpredictability and fuzziness. Human thinking is represented in fuzzy logic using mathematical notations. A proposition's truth value may be approximated using fuzzy logic, which makes use of inference rules and linguistic factors.

High transmission power, which is necessary to increase sensor node range and hence sensor node level, results in a significantly lower level of energy efficiency. As a result, network reachability, a key WSN property, is frequently low when WSNs are poorly connected; robust connection in the network costs a lot of power, which lowers WSN quality of service. The appropriate transmission power requirements of a sensor node necessitate thorough research. An optimization approach for attaining an energy-efficient clustering-based routing framework is also described in [7]. Total packets transmitted to base stations, network longevity, energy consumption, and dead nodes are all balanced out in this architecture. The framework takes advantage of particle swarm optimization. Particle swarm optimization relies on the swarm's collective intelligence. In recent years, the particle swarm optimization approach has been used to handle a wide range of optimization problems.

WSNs have a large number of small sensor nodes planted densely, and this has an impact on the overall performance, leading to errors in the reconstruction of the physical signal. Using a genetic algorithm technique, the original problem is divided into smaller subproblems. When it comes to handling complex optimization issues, genetic algorithms have been shown to be a highly effective method. Genetic algorithms begin with a population of randomly produced candidate solutions, in accordance with Darwinian evolutionary theory. Each person in the population is a possible solution to the problem since each one has a unique genetic code.

Development efforts throughout the first decade (2000–2010) are referred to as the "first era." Energy harvesting devices with improved fabrication methods and the creation of hybrid models for energy harvesting systems are the main characteristics of this age in terms of substantial advancement in this field.

The second period encompasses the years 2010–2015. Triboelectric nanogenerators, a variety of energy harvesting system designs, and a hybrid energy harvesting system model are a few of the most significant breakthroughs in the field of power generation technology.

With the introduction of Industry 5.0 in December 2015 and the trend toward producing self-powered IoT, energy harvesting technology has seen a substantial rise in interest.

As small-scale self-powered devices are created with system-level deployment in mind, this trend will be supported as energy collecting technologies are studied as a method to improve the current smart city and society environment. The signal from the energy harvester will be examined, using machine learning to predict the future environment. Metamaterials for machine learning energy harvesting are currently being investigated. The performance of energy harvesting systems is also being improved by new technology.

20.7 Consummation

Here, the EH-WSN energy optimization problem is seen from a broader perspective than in previous chapters. Network sensor nodes are expected to have their lifespans increased by this project. We go over each of the 11 key types of EH-WSN energy optimization solutions. For sensor nodes to use energy harvesting as a backup power source, we need to think about how to make the most efficient use of the energy harvested and how to improve the efficiency of the energy captured. Because there is room for improvement, a lot of work has been invested in improving the process of obtaining energy. New sources of power for the small sensor nodes should be explored, even if upgrading the energy gathering process poses a lot of challenges. Energy scarcity for tiny sensor nodes may be addressed with a hybrid solution that utilizes all three of WSN's primary power sources: environmental collection, batteries, and ultimately wireless power transmission. Although this is a promising concept, there is still more work to be done before a long-term WSN can be realized.

Acknowledgments

We thank our parents for their support and motivation.

References

[1] Li, Junling, and Danpu Liu. "An energy aware distributed clustering routing protocol for energy harvesting wireless sensor networks." In 2016 IEEE/CIC International Conference on Communications in China (ICCC), pp. 1–6. IEEE, 2016.

[2] Ueda, Tetsuro, Akira Idoue, and Eiji Utsunomiya. "Comparison of routing protocols for wireless sensor networks under battery-powered and energy harvesting conditions." In 2018 24th Asia-Pacific Conference on Communications (APCC), pp. 433–438. IEEE, 2018.

[3] Galmés, Sebastià. "Optimal routing for time-driven EH-WSN under regular energy sources." Sensors 18, no. 11 (2018): 4072.

[4] Mazunga, Felix, and Action Nechibvute. "Ultra-low power techniques in energy harvesting wireless sensor networks: Recent advances and issues." Scientific African 11 (2021): e00720.

[5] Khademi Nori, Milad, and Saeed Sharifian. "EDMARA2: a hierarchical routing protocol for EH-WSNs." Wireless Networks 26, no. 6 (2020): 4303–4317.

[6] Eu, Zhi Ang, and Hwee-Pink Tan. "Adaptive opportunistic routing protocol for energy harvesting wireless sensor networks." In 2012 IEEE international conference on communications (ICC), pp. 318–322. IEEE, 2012.

[7] Hao, Sheng, Yong Hong, and Yu He. "An Energy-Efficient Routing Algorithm Based on Greedy Strategy for Energy Harvesting Wireless Sensor Networks." Sensors 22, no. 4 (2022): 1645.

[8] Adu-Manu, Kofi Sarpong, Nadir Adam, Cristiano Tapparello, Hoda Ayatollahi, and Wendi Heinzelman. "Energy-harvesting wireless sensor networks (EH-WSNs) A review." ACM Transactions on Sensor Networks (TOSN) 14, no. 2 (2018): 1–50.

21

Various Simulation Tools for Wireless Sensor Networks

Hakan Koyuncu and Ashish Bagwari

CONTENTS

21.1 Introduction 313
21.2 Network Structure 314
21.3 Various Simulation Tools for Wireless Sensor Networks 316
    21.3.1 NS-2/NS-3 317
    21.3.2 OMNET++ 318
    21.3.3 J-Sim 318
    21.3.4 JiST/SWANS 319
    21.3.5 GloMoSim 320
    21.3.6 SHAWN 320
    21.3.7 OPNET Network Simulator 321
    21.3.8 SENSE 322
    21.3.9 VisualSense 322
    21.3.10 TOSSIM 324
    21.3.11 EmStar 324
        21.3.11.1 EmSim/EmCee 324
        21.3.11.2 EmView/EmProxy 325
        21.3.11.3 EmRun 325
    21.3.12 NetSim 326
    21.3.13 ATEMU 327
    21.3.14 PiccSim 327
21.4 Conclusion and Future Work 328
Acknowledgments 329
References 329

21.1 Introduction

With the development of wireless technology in recent years, formerly wired applications became wireless, and whole new application domains emerged. One of the most significant research areas that emerged in the context of wireless technology was mobile ad hoc networks (MANETs). Mobile ad hoc networks are self-configuring networks where the network architecture is constantly changing and the movement of the wirelessly connected nodes is not constrained.

Advancements in highly integrated and tiny devices, as well as improvements in energy accumulators, have led to the emergence of a new generation of hardware known as sensor nodes. One division of mobile ad hoc networks is the sensor node. The main function of sensor nodes is to carry out distributed sensing operations with the least amount of resources possible. Wireless sensor networks (WSNs) are designed to cooperate in order to accomplish the common goal of gathering, aggregating, and sending sensed data from a particular region by means of a number of hops to a particular position in the network for analysis and assessment. Analytically, the properties of sensor networks need a paradigm shift away from conventional computing models: New analytical approaches that mix network protocols, distributed computing, and traditional centralized network algorithms are required due to the constrained capacities of nodes, the absence of centralized control, and the constrained bandwidth in transmission between nodes.

Understanding complex networks may be improved by three essentially diverse methods: computer simulations, analytical methods, and physical research. It is challenging to create algorithms for sensor networks. Limited resources, fault tolerance, analysis of global behavior arising from local communication, decentralized collaboration, and energy efficiency are just a few of the difficulties that must be solved.

Theoretically, testing using real-world sensor networks should be a great way to show that a system is capable of achieving particular objectives in real-world settings. However, there are a number of issues with this strategy in actual use. First of all, operating and troubleshooting such systems can be challenging. This may help to explain why so few of these networks have really been put into use. Numerous sensor nodes make up most real-world systems, but future networks are expected to have hundreds or even millions of nodes. If sophisticated methods are employed to mimic a wide variety of factors, the numbers in the actual world may climb by several orders of magnitude.

The intricacy of this approach, however, conceals and ignores a different, far more significant issue: building extremely complicated simulation software for each sensor node is similar to building a functioning model for individual brain cells. The objective of a self-organizing, decentralized, and effective network involves more than simply a great number of sensor nodes, just as a brain is more than a collection of cells.

Instead, creating the ideal functional structure is the most crucial scientific endeavor for achieving the objective of sensor networks. Beyond the specific technological characteristics of the individual nodes, the research and development of these systems provide other computational hurdles. New demands for the capabilities of particular nodes may develop as our knowledge grows. Additionally, as miniaturization advances, new characteristics and parameters for microsimulation may be defined.

Several different forms of WSN simulation environment are now accessible. These simulation environments have a wide range of features and structures, including protocols and models. The goal of this project is to offer various modeling tools for wireless sensor networks.

21.2 Network Structure

Typically a WSN contains sink nodes, task management nodes, and sensor nodes. Figure 21.1 depicts the network architecture. WSN sensor nodes can be randomly distributed inside

Various Simulation Tools for WSNs

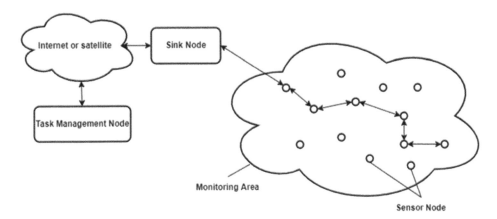

FIGURE 21.1
WSN architecture.

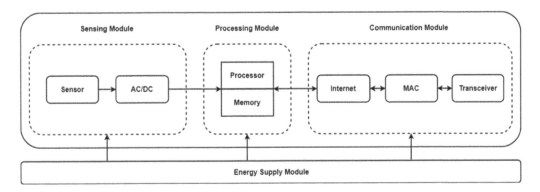

FIGURE 21.2
Node structure of WSNs.

the monitoring region, according to the application's needs. By self-organization, each randomly spread node becomes a WSN. The necessary data received by the node is delivered to the sink node via other nodes in the network, hop by hop. Information is subsequently sent to the task management node through a satellite or the internet. The data collected may be evaluated and interpreted by users.

A sensor node is the network's fundamental unit in WSNs. Figure 21.2 depicts the construction of a sensor node. Processing modules, sensor modules, energy supply modules, and communication modules make up the majority of a sensor node.

A sensor and an analog-to-digital converter (AC/DC) make up the sensing module. A sensor's function is to gather data by detecting things in the environment and altering it. The analog-to-digital converter's job is to convert analog signals to digital signals that may subsequently be sent to the processing module. One of two approaches can be used to implement the sensor module of the sensor node. One approach is to combine numerous sensors. This mode has the benefit of compact size and high integration, making it ideal for sensors with basic circuits, but it is limited in terms of adaptability and scalability. The alternative option is to use plug-ins to link different types of sensors to nodes. This model has strong scalability and can be used for sensors with complicated circuitry in a variety of ways. The sensor node's main component is the processing module. It is made up of two parts: a memory and a CPU. The major responsibility is to manage the entire node's activity. It is in charge of storing and processing information gathered by the node, as well as information transmitted through other nodes. The purpose of the communication module is to exchange control information, interact wirelessly with other sensor nodes, and distribute and receive information gathered by the nodes. One of the most critical parts of a node is the energy supply module. Its purpose is to utilize low-energy batteries to supply the node with all of the energy it requires to function.

The network protocol stack is another significant feature of WSNs. A data connection layer, a physical layer, a transport layer, a network layer, and an application layer make up a network protocol stack. The physical layer is in charge of modifying signals, generating carrier frequencies, and demodulating them. Error checking and media access are handled by the data connection layer. The network layer's job is to find and keep track of routes that allow nodes to interact with one another. The transport layer assures communication quality by supplying control data streams. The application layer's role is to arrange and supply data in accordance with varied needs.

WSNs use a cross-layer design strategy for their protocol stack, which includes a mobility management platform, a task management platform, and an energy management platform. The energy management platform's mission is to find out how to save energy while prolonging the network's lifespan at each protocol layer. The mobile management platform's role is to maintain track of the route between the sensor and the sink node, as well as to record and identify node movements. The function of the task management platform is to coordinate the work of various nodes based on various needs. These management solutions enable sensor nodes to interact more efficiently while using less energy, and they also provide multitasking and resource sharing.

21.3 Various Simulation Tools for Wireless Sensor Networks

It is quite expensive to construct a WSN testbed. Real-world tests on a testbed are expensive and time-consuming. Furthermore, because several factors influence the experimental outcomes at the same time, reproducibility is severely harmed. Isolating a single aspect is difficult. Furthermore, doing meaningful tests takes time. As a result, WSN simulation is critical for WSN development. Protocols, methods, and even new concepts may all be tested on a massive scale. Users can isolate distinct aspects using WSN simulators by adjusting settings. As a result, simulation is critical for studying WSNs, as it is the most prevalent method for field-testing novel applications and protocols. This has resulted in a recent surge in simulator development.

Obtaining strong findings from simulation research, on the other hand, is not an easy task. In WSN simulators, there are two important factors to consider: (1) the accuracy of the simulation models, and (2) the applicability of a specific tool for implementing the model. To get reliable results, you need a "correct" model based on strong assumptions. Precision and the need for fine details vs. performance and scalability are the main trade-offs. Several popular WSN simulators are discussed in this section.

21.3.1 NS-2/NS-3

NS-3 is preferred because of its open-source nature. NS-3 allows protocols to be redesigned, implemented, or tested to observe their efficiency [1]. NS-3 is not a continuation of NS-2. It is a completely new simulator developed from scratch. Good programming skills in C++ or Python are required to use NS-3. Concepts such as classes, objects, and structures are important as they are part of object-oriented programming. Users typically interact by developing a C++ or Python program that creates a collection of simulation models, which enters the main simulation loop, and exits when the simulation is complete. It is possible to simulate complex and simple network scenarios. Although it is permitted to program scripts in Python, it is not recommended due to insufficient support.

NS-3 Wireless Simulation creates a dynamic library of past events stored in a sorted event list, and allows packets to be sent and received and timers to be used by the event model. The network animator (NAM) of NS-2 does not show packet flow in the wireless component of a wired and wireless network. However, with NS-3 there are no such problems. It is possible to simulate either a purely wireless network with only wireless nodes or a network with both wired and wireless nodes. With NS-3, groups of nodes can work together to form a cooperative wireless sensor network. Each node has processing capabilities, numerous memory types, a RF transceiver, a power supply, and the ability to accommodate various sensors and actuators. NS-3 allows WSNs to incorporate a gateway that enables wireless communication with the wired world as well as with dispersed nodes. The choice of wireless protocol depends on the application requirements. NS-3 allows WPAN, IEEE 802.15.1, WLAN, LAN, GSM, CDMA, TDMA, FDMA, 5G, and Zigbee networks to be simulated. Figure 21.3 depicts the user interface of NS-2 and NS-3.

New features of NS-3 include an execution environment for implementation code, which was not available in NS-2 (so that the simulator could be used to run real implementation code). NS-3 has a smaller abstraction base than NS-2, which brings it more in line with the way real systems are built. Some issues from NS-2 have been fixed in NS-3, such as effectively supporting different types of node interfaces. NS-3 is well maintained, with an interactive user mailing list, while NS-2 is poorly maintained and has not had significant code changes in over a decade.

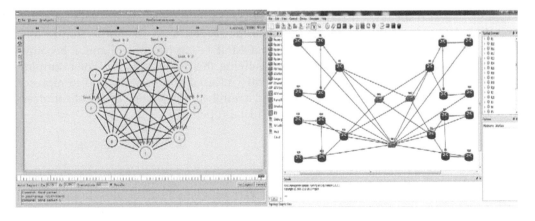

FIGURE 21.3
NS-2 and NS-3 software interface.

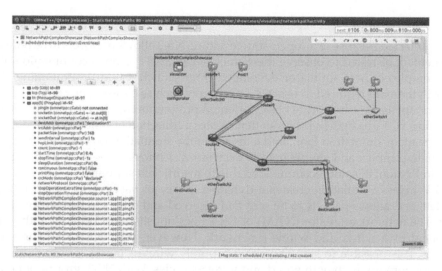

FIGURE 21.4
OMNET++ graphical user interface.

21.3.2 OMNET++

A common framework and simulation tool for C++ networks is called OMNeT++. OMNeT++ utilizes model frameworks to enable sensor and wireless ad hoc networks, Internet protocols, photonic networks, and energy modeling. OMNeT++ offers a component framework for models. The NED high-level language is used to combine smaller parts and structures into modules that were originally built in the C++ programming language. The INET framework serves as the universal protocol model framework for the OMNeT++ simulation tool. The INET framework includes the web stack in addition to other features and protocols. The key characteristics of OMNeT++ are that it is a data graph program with a simulation core and two different user interfaces, a graphical user interface and a command line user interface. Small- to large-scale wireless sensor networks may be simulated using the OMNeT++ program.

OMNET++ also includes a robust GUI framework with support for animation, tracing, and debugging, as shown in Figure 21.4. Its biggest flaw, in comparison to other simulators, is the lack of protocols in its library. OMNET++, on the other hand, is becoming a common tool, and fresh contributions are solving the model scarcity. For example, a mobility framework for OMNET++ was recently released and may be used to represent WSNs. The Consensus Project has also released a number of new ideas for localization and MAC protocols for WSN using OMNET++, which is open-source [2].

21.3.3 J-Sim

J-Sim is a discrete event network simulator that makes use of Java programming language classes to generate its objects, as seen in Figure 21.5. J-Sim is one of the most popular simulation programs, providing pre-built wireless sensor network packets, and is simple to use. Additionally, real-time processes can be simulated. With funding from the NSF, Ohio University created this Java-based simulator, which allows users to create quantitative numerical models and compare them to experimental reference data.

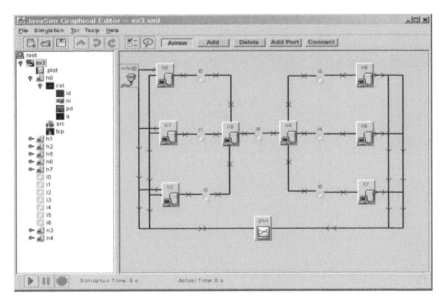

FIGURE 21.5
J-Sim user interface.

J-Sim is an autonomous component architecture (ACA)-based component-based simulation environment [3]. In order to select the best choice and carry it out in a given circumstance, autonomous systems can utilize their artificial intelligence to make their own judgments, learn the appropriate behaviors through time, and carry out a given mission in previously undefined settings. The fundamental component structure creates a simulation environment that is most like the actual world because it enables the simulation's other components to function independently.

The value of J-sim is the extensive range of protocols it supports, which includes a WSN simulation framework with a detailed model of WSNs and implementations of WSN algorithms for localization, routing, and data dissemination. J-sim models are interchangeable and reusable, giving them a wide range of applications. It also comes with a Jacl scripting interface and a graphical user interface library for tracing, debugging, and animation. J-sim promises to be able to grow to a similar number of wireless nodes (about 500) as NS-2, but with two orders of magnitude less memory usage and 41 percent less execution time.

21.3.4 JiST/SWANS

JiST/SWANS is a discrete event simulation framework with a Java bytecode-based simulation engine [4]. The models are implemented and built using Java. After that, the bytecodes are recreated in order to offer simulation semantics. After that, they are executed on a standard Java virtual machine. This method, like NCTUns 2.0 and UNIX, allows the simulation to run on existing Java software without any modifications. The JiST tool's primary shortcoming is the shortage of protocol models. SWANS, an ad hoc network simulator developed on top of the JiST engine with limited protocol compatibility, is already available. An event logger is the sole graphical aid. Jython is the scripting language. JiST promises to grow to networks of 106 wireless nodes while outperforming NS-2

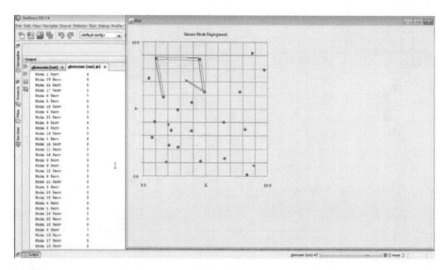

FIGURE 21.6
GloMoSim user interface.

and GloMoSim by two and one orders of magnitude, respectively, in terms of execution time. In terms of event throughput and memory usage, it outperforms GloMoSim and NS-2 despite being developed in Java.

21.3.5 GloMoSim

The term GloMoSim refers to the Global Mobile Information System Simulator. GloMoSim can comprehensively simulate satellite networks, wireline communications systems, and mobile ad hoc network environments as shown in Figure 21.6. It was developed at the Parallel Computing Laboratory at Cornell University. GloMoSim was developed using the parallel discrete event simulation capability of Parsec, a parallel programming language [5].

GloMoSim is a Parsec-based simulation environment for wireless networks. Parsec is a C-based simulation language with semantics for constructing simulation entities and exchanging messages across a wide range of parallel systems. It has been demonstrated that it can grow to 10,000 nodes via parallelization. It has been used to test several WSN protocol concepts. sQualnet, a WSN developer kit, was introduced as a new capability to the simulation software.

21.3.6 SHAWN

Shawn is a sensor network discrete event simulator. It is very quick due to its tremendous customizability, yet it may be tweaked to whatever precision is necessary by the simulation or application. There are various applications for simulations, and many different simulators have been created. Each of them focuses on a certain application area where it shines. As seen in Figure 21.7, Shawn does not make an effort to compete with other simulation environments in the field of network stack modeling. Instead, the basic idea is to replace low-level effects with interchangeable and abstract models in order to quickly apply the simulation to huge networks while keeping the study issue in mind.

Various Simulation Tools for WSNs

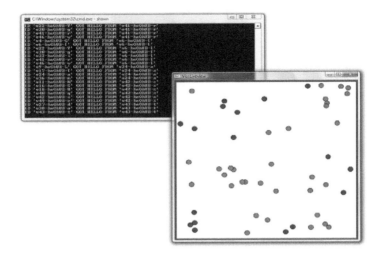

FIGURE 21.7
Shawn simulation visualization tool.

Instead of explicitly simulating phenomena, Shawn's major technique is to employ abstract models and their application to mimic the impacts of phenomena. Shawn models the impacts of the MAC layer rather than replicating the complete MAC layer together with the radio propagation model (e.g., packet delay, loss, and corruption). This affects the simulations in a number of ways: Since such models are often constructed very rapidly, they gain better reliability and information, as well as increased performance.

As long as the enormous power of the macrosystem is focused on unaffected functions, it is possible to support a high number of sensor nodes simultaneously by reducing the low-level parameter architecture, which enables quick simulation to take on the role of time-consuming computations. Shawn must thus be adjusted to the issue by concentrating on the crucial factors in the given situation. On basic PC hardware, Shawn has conducted simulations including more than 100,000 nodes.

Shawn permits several iterations of development when the programmer has full access to the communication diagram, including each node and the data that goes with it [6]. As a result, it is simpler to develop some components as centralized diagram methods and save conversion to a distributed variation until later.

21.3.7 OPNET Network Simulator

A popular network simulator is Opnet. An object-oriented application called Opnet Modeler is used to model communication networks and systems. A visual simulation environment is provided by the object-oriented application OPNET Modeler for the modeling of communication systems and networks. Using the discrete event simulation approach, behavioral and performance analyses of the modeled systems are carried out. Hierarchical modeling layers make up OPNET. It may be used with UNIX or Windows NT/2000/XP.

OPNET is in charge of gathering data, modeling the system, running simulations, and analyzing the outcomes. It contains a large collection of models for switches, routers, and network protocols, including ATM, TCP/IP, and MPLS from manufacturers like 3Com, Cisco, and Bays Network. The primary benefit of OPNET is the capability to use its editors

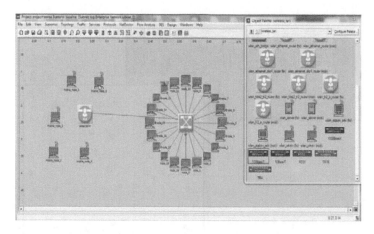

FIGURE 21.8
OPNET user interface.

to build models for new protocols and products and add them to the model library [7]. A tool for event-based simulation is Opnet Models Suite. Network protocols and fixed commercial networks are included. Figure 21.8 shows the OPNET graphical user interface.

21.3.8 SENSE

A sensor network simulator that is effective, powerful, and simple to use is called SENSE (Sensor Network Simulator and Emulator). The three primary criteria in its creation are extensibility, reusability, and scalability. High-level users, network developers, and component developers are separated.

The physical layer of the component memory SENSE may already access application components such as IEEE 802.11, AODV, DSR, SSR, SHR, battery models, and power supply models [8]. SENSE does not yet seem to have additional functions, including a visualization tool enabling a graphical examination of network activities. In 2004, the potent sensor network simulator (SENSE) made its debut. The most recent version is SENSE 3.0.3. SENSE is a standalone CompC++ (component-based extension of C++) event simulator built on COST. It operates in a manner akin to J-Sim. For wireless sensor networks, SENSE offers battery models, MAC layer protocols including IEEE 802.11 with DCF and NullMAC, as well as applications, networks, and physical layer models. SENSE has SSR protocols including SSR, SHR, and Self-Selective Reliable Path (SRP). SENSE only has a text-based user interface. G-SENSE was developed with graphical tools to address this, as seen in Figure 21.9.

21.3.9 VisualSense

Based on Ptolemy II, VisualSense is a modeling and simulation tool for wireless and sensor networks [9]. A thorough description and analysis of communication channels, ad hoc and network protocols, sensors, localization algorithms, media access control systems, and energy usage of sensor nodes are among the requirements for modeling wireless networks. The creation of such models with components was made as easy as possible with the help of the modeling framework that was created. It allows for the actor-oriented definition of network nodes, wired subsystems, and physical media including acoustic channels and

Various Simulation Tools for WSNs 323

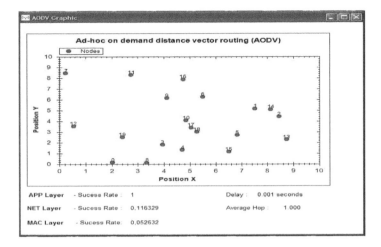

FIGURE 21.9
G-SENSE, a graphical user interface for SENSE.

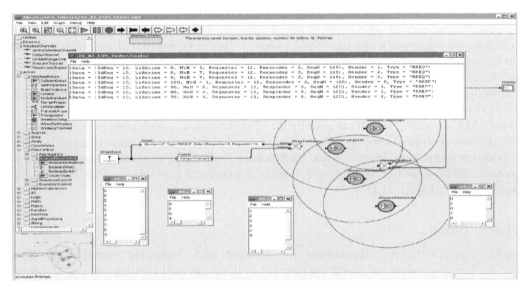

FIGURE 21.10
VisualSense graphical user interface.

wireless communication channels. A collection of base classes for building channels and sensor nodes, a library of subclasses that offer particular channel and node models, and an expandable visualization framework make up the software architecture. By subclassing the base classes and defining behavior in Java, or by utilizing one of Ptolemy II's modeling environments to build composite models, user-defined nodes may be constructed. By subclassing WirelessChannel's base class and implementing the Ptolemy II model's connecting mechanism, custom channels may be made. Ptolemy's goal is to enable academics to exchange models of various sensor network components and create models that include complicated issues from several angles. Figure 21.10 depicts the graphical user interface in detail.

21.3.10 TOSSIM

TOSSIM is a TinyOS bit-level discrete event simulator, or, to put it another way, an emulator [10]. One event is formed for each bit received or sent, instead of a single event per packet. For nesC model execution, TOSSIM is a TinyOS/MICA hardware simulator. TOSSIM may also act as a hardware emulator by converting hardware interrupts into discrete events. A simulated radio model is also included with TOSSIM. To conclude, TOSSIM is a high-fidelity sensor node network emulator based on TinyOS/MICA.

TOSSIM is a tool for researching TinyOS and its applications, not for evaluating new protocols' performance. As a result, one big disadvantage exists: each sensor node must execute the same code. As a result, TOSSIM's applicability to heterogeneous sensor applications is at best limited. TOSSIM is incompatible with a large number of programs. TOSSIM has been shown to be capable of simulating tens of thousands of nodes in a sensor network. Bit-level granularity, which diminishes as traffic grows, limits the capacity to mimic a higher number of sensor nodes. Because channel sampling is also mimicked at the bit level, the usage of a CSMA protocol adds substantial overhead, especially when compared to a TDMA-based MAC approach.

In TOSSIM, wireless communications for every sensor node within a one-hop radius of a single sensor node are intended to be error-free. To put it another way, every bit transmitted over the network is correctly received. During network communication, there is no cancellation and all signals are of equal strength. There is no signal cancellation and collisions are expressed using the CSMA protocol. This indicates that distance has no effect on signal intensity. The overlap of the signals may be heard by sensor nodes in the network. Each node has the same transmission range, which is a disk with a radius of r.

Through the location interface, TOSSIM makes each sensor node's virtual location available to that sensor node. This is accomplished by changing the value of each sensor node's "fake" ADC channels, which are read by the "FakeLocation" component in order to establish the sensor node's virtual position. This is intended to be used in place of a real location service for mimicking TinyOS apps. Each sensor node has a timer that may be used to create periodic activities like a sleep schedule. A graphical user interface for TOSSIM can be observed in Figure 21.11.

21.3.11 EmStar

EmStar is a software environment for creating and deploying complicated WSN applications on embedded 32-bit microserver systems, as well as integrating with Motes networks. EmStar consists of message-passing IPC primitive libraries, tools for simulation, emulation, and visualization of actual and simulated systems, and services for networking, sensing, and time synchronization. While the simplicity of use and versatility were prioritized over efficiency throughout EmStar's development, the increased overhead has not posed a problem for any current applications [11]. EmStar includes a number of tools and services that are relevant to the development of WSN applications. Below is a brief explanation of various tools and services, with some implementation detail.

21.3.11.1 EmSim/EmCee

Transparent simulation at varying degrees of precision is necessary for the creation and deployment of big systems. The levels of accuracy offered by EmSim and EmCee are various. EmSim creates a virtual environment with many storage nodes to imitate radio

Various Simulation Tools for WSNs

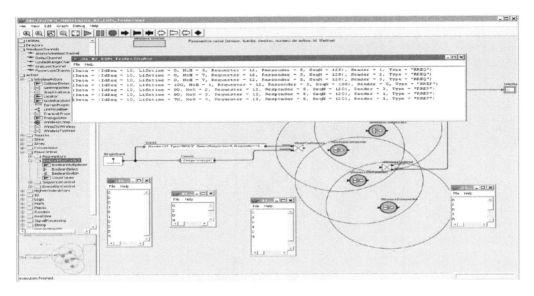

FIGURE 21.11
TOSSIM user interface.

and sensor networks. EmCee makes use of actual low-power radios as opposed to EmSim, which uses a simulated channel. These simulations hasten the debugging and development processes. While field emulation gives a knowledge of environmental dynamics prior to actual deployment, pure replication enables logically sound code. Using simulation and/or emulation techniques can greatly minimize the debugging work for a deployed system. In other words, switching between various environments for development and debugging is considerably simpler because they all utilize the same EmStar settings and source code compared to an installed system.

21.3.11.2 EmView/EmProxy

EmView is a visual representation of EmStar systems, shown in Figure 21.12. Developers may easily construct "plugins" for new apps and services because of the flexible design. EmView connects to both actual and simulated nodes via the UDP protocol. EmView can capture system dynamics in real-time since the protocol operates according to the best-effort rule. EmProxy is a proxy that works on a node or in a simulation and answers to EmView queries. EmProxy continuously monitors the node's status and responds to requests for information in real time.

21.3.11.3 EmRun

EmRun is in charge of initiating, terminating, and resuming the operational services for EmStar. EmRun launches the systems in order of dependence when EmStar services are "wired" together to optimize parallelism.

EmRun creates a communication channel with each of its child applications when a new computer is turned on so that it may keep track of their health, restart halted or dead processes, shut down the machine appropriately, and be informed when the installation is

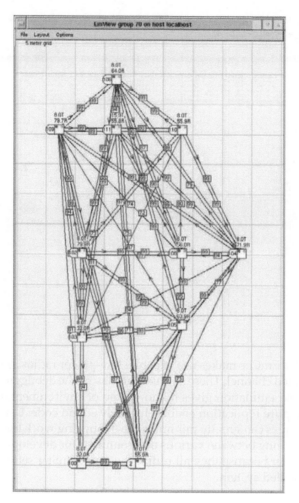

FIGURE 21.12
EmView, the EmStar visualizer.

finished. If they are runtime-modifiable in-memory log rings, EmStar services generate log messages that EmRun centralizes and makes available to participating clients.

21.3.12 NetSim

NetSim is the most widely used network simulation program for simulation and protocol modeling, development and network research, and security applications [12]. It provides unrivaled depth, power, and flexibility for analyzing computer networks. Its drag-and-drop capabilities for devices, connections, apps, etc. makes it simple to use. Netsim can scale to a network with hundreds of nodes. It supports all currently used protocols, including Fast and Gigabit Ethernet, IP and Routing, TCP, MANET, and radio, and it also supports multi-tenancy, which enables the connection of numerous sources, destinations, clients, and servers. The NetSim emulator is used and trusted by hundreds of military and space communications companies, research institutions, and government agencies.

Various Simulation Tools for WSNs　　327

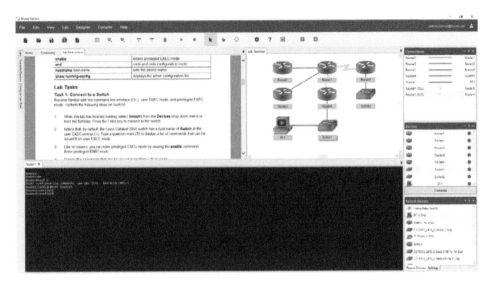

FIGURE 21.13
NetSim user interface.

NetSim is a network simulation program for creating network scenarios, designing protocols, modeling traffic, and analyzing network performance. Users may investigate the behavior of a network by experimenting with different network settings. NetSim can simulate 5G, 802.11, 802.22, LTE/LTE–ADV, MANET, IOT/WSN, VANETS, satellite communications, and software-defined networks. A NetSim simulation tool interface can be seen in Figure 21.13.

21.3.13 ATEMU

ATEMU is a finer granularity sensor network simulator. ATEMU's mission is to bridge the gap between real-world sensor network deployment and simulations. It uses a hybrid approach, simulating both the command-by-command functioning of individual sensor nodes and their actual wireless broadcast interactions. The ability to simulate a sensor network which is heterogeneous is one of ATEMU's distinctive features [13]. Many applications, as well as a complete sensor network based on multiple hardware platforms, may be exactly replicated on the MICA2 platform utilizing ATEMU. It is also explained how to use XATDB, an ATEMU front-end debugger/GUI. Without needing to invest in real sensor node hardware, XATDB is an excellent teaching tool for understanding how sensor nodes and sensor networks work. The software has previously been carefully tested and debugged on an accurate platform if real hardware is used, thanks to ATEMU's accuracy and emulation capabilities. This makes it much easier for the sensor network community to evaluate the performance of different algorithms and protocols in the real world. An ATEMU user interface can be seen in Figure 21.14.

21.3.14 PiccSim

PiccSim Toolchain streamlines wireless networked control system development, simulation, and deployment. PiccSIM is a Matlab/Simulink and ns-2-based simulation framework

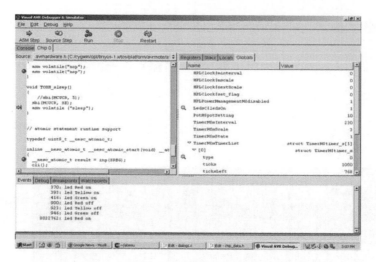

FIGURE 21.14
ATEMU user interface.

for (wireless) networked control systems [14]. "Platform for integrated communications and control design, simulation, implementation, and modeling" is what PiccSIM stands for. PiccSIM's purpose is to offer a complete set of tools for designing, modeling, and implementing wireless control systems. It is a networked control systems (NCS) or wireless network control systems (WNCS) research co-simulation tool. PiccSIM replicates both the control system and the network at the same time. Combining them into a single tool has the benefit of making it simple to investigate all areas of NCS control and communication, as well as their relationships. It is more efficient to combine well-known and powerful tools like Simulink and ns-2 into a single toolset than to use many programs to study different topics. Researchers may use PiccSIM to analyze all layers and linkages of large, sophisticated networked control systems. The PiccSIM toolchain features a graphical user interface for planning and modeling the control system and the network in addition to NCS simulation. This makes it easier to handle both simulators. The toolchain may also use the Matlab real-time workshop and target language compiler to turn the simulation model into C code, which can then be executed on the node hardware. Without any extra programming, the simulated system may be evaluated in practice. A NS-2 script generator for PiccSim can be seen in Figure 21.15

21.4 Conclusion and Future Work

In this chapter, we present various simulation tools for wireless sensor networks. The aim of this chapter is to give a general idea of the various simulations being used generally. In the first part of this chapter, we discuss what a wireless sensor network is. Then, wireless sensor network architecture and node structure are discussed. Finally various simulation tools – NS-2/NS-3, OMNET++, J-Sim, JiST/SWANS, GloMoSim, SHAWN, OPNET Network Simulator, SENSE, VisualSense, TOSSIM, and EmStar – are presented with their capabilities. The chapter aims to help people choose from various simulations

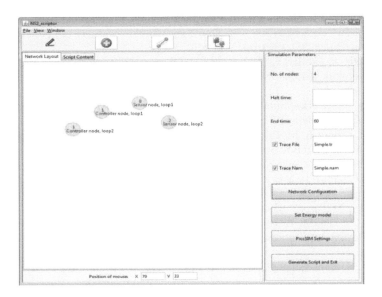

FIGURE 21.15
PiccSim user interface.

tools for their different needs. In most cases, general simulators lack some elements that are required for particular simulations in WSNs. Specialized simulators with more complete features, on the other hand, may perform better. Choosing alternative simulation tools for WSNs is more efficient and effective depending on the objectives.

Acknowledgments

We thank our parents for their support and motivation.

References

[1] P. Rajankumar, P. Nimisha, and P. Kamboj, "A comparative study and simulation of AODV MANET routing protocol in NS2 & NS3," *2014 International Conference on Computing for Sustainable Global Development (INDIACom)*, 2014, pp. 889–894.

[2] A. Varga, "OMNeT++," in K. Wehrle, M. Güneş, and J. Gross, (eds.), *Modeling and Tools for Network Simulation*, pp. 35–59. Berlin and Heidelberg: Springer, 2010.

[3] A Sobeih et al., "J-Sim: a simulation and emulation environment for wireless sensor networks," in *IEEE Wireless Communications*, vol. 13, no. 4 (2006), 104–119.

[4] F. Kargl and E. Schoch. "Simulation of MANETs: a qualitative comparison between JiST/SWANS and ns-2," in *Proceedings of the 1st International Workshop on System Evaluation for Mobile Platforms (MobiEval '07)*, 2007, pp. 41–46.

[5] X. Zeng, R. Bagrodia, and M. Gerla, "GloMoSim: a library for parallel simulation of large-scale wireless networks," in *Proceedings: Twelfth Workshop on Parallel and Distributed Simulation (PADS '98)*, 1998, pp. 154–161.

[6] S.P. Fekete, A. Kroller, S. Fischer, and D. Pfisterer, "Shawn: the fast, highly customizable sensor network simulator," in *2007 Fourth International Conference on Networked Sensing Systems*, 2007, p. 299.

[7] X. Chang, "Network simulations with OPNET," in *WSC '99: 1999 Winter Simulation Conference Proceedings. "Simulation – A Bridge to the Future,"*, 1999, vol.1, pp. 307–314.

[8] G. Chen, J. Branch, M. Pflug, L. Zhu, and B. Szymanski, "SENSE: a wireless sensor network simulator," in B.K. Szymanski and B. Yener (eds.), *Advances in Pervasive Computing and Networking*, pp. 249–267. Boston, MA: Springer, 2005.

[9] V. Roselló, J. Portilla, Y.E. Krasteva, and T. Riesgo, "Wireless sensor network modular node modeling and simulation with VisualSense," in *2009 35th Annual Conference of IEEE Industrial Electronics*, 2009, pp. 2685–2689.

[10] P. Levis, N. Lee, M. Welsh, and D. Culler, "TOSSIM: accurate and scalable simulation of entire TinyOS applications," in *Proceedings of the 1st International Conference on Embedded Networked Sensor Systems (SenSys '03)*, 2003, pp. 126–137.

[11] L. Girod, N. Ramanathan, J. Elson, T. Stathopoulos, M. Lukac, and D. Estrin. "EmStar: a software environment for developing and deploying heterogeneous sensor-actuator networks," *ACM Transactions on Sensor Networks*, vol. 3, no. 3 (2007), 13–es.

[12] P. Nayak and P. Sinha, "Analysis of random way point and random walk mobility model for reactive routing protocols for MANET using NetSim simulator," in *3rd International Conference on Artificial Intelligence, Modelling and Simulation* (AIMS), 2015, pp. 427–432.

[13] J. Polley, D. Blazakis, J. McGee, D. Rusk, and J.S. Baras, "ATEMU: a fine-grained sensor network simulator," in *2004 First Annual IEEE Communications Society Conference on Sensor and Ad Hoc Communications and Networks*, 2004, pp. 145–152.

[14] T. Kohtamaki, M. Pohjola, J. Brand, and L.M. Eriksson, "PiccSIM Toolchain – design, simulation and automatic implementation of wireless networked control systems," in *2009 International Conference on Networking, Sensing and Control*, 2009, pp. 49–54.

Index

5G, 68, 69, 74, 76

A

Active state, 139, 157, 229
Additive white Gaussian noise, 26, 28
Advanced Encryption Standard (AES), 208, 211
Agent-based approach (ABA), 61
Agriculture, 263, 274
Anechoic chamber, 44, 45
Area spectral efficiency, 245, 250
Artificial intelligence, 67, 319
Asymmetric digital subscribers lines (ADSL), 69–70

B

Beam Division Multiple Access (BDMA), 12, 14
Body Area Network (BAN), 62, 206
Building maintenance, 63, 211

C

Carrier 4, 7, 8
Cellular users (CUs), 244, 254
Central processing unit (CPU), 99, 137, 260, 282, 289, 302
Cluster head, 138, 168, 296
Clustered network, 296, 297
Code division multiple access (CDMA), 107
Cognitive cycle, 93
Cognitive radio (CR), 92, 94
Column Access Strobe (CAS), 145
Communication channel, 322, 325
Computational complexity, 22, 30, 36
Computational complexity reduction ratio (CCRR), 23
Concurrent power and information transfer (CPIT), 22
Confidentiality, integrity, authentication and availability (CIAA) 199
Conventional antennas, 15
Coplanar waveguide (CPW), 41
Coverage, 5, 13, 66, 76, 128, 149
Coverage probability, 229, 245
Cross-layer design, 98, 316
Cumulative distribution function (CCDF), 30, 33, 36

Cyclic-prefix (CP), 22

D

Data aggregation, 139, 212
Data Encryption Standard (DES), 7, 211
Data integrity, 268, 281
Data rate, 39, 50, 250, 254, 287
Data security, 67, 131, 191, 273
Defense Advanced Research Projects Agency (DARPA), 84
Definition video, 22, 67
Denial-of-service (DoS) attack, 200, 201
Dense area, 244, 249, 254
Dielectric substrate, 14, 19, 41, 50
Digital audio broadcasting, 73
Discrete event simulation, 319, 320
Downlink, 74, 245
DRAM, 112, 145, 205, 217
Dual band, 12, 19, 40, 41
Duty cycle, 97, 142, 146
Dynamic bandwidth allocation, 98, 102

E

E-health, 58
EH-WSNs, 305, 306, 308
Electric batteries, 175
Embedded system, 76, 182
Encryption, 7, 9, 268, 274, 302
Energy consumption, 110, 115, 137, 145, 154
Energy efficiency, 178, 200
Energy harvesting, 26, 58, 89, 146, 211, 294
Energy management, 110, 176, 316
Energy monitoring, 305
Energy storage, 177
Envelope correlation coefficient (ECC), 46
Event driven, 88, 223

F

Fading, 8, 65, 66
Far field, 17, 44, 76
Fast Fourier Transform (FFT), 26, 72
FBMC, 12, 22, 36
Field station, 60
Firmware control, 177
Fog computing, 265, 299

331

Frequency, 66, 69, 70, 74, 77, 92
Fuzzy, 139, 310

G

Gain, 16, 26, 36, 45
GA-LPTS, 28, 29, 33
General Packet Radio Service, 283
Genetic algorithm, 36, 310
Geographic Information System (GIS), 206
Global Positioning System (GPS), 88, 100
Green communication, 305
GSM, 67, 219, 317
Guard band, 69, 73

H

Hardware control, 176
Health care, 121, 173, 299
HFSS, 16, 18
Homogeneous Poisson point process (HPPP) 246
Hopping, 6
Hybrid energy harvesting, 175, 310

I

IEEE 802.15.4, 60, 89, 219
IIoT, 187, 298
Interference, 93, 154, 158, 201, 245
Interference measurements, 308
International Telecommunication Union (ITU), 46, 275
Internet of things, 120, 127, 184, 244

L

Latency, 12, 14, 18
Layered network, 295
Leakage gating, 142
Link quality measurements, 307
Local area networks, 5, 71, 102, 218
Long distance communication, 144, 266
Long term evolution (LTE), 106
Low energy adaptive clustering hierarchy (LEACH), 90, 296

M

MAC protocol, 64, 202, 306
Machine-to-machine communication, 74
Mean effective gain (MEG), 48
Message driven, 138, 149
Microcontroller, 60, 88, 177, 182, 264, 288

Micro electro mechanical system (MEMS) 84, 263
Microstrip patch antenna, 13, 18
Microwave, 12, 41, 283
Mobile data collector, 90
Mobility, 115, 138, 224
Modulation, 69, 71, 73, 106, 156
Multiple input multiple output (MIMO), 198

N

National Institute of Standards and Technology (NIST), 59
Network level, 133, 155, 156, 167
Network lifetime, 110, 202, 228, 230, 294
Network parameters, 164, 300
Network protocol, 98, 314, 321
Network simulator, 61, 318, 322
New radio (NR), 39, 77
Next generation networks, 297, 303
NMOS transistor, 141
Node contribution, 228, 232
Non-functional area (NFA), 243
Non-line-of-sight, 7
Non-orthogonal multiple access, 22, 28
Non-rechargeable batteries, 86, 144, 175

O

OFDMA, 71, 74
Optimization, 33, 99, 112, 183, 212, 288, 295
Orthogonal frequency division multiplexing, 22
Outage probability, 244, 248

P

Partial transmit sequence, 22, 33
Peak to average power ratio, 22
Phase shift keying, 7, 8
Photovoltaic energy, 175
PHY layer, 106, 114
Plexiglass, 40
Power and information transfer, 30
Power conservation, 141, 145, 152
Power consumption, 152, 156, 159
Power efficient routing protocols, 160
Power management, 159, 167
Power spectral density, 23, 39
Power waste, 158
Precision agriculture, 60, 63, 206, 275
Primary user, 76, 93
Priority, 93, 185, 229, 230
Probability density function (PDF), 247
Protocol, 236, 260, 269, 275

Index

Q

Quadrature amplitude modulation, 7, 69
Quality of service (QoS) 7, 68, 106, 267, 294, 310

R

Radiation, 9, 15, 17, 40
Radiation intensity, 15, 16, 40, 41, 44
Radiation pattern, 17, 40, 41, 44
Radio, 3, 7, 13, 18, 39, 86, 142, 268
Ramifications, 122
Rayleigh fading channel, 106, 246
Real time clock (RTC), 178
Received signal strength indicator (RSSI), 205
Reception, 110, 157, 175, 283
Rechargeable batteries, 86, 110, 144, 306
Residual energy, 139, 146, 307
Resource allocation, 33, 99
RFID, 5, 7, 299
Rogers Duroid 5880, 41, 50
Routing protocols, 89, 152
Row Access Strobe (RAS), 145

S

Secondary users, 93, 107
Security, 12, 22, 36, 68, 267, 273
Security and safety, 267
Sensing range, 68, 74, 123
Sensor, 127, 129, 137
Sensor node, 138, 145, 152, 175, 201
Sensor technology, 62, 211
Session key agreement, 277
Signal to interference noise ratio (SINR), 246
Signal to noise ratio, 250, 309
Simulation results, 16, 160, 164, 245
Sink node, 283, 315
Sleep state, 138, 148, 162
Smart cities, 182, 192, 245
Smart grid, 192, 265, 303
Social impact, 120, 128, 130
Social networking, 22
Software level, 157, 159
Sound Surveillance System (SOSUS), 84
Spectral efficiency, 22, 23, 30, 77, 243, 249
Spectrum, 6, 14, 41, 197, 300
Spectrum sensing, 76, 309
S-plane, 17
Standard, 69, 77, 104, 188, 211, 267, 289
Structural health monitoring, 89, 91
Sub-6GHz, 13
Sub-Channel, 22, 26
Subscriber stations, 74
Surface current distribution (SCD), 44
Susceptance, 15

T

Target coverage, 227, 230, 233
TCP/IP Model, 208
Thermal energy, 175
Time division multiple access (TDMA), 24
Time of arrival (TOA), 61
Timer driven, 138, 149
Total active reflection coefficient (TARC), 48
Transceiver, 154, 174
Transmission Control Protocol, 104, 112
Transmission process, 199, 262
Transport layer, 105, 276, 316

U

Unmanned Aerial Vehicle (UAV), 245
Uplink Phase, 27

V

Vibrational energy, 175
Virtual private network (VPN), 210, 219
Voltage regulator, 141, 142
Voltage Standing Wave Ratio (VSWR), 17, 44

W

Wavefront, 17
Wearable Wireless Body Area Network (WWBAN), 62, 211
WGAN, 219
Wi-Fi Protected Access (WPA), 207
Wireless, 50, 58, 62, 68, 76, 260, 264
Wireless Access Standards, 6
Wireless communications, 4, 8, 68, 74
Wireless local area network, 219
Wireless networks, 57, 75, 104, 270, 274
Wireless power transfer, 74, 78, 174
Wireless sensor network, 176, 178, 183, 198, 216
Wireless wide area network, 219
WiseMAC, 177
WSN architecture, 212, 265

Z

ZigBee, 89, 220, 244, 317

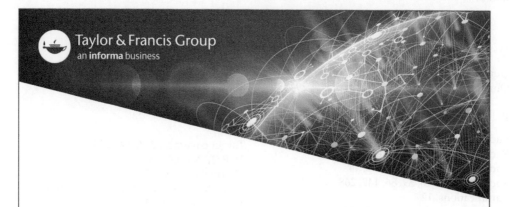